WITHDRAWN

W9-AOE-109

ALSO BY MICHAEL BENSON

Otherworlds: Visions of Our Solar System

Cosmigraphics: Picturing Space Through Time

Planetfall: New Solar System Visions

Far Out: A Space-Time Chronicle

Beyond: Visions of the Interplanetary Probes

SPACE ODYSSEY

STANLEY KUBRICK, ARTHUR C. CLARKE, AND THE MAKING OF A MASTERPIECE

MICHAEL BENSON

Mount Laurel Library
100 Walt Whitman Avenue
Mount Laurel, NJ 08054-9539
856-234-7319
www.mountlaurellibrary.org

Simon & Schuster
New York   London   Toronto   Sydney   New Delhi

Simon & Schuster
1230 Avenue of the Americas
New York, NY 10020

Copyright © 2018 by Michael Benson

All rights reserved, including the right to reproduce this book or portions thereof in any
form whatsoever. For information, address Simon & Schuster Subsidiary Rights Department,
1230 Avenue of the Americas, New York, NY 10020.

First Simon & Schuster hardcover edition April 2018

SIMON & SCHUSTER and colophon are registered
trademarks of Simon & Schuster, Inc.

For information about special discounts for bulk purchases,
please contact Simon & Schuster Special Sales at
1-866-506-1949 or business@simonandschuster.com.

The Simon & Schuster Speakers Bureau can bring authors to your live event. For more
information or to book an event, contact the Simon & Schuster Speakers Bureau at
1-866-248-3049 or visit our website at www.simonspeakers.com.

Interior design by Silverglass

Manufactured in the United States of America

10   9   8   7   6   5   4   3   2   1

Library of Congress Cataloging-in-Publication Data is available.

ISBN 978-1-5011-6393-7
ISBN 978-1-5011-6395-1 (ebook)

PHOTO CREDITS

Throughout: 2001: A SPACE ODYSSEY and all related characters and elements © & ™
Turner Entertainment Co. (s17): 95, 121, 122, 142, 153, 163, 166, 173, 174, 175, 178, 179
(photo illustration), 186, 191, 195, 198, 205, 211, 217, 220, 232, 270, 324–325, 327, 387, 398;
Photographer unknown unless otherwise specified. Courtesy Hector Ekanayake and Rohan
De Silva: 18; Courtesy Kavan Ratnatunga, from his private collection: 22; Estate of Roger
Caras: 40; Courtesy Christiane Kubrick and Jan Harlan: 48, 68, 416; Courtesy Tony Frewin:
109; Courtesy Doug Trumbull: 132, 253, 257, 296, 311, 388; Dmitri Kasterine, © & ™ Turner
Entertainment Co. (s17): 169, 202; Hayley Mills, Courtesy Andrew Birkin: 226, 245; Andrew
Birkin, Courtesy Andrew Birkin 275, 277; Courtesy Freeborn Estate: 234; Keith Hamshere;
Courtesy Dan Richter: 250; Courtesy Dan Richter: 282, 301, 317; Courtesy Con Pederson:
321, 361; Courtesy Anne McDonigle: 335; Doug Trumbull, Courtesy Bruce Logan: 342;
Robert U. Taylor, Courtesy Doug Trumbull: 343; Courtesy Hector Ekanayake and Rohan De
Silva: 432 Insert: © & ™ Turner Entertainment Co. (s17): 1; Andrew Birkin, Courtesy Andrew
Birkin: 2, 3; FX team, © & ™ Turner Entertainment Co. (s17): 4; John Alcott, © & ™ Turner
Entertainment Co. (s17): 5; "2001" FX team, © & ™ Turner Entertainment Co. (s17): 6, 9,
13, 15; Geoffrey Unsworth, "2001" FX team, © & ™ Turner Entertainment Co. (s17): 7, 8,
10, 14, 16, 17, 18; Geoffrey Unsworth, © & ™ Turner Entertainment Co. (s17): 11; John Jay/
mptvimages.com: 12

In memory of my father,
Raymond E. Benson,
November 2, 1924—November 12, 2017

Politics and religion are obsolete. <u>The time has come for science and spirituality</u>.

—VINOBA BHAVE

Quoted by Arthur C. Clarke in his keynote address to the American Astronautical Society's fifth Robert Goddard Memorial Symposium in Greenbelt, Maryland, on March 14, 1967. Clarke erroneously attributed the statement to Indian prime minister Jawaharlal Nehru, who frequently quoted Bhave. The last sentence was underlined by Stanley Kubrick in his copy of Clarke's address.

CONTENTS

MAJOR CHARACTERS

IN ORDER OF APPEARANCE

Stanley Kubrick—film director and producer

Arthur C. Clarke—futurist, science fiction writer, essayist

Christiane Kubrick—artist, actress, Stanley's wife

Carl Sagan—astronomer, planetary scientist; later, best-selling author

Mike Wilson—Clarke's partner in the 1950s and 1960s

Roger Caras—publicist, VP of Stanley Kubrick's two companies, Hawk Films and Polaris Productions; later, a leading animal rights advocate

Scott Meredith—Clarke's New York literary agent

Hector Ekanayake—Clarke's assistant and later, business partner

Ray Lovejoy—Kubrick's assistant, lead film editor

William Sylvester—actor, played presidential science advisor Heywood Floyd

Con Pederson—visual effects supervisor

Doug Trumbull—visual effects supervisor

Robert Gaffney—cinematographer, advisor to Kubrick, did second-unit aerial shots

Louis Blau—Kubrick's Los Angeles lawyer and close associate

Wally Gentleman—initial director of visual effects

Douglas Rain—Canadian actor, voice of HAL-9000 computer

Fred Ordway—technical and scientific consultant

Harry Lange—graphic artist and production designer

Robert O'Brien—MGM president and CEO

Keir Dullea—actor, played mission commander Dave Bowman

Gary Lockwood—actor, played second-in-command Frank Poole

Victor Lyndon—associate producer

Tony Masters—lead production designer

Bob Cartwright—initial set decorator

Tony Frewin—assistant to the director

Ernie Archer—production designer, assistant to Masters

Wally Veevers—visual effects supervisor

Brian Johnson—special effects assistant

Robert Beatty—actor, played lunar base commander Ralph Halvorsen

Geoffrey Unsworth—director of photography

Derek Cracknell—first assistant director

Kelvin Pike—camera operator

David de Wilde—first assistant film editor

John Alcott—assistant to Unsworth, director of photography for Dawn of Man sequence

Bryan Loftus—special photographic effects

Andrew Birkin—assistant to the director

Stuart Freeborn—makeup artist

Dan Richter—mime, played lead man-ape, Moonwatcher

Bill Weston—stuntman

Tom Howard—visual effects supervisor

Pierre Boulat—still photographer, desert landscapes for Dawn of Man sequence

Colin Cantwell—special photographic effects

Jan Harlan—Kubrick's brother-in-law and informal music advisor

CHAPTER ONE

PROLOGUE: THE ODYSSEY

The very meaninglessness of life forces
man to create his own meanings.

—STANLEY KUBRICK

The twentieth century produced two great latter-day iterations of Homer's *Odyssey*. The first was James Joyce's *Ulysses*, which collapsed Odysseus's decade of wandering down to a single city, Dublin, and a seemingly arbitrary day, June 16, 1904. In *Ulysses*, the role of Ithaca's wily king was played by a commoner, Leopold Bloom—a peaceable Jewish cuckold with an uncommonly fascinating inner life, one the author effectively allowed us to hear. Serialized from 1918 to 1920, it was published in full in 1922.

The other was Stanley Kubrick and Arthur C. Clarke's *2001: A Space Odyssey*, in which the islands of the southeastern Mediterranean became the solar system's planets and moons, and the wine-dark sea the airless void of interplanetary, interstellar, and even intergalactic space.

Shot in large-format panoramic 65-millimeter negative and initially projected on giant, curving Cinerama screens in specially modified theaters, *2001* premiered in Washington, DC, on April 2, 1968, and in New York City the following day. Produced and directed by Kubrick and conceived in collaboration with Clarke, one of the leading authors of science fiction's "golden age," the film was initially 161 minutes long. Following a disastrous series of preview and premiere screenings, the director cut it down to a leaner 142 minutes.

Where Joyce's strategy had been to transform Odysseus into a benevolently meditative cosmopolitan flaneur, and to reduce ten years of close calls and escape artistry to twenty-four hours in proximity of the River Liffey, Kubrick and Clarke took the opposite approach. Deploying science as a kind of prism, which during the nineteenth and twentieth centuries entirely transformed our sense of the size and duration of the universe, they vastly expanded Homer's spatiotemporal parameters. *2001: A Space Odyssey* encompassed four million years of human evolution, from prehuman Australopithecine man-apes struggling to survive in southern Africa, through to twenty-first-century space-faring Homo sapiens, then on to the death and rebirth of their Odysseus astronaut, Dave Bowman, as an eerily posthuman "Star Child." In the final scene, the weightless fetus returns to Earth as Richard Strauss's 1896 composition *Thus Spoke Zarathustra* pounds cathartically on the soundtrack.

In *2001: A Space Odyssey*, the meddlesome gods of the ancients have become an inscrutable, prying alien super-race. Never seen directly, they swoop down periodically from their galactic Olympus to intervene in human affairs. The instrument of their power, a rectangular black monolith, appears at key turning points in human destiny. First seen among starving man-apes in a parched African landscape at the "Dawn of Man," *2001*'s totemic extraterrestrial artifact engenders the idea among our distant ancestors of using weaponized bones to harvest the animal protein grazing plentifully all around them. This prompting toward tool use implicitly channels the species toward survival, success—and, eventually, technologically mediated global domination.

After vaulting into that happy future in a match cut that has deservedly acquired the reputation of being the single most astonishing transition in cinematic history, *2001* leads us to understand that a lunar survey team has discovered another monolith, this one seemingly deliberately buried under the surface of the Moon eons before. When excavated and hit by sunlight for the first time in millions of years, it fires a powerful radio pulse in the direction of Jupiter—evidently a signal, warning its makers that a species capable of space travel has arisen on Earth. A giant spacecraft, *Discovery*, is sent to investigate.

While parallels with *The Odyssey* aren't as thoroughly woven into the structure of *2001* as they are in *Ulysses*, they certainly exist. Seemingly prodded into action by flawed programming, a cyclopean supercomputer named HAL-9000—represented by an ultracalm disembodied voice and a network of individual glowing eyes positioned throughout *Discovery*—goes bad and kills off most of the crew. The sole surviving astronaut, mission commander Dave Bowman, then has to fight the computer to the death. Apart from dueling a cybernetic Cyclops, Bowman's name references Odysseus, who returns to Ithaca, strings the bow of Apollo, shoots an arrow through twelve axe shafts, and proceeds to slaughter his wife's suitors. A *nostos*, or homecoming, was as necessary to Kubrick's and Clarke's *Odyssey* as it was to Homer's.

Much like Joyce and in keeping with their expansive vision, *2001*'s authors took parallels with Homer as a starting point, not final word. When they began work in 1964, one initial motivation was to study the universal structures of all human myths. They were aided by Joseph Campbell's magisterial study *The Hero with a Thousand Faces*, which provided them with a template for the conscious creation of a new work of mythology. Early in their collaboration, Kubrick quoted a passage to Clarke concerning the universal rite of passage of any mythological hero, which Campbell suggested invariably encompasses "*separation–initiation–return.*" This tripartite structure "might be named the nuclear unit of the monomyth," Campbell wrote—a term he borrowed from Joyce, who'd coined it in his last major work, *Finnegans Wake*.

Campbell's research helped Kubrick and Clarke delve into the archetypal workings of human mythological yearnings, expanding that template to encompass not just one story and hero, and not even just one species, but rather the entire trajectory of humanity—"from ape to angel," as Kubrick put it in 1968. In this, they also overtly referenced Friedrich Nietzsche's 1891 philosophical novel *Also sprach Zarathustra*, with its concept of mankind as merely a transitional species—sentient enough to understand its animal origins but not yet truly civilized. It was an idea both could get behind, Clarke with his innate optimism about human possibilities, and Kubrick with his deeply ingrained skepticism. It was this seemingly contradictory mesh of worldviews that gave *2001: A Space Odyssey* its exhilarating fusion of agnosticism and belief, cynicism and idealism, death and rebirth.

In Clarke, Kubrick had found the most balanced and productive creative partnership of his career. While the director made all the critical decisions during the film's production, the project started out—and in important ways remained—a largely equal collaboration between two very different, singularly creative characters. Like Joyce, both were expatriates, with the Kubrick family finally settling for good in England during the making of *2001*, and Clarke being a resident of Ceylon—later Sri Lanka—from 1956 until his death in 2008.

At *2001*'s release in 1968, Kubrick was thirty-nine, the same age as Joyce when *Ulysses* was being serialized. He was at the pinnacle of his abilities, having already made two of the twentieth century's great films. Each was a devastating indictment of human behavior as expressed through the military mind-set. Released in 1957, *Paths of Glory* served as a comprehensive indictment of the hypocrisy of the French general staff during World War I—though its meanings were by no means limited to any one army or conflict. And his 1964 satire *Dr. Strangelove*, written in collaboration with Peter George and Terry Southern, cut to the core of the Cold War nuclear arms race, equal parts savage critique and caustic black comedy. A resounding critical and commercial success, it set the stage for the large-scale studio support necessary to realize *2001*.

Kubrick's method was to find an existing novel or source concept and adapt it for the screen, always stamping it with his own bleak— but not necessarily despairing—assessment of the human condition. A self-educated polymath, he was in some ways the ultimate genre director, switching virtuosically between established cinematic categories and forms with a restless analytical intelligence, always transcending and expanding their boundaries. During his career, he reinvented and redefined the film noir heist film, the war movie, the period costume feature, the horror flick, and the science fiction epic, each time transforming and reinvigorating the genre through extensive, time-consuming research followed by an uncompromising winnowing away of clichés and extraneous elements.

Kubrick treated every film as a grand investigation, drilling down into his subject with a relentless perfectionist's tenacity as he forced it to yield every secret and possibility. Once he'd decided on a theme, he subjected it to years of interrogation, reading everything and exploring all aspects before finally jump-starting the cumbersome filmmaking machinery. Having concluded his preproduction research, he directed his pictures with all the authority of an enlightened despot. Following a stint as hired-gun director on *Spartacus* in 1960, he conceived of a personal kind of slave revolt, never again working on a project he didn't produce himself. While in practice, studios such as MGM footed the bills and exerted some influence, this gave him near-complete artistic independence. (Still *Spartacus*, which Kirk Douglas both produced and starred in, marked Kubrick's definitive induction into big-budget Hollywood filmmaking. The picture, which dramatized the bloody trajectory of a Thracian gladiator as he led a successful uprising against Rome, won four Oscars and a Golden Globe award for Best Motion Picture Drama.)

As the ne plus ultra example of Kubrick's methods, *2001: A Space Odyssey* wasn't just rooted in extensive preproduction fieldwork, it continued throughout—an uninterrupted, well-funded research project spanning its live-action filming and extending across its postproduction as well (which, given the importance of its visual effects, was actually production by another name). All the while, the director and his team pioneered a variety of innovative new cinematic techniques. Highly unorthodox in big-budget filmmaking, this improvisatory, research-based approach was practically unheard of in a project of this scale. *2001* never had a definitive script. Major plot points remained in flux well into filming. Significant scenes were modified beyond recognition or tossed altogether as their moment on the schedule arrived. A documentary prelude featuring leading scientists discussing extraterrestrial intelligence was shot but discarded. Giant sets were built, found wanting, and rejected. A transparent two-ton Plexiglas monolith was produced at huge expense and then shelved as inadequate. And so forth.

Throughout, Kubrick and Clarke remained locked in dialogue. One strategy they'd agreed on in advance was that their story's metaphysical and even mystical elements had to be earned through absolute scientific-technical realism. *2001*'s space shuttles, orbiting stations, lunar bases, and Jupiter missions were thoroughly grounded in actual research and rigorously informed extrapolation, much of it provided by leading American companies then also busy providing technologies and expertise to the US National Aeronautics and Space Administration, or NASA. In late 1965, George Mueller, the czar of the Apollo lunar program, visited *2001*'s studio facilities north of London. Apollo was then still flight-testing unmanned launch vehicles, while NASA launched the precursor Gemini program's two-man capsules in an ambitious series of Earth-orbiting missions. After touring the film's emerging sets and viewing detailed scale models of its centrifuges and spacecraft, the man in charge of landing men on the Moon and returning them safely to Earth—the ultimate Odyssean voyage yet accomplished by the species—was impressed enough to dub the production "NASA East."

Clarke was fifty when *2001* came out. When Kubrick first contacted him early in 1964, he had already enjoyed an exceptionally prolific career. Best known as a formidably imaginative science fiction novelist and short-story writer, he was also a trenchant essayist and one of the twentieth century's leading advocates of human expansion into the solar system. Apart from his fictional and nonfictional output, he had played a noteworthy role in the history of technology. Clarke's 1945 paper on "extraterrestrial relays," published in the British magazine *Wireless World*, proposed a global system of geostationary satellites, which, he argued, would revolutionize global telecommunications. While some of the ideas he presented had already been in circulation, he synthesized them impeccably, and the paper is regarded as an important document of the space age and the information revolution.

Clarke's fictions were greatly influenced by the work of British science fiction novelist Olaf Stapledon (1886–1950), whose seminal *Last and First Men* and *Star Maker* encompassed multiple phases of human evolution across vast timescales. Clarke's early novels *Childhood's End* (1953) and *The City and the Stars* (1956) likewise encompassed sweeps of time so expansive that monumental civilizational changes could be examined in great detail. Still considered his best work, *Childhood's End* closed with the human race being shepherded through an accelerated evolutionary transformation by a seemingly benevolent alien race, the "Overlords." In it, humanity is depicted as obsolete—destined for replacement by a telepathically linked successor species composed, oddly, of children. Clarke's strange vision of mankind outgrowing its childhood was also influenced directly by the great Russian rocket scientist and futurist Konstantin Tsiolkovsky, who, in an essay published in 1912, stated, "Earth is the cradle of the mind, but humanity can't remain in its cradle forever." As the central utopian credo of the space age, Tsiolkovsky's pronouncement would find direct expression in *2001*'s final scenes.

As with *Ulysses*, *2001* was initially greeted with varying degrees of incomprehension, dismissal, and scorn—but also awed admiration, particularly among the younger generation. Its first screenings were a harrowing ordeal, with audience reactions at the New York premiere including boos, catcalls, and large-scale walkouts. Most of the city's leading critics dismissed the film, some in personal and humiliating terms. And as with Joyce, some of Kubrick's and Clarke's peers went out of their way to disparage the film. Russian director Andrei Tarkovsky, possibly the greatest filmmaker of the twentieth century, found *2001* repellant. Calling it "phony on many points," he argued that its fixation on "the details of the material structure of the future" resulted in a transformation of "the emotional foundation of a film, as a work of art, into a lifeless schema with only pretensions to the truth." Soon after its release, Clarke's friend and fellow science fiction writer Ray Bradbury wrote a negative review decrying *2001*'s slow pace and banal dialogue. He had a solution, though: it should be "run through the chopper, heartlessly."

In retrospect, these initial waves of hostility and incomprehension can be understood as a result of the film's radical innovations in technique and structure—another similarity to *Ulysses*. They were followed by grudging reappraisals, at least on the part of some, and a dawning understanding that a truly significant work of art had materialized. *2001: A Space Odyssey* is now recognized as one of the exceedingly rare works that will forever define its historical period. Put simply, it changed how we think about ourselves. In this way, too, it easily withstands comparison to James Joyce's masterpiece.

In both of these modernist Odysseys, audiences were asked to accept new ways of receiving narrative. While Joyce didn't invent stream of consciousness and interior monologue as literary devices, he brought them to unprecedented levels of proficiency and complexity. Likewise, Kubrick didn't create oblique auteurist indirection and dialogue-free imagistic storytelling—but by transposing it into the science fiction genre and setting it within such a vast expanse of space and time, he effectively kicked it upstairs. *2001* is essentially a nonverbal experience, one more comparable to a musical composition than to the usual dialogue-based commercial cinema. An art film made with a Hollywood blockbuster budget, it put audiences in the unaccustomed position of "paying attention with their eyes," as Kubrick put it.

Joyce's tidally impressionistic portrait of provincial Dublin allowed us to sample previously inaccessible internal currents of human thought and feeling. Kubrick's and Clarke's *2001: A Space Odyssey* presented a disturbing vision of human transformation due to technology, positioning all our strivings within a colossal cosmic framework and evoking the existence of extraterrestrial entities so powerful as to be godlike. Each was highly influential, with innumerable successor works striving to equal their philosophical breadth and technical virtuosity. Neither has yet been surpassed.

· · ·

My own lifelong engagement with *2001* started in the spring of 1968 at the age of six. My mom, a confirmed Clarke fan, took me to an afternoon matinee within weeks of the film's premiere. Whether it was in Washington (where we then lived) or New York (as I remember it) is unclear. While I was already excited by the jump into space as then best represented by the Apollo program—which had already launched two of its towering Saturn V Moon rockets on unmanned test flights—it was no preparation for my first exposure to such a powerfully ambiguous, visually stunning work.

At six, of course, your receptors are about as open as they'll ever be, and I consider myself fortunate to have seen the film at that age. The Dawn of Man prelude was both riveting and disturbing, and the mysterious appearance of the monolith, accompanied by the unholy keening of György Ligeti's *Requiem*, reverberated in my childish imagination with almost overpowering overtones of mystery, wonderment, and horror. The ecstatic discovery by the lead man-ape that a heavy bone could be used as a weapon, which Kubrick conveyed with a wordless cinematic assurance, needed no explanation and didn't even require conscious understanding. It spoke in its own language; as with much of the rest of the film, the authority and power of the images themselves didn't necessitate literal comprehension.

The lunar, spaceflight, and space-walk scenes were mesmerizing. The effects of zero gravity on the human body were conveyed with utterly convincing realism. Bowman's methodical lobotomization of HAL couldn't have been more disturbing and frighteningly strange. And the film's abstract "Star Gate" sequence, which led to Bowman's multistage transformation into an elderly man, seen on his deathbed in a phantasmagorical hotel room—and then finally his transformation into that ethereal floating fetus—was spellbinding.

Much of it was also incomprehensible, however, and afterward, I trailed my mom across the pavement of whichever city it may have been, exhausted by a surfeit of wonderment and squinting in the dazzling late-afternoon sunlight. "But what did it *mean*?" I wailed. "I don't know!" she replied, to her great credit. Mom was always honest with me—and is to this day.

Much later, I grew to understand that *2001*'s power over me then and thereafter came about at least partially due to personal circumstances. As the child of foreign service parents, I had already lived in Belgrade, Yugoslavia, as well as Hamburg, West Germany—two countries that no longer exist. Although an American by birth, my world was *the* world, and my nascent identity was global by default. What I'm getting at is that I believe that even at the age of six, I grasped, in a preconscious, precocious kind of way, that we live in a complex world surrounded by multiple contradictory cultures, perspectives, and ways of being. Unfortunately, an unwanted side effect of changing countries so regularly can be a sense of not belonging to any one place—the so-called third-culture syndrome of the expat kid.

It would risk both cliché and oversimplification to suggest that *2001: A Space Odyssey* helped me be at home in the world. But there's no doubt that as with any major work of art that has a decisive impact on a person, it did so for a very good reason. As the years passed, I saw the film again many times, and it always struck me as an extraordinarily prescient account of the human situation in a mystifyingly grand, seemingly indifferent cosmos. The visual magnificence and uncompromising artistic integrity of Kubrick's and Clarke's achievement made the accuracy of its individual details beside the point. In the decades since the film's release, paleoanthropologists have largely discredited its depiction of the transition from vegetarian man-apes to carnivorous killers, which was based on the work of paleontologist Raymond Dart. And, at least to date, we don't have credible evidence of even microbial life beyond Earth, let alone superpowered extraterrestrial interlopers with an uncanny interest in our evolutionary progress. (Of course, the surest sign that intelligent life exists out there may well be that it *hasn't* come here—as Clarke used to joke.)

And, unfortunately, *2001*'s vision of the Moon and planets being colonized by human beings simply hasn't come true—or, at least, not nearly in the way its authors had envisioned. The film was made when NASA's budget was at its peak, so their extrapolation is understandable. (Clarke even predicted to Kubrick, "This is the last big space film that won't be made on location.") In fact, no human being has ventured beyond low Earth orbit since the return of the last Apollo crew from the Taurus-Littrow valley on the Moon in 1972, just four years after the film's release. Since then, true space exploration has been conducted exclusively by automated spacecraft. But even if we consider HAL's attempts to kill off *Discovery*'s crew and continue on to Jupiter without pesky human interference as predictive of these circumstances (and it's a valid interpretation), the film's deviations from literal accuracy are beside the point. Like Joyce—like Homer himself—*2001*'s authors were crafting a *story*. As a fiction, it created its own reality and demands to be viewed within that frame.

In any case, Kubrick's and Clarke's portrayal of twenty-first-century humanity suspended within an evolutionary trajectory spanning millions of years, and their placement of that story within a universe potentially filled with ancient civilizations—not to mention their depiction of human beings as desensitized parts of the machinery they've created, and their evocation of an artificial intelligence brought into being through human genius, yet driven through human error into conflict with its makers—it all had, and still has, a perspicacious, even ominous ring of truth to it. Ask not for whom the monolith tolls.

■ ■ ■

I never met Stanley Kubrick, though I've had the privilege of spending many entertaining hours in discussion with his widow, Christiane, at their impressive estate and manor in Childwickbury, a hamlet north of London. I did get to know Arthur Clarke during the last decade of his life, and visited him three times in Sri Lanka, the last with my family in tow. When I first met him, in the year 2001, no less, he was already confined to a wheelchair due to a progressive neurological ailment called post-polio syndrome, but I found him to be an alert, cheerful, wickedly humorous presence, always up for an in-depth discussion and gratifyingly willing to mobilize a small motorcade and show me around the southern part of the island. We discussed *2001: A Space Odyssey* at some length, and though most of what he had to say he'd already published in one form or another, he would occasionally come up with an unexpected glint of insight that has proven valuable in the writing of this book. It was Clarke, for example, who told me about Kubrick's instant antipathy to his friend Carl Sagan—something I might not have known otherwise.

At one of our first meetings, I had the temerity to ask who had written perhaps *2001*'s most powerful scene: the one where Dave Bowman, having forcibly gained reentry to his ship, proceeds to HAL's Brain Room and deprograms the computer. "Who do you *think* wrote it? I did!" he boomed in mock indignation. While I accepted his statement, the reason I asked—and I told him this—was that the sequence has a chilly intensity that I recognized as more Kubrickian than Clarkean. In fact, as with any good collaboration, the truth lies somewhere in between. While Clarke appears to have conceived of the scene up to a point—or at least put forward the Cartesian proposition that an artificial intelligence is alive and therefore can be hurt—Kubrick did, in fact, write it, as he did most of *2001*'s dialogue. Of which there isn't much: the film has less than 40 minutes of spoken words in its 142 minutes of running time. You could say that in this case, Kubrick handled the lyrics and Clarke the tune—something the latter wasn't particularly prepared to adjudicate more than three decades later to a whippersnapper such as myself, understandably enough.

Of course, this presents a paradox. Clarke was the writer and Kubrick the filmmaker, so one might be excused for assuming that what words there are in *2001* must have chattered from his portable typewriter at some point between 1964 and 1968. Not so. Almost every scene was rewritten multiple times by the director during live-action production, which—if we exempt the wordless Dawn of Man sequence—extended for just over six months, from late December 1965 to mid-July 1966. (The prehistoric prelude was shot in the summer of 1967.) And throughout the film's production, but in particular during editing, Kubrick's instinct was to remove as much verbal explication as possible in favor of purely visual and sonic cues. Much to the consternation of his collaborator, this included Clarke's voice-over narrations, which were originally intended to frame the story.

Kubrick thereby lifted away what would have been a superstructure of overtly stated truths. He did so without necessarily losing them altogether; they were now implicit rather than explicit. The result was a masterwork of oblique, visceral, and intuited meanings. *2001*'s conscious deployment of a mythological structure, its insistence on first-person experiential cinema, and the inherent opacity of its "true" messages permitted every viewer to project his or her own understandings on it. It's an important reason for the film's enduring power and relevance.

Finally, *2001: A Space Odyssey* is about our situation as creatures conscious of our own mortality, aware of inherent limitations to our imaginations and intellectual capacities, and yet perpetually striving for more exalted states and higher planes of being. And that's where it best reveals itself as a profoundly collaborative work. While obviously Kubrick's film, it's Clarke's as well, and represents a grand synthesis of themes the writer had been working on for decades.

These include the rebirth of the species into a transcendent new form. While it took Kubrick's brilliance to recognize it and make it so, it's no accident that the single most optimistic vision in his entire body of work—*2001*'s Star Child—was Clarke's idea. The alliance between these two gifted, idiosyncratic men during the four years it took to bring *2001: A Space Odyssey* to the screen required great patience and sensitivity on both sides. It was the most consequential collaboration in either of their lives.

CHAPTER TWO

THE FUTURIST

> *For every expert, there is an equal and opposite expert.*
>
> —ARTHUR C. CLARKE

Things were not going well at the Ceylon Astronomical Association. At their meeting on March 19, 1964, Herschel Gunawardene had been, shall we say, less than complimentary about the work that Chandra de Silva had done—for free, mind you—in composing the *Quarterly Bulletin* and getting it into publishable shape. For one, he'd supplied only a percentage of the ream of paper required to print the *Bulletin*—the exact number being subject to some rather heated debate—with the result that Harry Pereira, the vice president of the council, could produce only thirty-five or so copies of it. This was a problem, of course: eighty was the acceptable number of copies, something well established and ratified at council meetings. Adding insult to injury, Herschel insisted that he *had*, in fact, supplied the full ream, insinuating that it must be Harry's fault in some way—either by going to a substandard print shop filled with amateurs, or losing it through some slipup, or via some other wastage that only Harry would know.

For his part, Harry had an entirely different understanding of events. No way was the wastage more than, say, 15 percent, and certainly not the 50 percent of the twenty-five score sheets that Herschel had allegedly supplied. Furthermore, the print shop, when questioned, had stated in no uncertain terms that it had *not* received a full five hundred sheets; the package, rather, had been broken. In Harry's view, for these and other reasons, it was manifestly clear that Herschel was not a fit person to look after and account for Association funds and material, and that his behavior, in attempting to assign blame to Harry and Chandra, was—to quote Harry—quite rumbustious and irresponsible and most unpalatable.

It didn't end there, either. For too long now, Herschel had failed to read the minutes of the council's meetings, the excuse being poor attendance. Most likely, in Pereira's view, he wasn't bothering to maintain them in the first place. And Harry couldn't help but notice that, concerning Herschel's role in the *Bulletin*'s production—evidently a highly dubious role to begin with—he'd seen it fit to insert an appreciation of his, Herschel's, own recent *Booklet on Telescope Making*! Clearly this was a breach of privilege, to say the least. To quote Pereira again, Herschel still had not attained the caliber of an Arthur Clarke to imagine he could use the Association as a platform for personal gain or glory, and if, in fact, he had not received the concurrence of the Association, his conduct must be considered reprehensible.

If we add the irascible Herschel's constant interruptions of Pereira—who, after all, was the vice president, and therefore had every right to chair meetings in the president's absence—the situation was plainly intolerable. And so Pereira, with the provision that it came entirely without any admission of personal guilt in the matter, felt compelled to enclose a sum of five Ceylon rupees, which he believed was the approximate value of the vanished percentage, whatever it might be, of the paper in question. In any case, he was—to quote him again—at best a pseudo–vice president of a spurious council, and was therefore resigning with immediate effect.

The president of the Ceylon Astronomical Association, Arthur C. Clarke, leaned back in his chair and permitted himself a small sigh. He'd only skimmed the letter—he'd been distractedly catching up on yesterday's mail as he waited for today's—but he'd read enough, seated at the desk of his study on Gregory's Road, to indulge himself in a distracted meditation on the differences between interpersonal politics in the British Interplanetary Society—of which he'd been chairman twice from 1946 to 1953—and here in the port city of Colombo, Ceylon's capital.

They certainly existed in the former, of course. The process of preparing *its* journal for print could occasionally lead to certain tensions and misunderstandings. But unless he was letting nostalgia distort his memory, disagreements in the BIS were on the whole less smolderingly resentful, less poisonously conspiratorial—more like parliamentary jousting, really, with a "hear, hear" and a bit of sarcasm—than on this teardrop-shaped island suspended at the mouth of the Bay of Bengal. More Question Time, in other words, than making sausage.

Clarke remembered revising his essay "The Challenge of the Space-ship," first delivered as a lecture at St. Martin's Technical College on Charing Cross Road in the fall of 1946, then published in one of the first postwar editions of the BIS journal. He'd taken it in, after it had been neatly typed by Dot, to deliver personally to founding editor Phil Cleator, whom he'd always seen eye to eye with. As usual his brother's first wife had done an excellent job, flawlessly retyping his scribbled-over copy, and he'd managed to get a seat on the tube, open the folder, and reread parts of the pristine typescript with no small degree of pride. He'd really pulled it off with this one. It was a kind of mission statement or manifesto for a space age then only notional.

> The urge to explore, to discover, to "follow knowledge like a sinking star," is a primary human impulse that needs, and can receive, no further justification than its own existence. The search for knowledge, said a modern Chinese philosopher, is a form of play. If this be true, then the spaceship, when it comes, will be the ultimate toy that may lead mankind from its cloistered nursery out into the playground of the stars. . . . This then is the future that lies before us, if our civilization survives the diseases of its childhood.

Alas, the Ceylon Astronomical Association was no British Interplanetary Society, and never would be, and neither could its *Bulletin* compete with the *Journal of the British Interplanetary Society*. Opening a small drawer in his desk, Clarke deposited Harry's letter with its rumpled five-rupee note in a folder marked "CAA," and turned as his assistant Pauline entered with the mail. Clarke was more than usually alert when the post arrived these days—a kind of compound of trepidation and excitement.

For one, he'd been dealing with a cascade of bad news. His estranged American wife had finally found competent lawyers, who'd gone after Clarke's liquid assets and future income in the United States. Although they'd separated within a few months of getting married, they had never formalized divorce proceedings more than a decade before—a situation he regretted more with every mail delivery. With some scrambling, he'd succeeded in diverting incoming royalties to his UK company, Rocket Publishing, but there was no doubt that Marilyn had put him in a serious bind. Her lawyers had sued him in a New York court for $22,000 arrears of maintenance—about $175,000 in today's dollars—and they'd managed to seize and freeze his American account.

The United States was by far his largest source of income—he was due at the offices of Time-Life in New York later that very month, in fact, for the editing of his upcoming book, *Man and Space*—and although dollars went much further in Ceylon than they did in America, his financial needs had an uncanny way of matching his total income and even extending into his net worth, wherever he lived.

In this, he had a good deal of help from his partner Mike Wilson, whom he could hear on the other side of the door, pontificating rather loudly to a gathered group of Ceylonese filmmakers as he celebrated the release of his second feature film. "It's about boat building and boat racing," Clarke had written a friend, Major R. Raven-Hart, "with asides on the local scene and cracks at Western-oriented culture-vultures. It's called *Getawarayo*, which means, as far as I can gather, much the same as *The Wild Ones*." What he hadn't said, because Raven-Hart would have assumed this, was that all the boats built to illustrate the boat building had been built with Clarke's funds, not to mention the cameras, sound gear, film stock, processing, catering, and salaries. Even in Ceylon, film production wasn't cheap. Still, with luck, they would earn at least some of it back from ticket sales.

Meanwhile, the mismanaged yet voracious Ceylonese tax authorities, operating from an incoherent pastiche of regulations, were effectively holding Clarke to ransom before they would allow him to leave for New York. Foreign residents always had to brace themselves for a potential collision with tax officials when planning a trip, and Clarke, a

Arthur C. Clarke and his partner Mike Wilson with
diving gear in Ceylon in the late 1950s.

resident since 1956, was known to be a man of some means—it was no
good explaining that he'd spent almost all of it on Mike's film projects,
oceangoing dive boats, Land Rovers, scuba gear, air compressors, engine parts, steaks, women, booze, funny cigarettes, and hospital bills.

While his captive-expat state of tax bondage largely had to be dealt
with in person—by showing up in the sweltering halls of the tax office, pleased local barrister in tow, for another exasperating bargaining
session—Clarke still braced himself mentally for more bad news every
time the mail arrived, on account of Marilyn's machinations. And yet
there was more than the normal amount of anticipation as well. Because the game was afoot.

For example, just in the last month, there had been circuitously indirect indications that a young New York director—something of a wun-
derkind, really, by the name of Stanley Kubrick—was interested in talking
to him. And recently Clarke had struck up a rather interesting correspondence with an adjunct professor at Harvard University, an astronomer
named Carl Sagan. Like Clarke, Sagan was intrigued by the prospect of

alien intelligence. Both thought it was certainly out there, somewhere among the stars. Sagan had the rare ability to pontificate about the subject in leading scientific journals without embarrassing himself.

There had been something refreshing about his recent paper in the journal *Planetary and Space Science* on direct contact between galactic civilizations—subject matter that until just recently had been exclusively the preserve of science fiction, here couched in the cool language of a scientific paper. Sagan was only thirty, almost two decades younger than Clarke. He was a complete unknown, and yet he wrote with a cool analytical authority as he advanced the argument that any estimate of the number of advanced technical civilizations on planets of other stars depends on our knowledge of star formation, the frequency of favorably situated worlds, the probability of life originating in the first place—let alone intelligence—and the lifetimes of those civilizations. It was a remarkably assured performance, particularly given the riskiness of the subject.

"Those parameters are poorly known," Sagan had written with considerable understatement. Such back-of-the-envelope calculations had been worked out before, of course, most famously by astronomer Frank Drake as an analytical tool for the first meeting in 1961 of SETI, the Search for Extraterrestrial Intelligence—a meeting that Sagan had attended. After a lot of debate, the SETI group had produced a figure of between a thousand and a hundred million civilizations in the Milky Way alone—a rather wide spread, to be sure, but on the other hand, either figure would be extraordinary. In his paper, Sagan had attempted to refine this, working up an estimate of something on the order of 10^6 extant advanced technical civilizations in the Milky Way alone—a round million—and had even hazarded that the most likely distance to the nearest such community was several hundred light years. Rather close, given the 180,000-light-year diameter of the galaxy.

But it was section three of the paper, "Feasibility of Interstellar Spaceflight," that had really gotten Clarke's attention. After pointing out that if we relied on radio signals to do the job, it would take a thousand or more years for even the most simple exchange of views between galactic civilizations—say, on the inclement weather—Sagan had outlined the difficulty of being on the same wavelength in the first place. Quite literally: the question of what signal frequency to use, after all, was hardly a given even among scientists disposed to consider such questions.

"Finally," Sagan had written, brushing aside the equations for an exhilarating moment, "electromagnetic communication does not permit two of the most exciting categories of interstellar contact—namely, contact between an advanced civilization and an intelligent but pretechnical society, and the exchange of artifacts and biological specimens among the various communities." Clarke found himself nodding in approval at the sentence. He had also been advancing the possibility that contact with an advanced extraterrestrial civilization may have occurred within recorded history.

Actually, Clarke had already mooted most of these thoughts in both fiction and nonfiction, and it's easy to imagine the writer reaching for his IBM Selectric soon after arriving at Sagan's conclusion.

It is not out of the question that artifacts of these visits still exist, or even that some kind of base is maintained (possibly automatically) within the solar system to provide continuity for successive expeditions. Because of weathering and the possibility of detection and interference by the inhabitants of the Earth, it would be preferable not to erect such a base on the Earth's surface. The Moon seems one reasonable alternative.

But in fact, he'd first read a summary of Sagan's automatic lunar base idea in a review written by his friend, the prolific science fiction writer Isaac Asimov, in the September 1963 *Magazine of Fantasy and Science Fiction*. It's what had prompted him to write the astronomer in the first place. If there was one thing Clarke didn't like, it was not being cited when an idea he'd come up with first saw the light of day under somebody else's name—particularly in a scientific journal. Accordingly, that November, he'd written Sagan his first letter.

I was particularly interested in your suggestion that there might already be an automatic base in the solar system. I developed this idea in a short story called "The Sentinel" . . . The analogy I used here was that of the fire alarm, and I suggested that since advanced races would only be interested in a species that had reached a fairly high level of technology, they would put their station on the Moon so that it would not react until we got there.

Clarke found it entertaining, and also redeeming—yet somehow simultaneously disturbing—that this concept, which he had indeed come up with first, was being mooted these days in credible scientific journals rather than in pulp magazines with garish pictures of bikini-clad damsels in distress being rescued by caped crusaders from bearded bad guys wielding bullwhips. "The Sentinel" had been written in 1948 for a BBC story competition, where it failed to even place—he sometimes wondered what *did* win—and then had subsequently seen print in the sole issue ever published of the pulp rag *10 Story Fantasy*. Cover story: "Tyrant & Slave-Girl on Planet Venus." Twenty-five cents, please.

"The Sentinel" had subsequently been published in two of Clarke's anthologies. One had to wonder if scientists were taking material they'd read in their teenaged years—or even last Thursday—and giving it a wash, rinse, and scaffolding of equations before printing it under their own name in top journals. To his credit, though, young Sagan had responded almost immediately with a warm letter referencing two of Clarke's works of nonfiction, *The Exploration of Space* and *Interplanetary Flight*, which had provided him "with some stimulation towards my present line of work."

As for the Kubrick approach, so far it had been frustratingly inconclusive. First he'd received a tantalizing cable on February 17 from his New York friend Roger Caras, a publicist with Columbia Pictures. He'd first met Roger when French undersea explorer Jacques Cousteau introduced them at the Boston launch of Cousteau's 1953 book *The Silent World*, and they'd hit it off immediately. Caras had conveyed that Kubrick was interested in working with Clarke, but thought he was "a recluse." At this, Clarke had laughed hollowly—if the director only knew that he was in effect living in a small beatnik commune comprised of filmmakers, writers, hangers-on, secretaries, servants, boy-

Clarke's short story "The Sentinel" was first published in 1951.

friends, and girlfriends. A gay, bisexual, straight, Anglo, Asian group ensconced in a sunny bungalow in the capital of this former English colony. If a recluse, he'd failed spectacularly.

In any case, he'd sat up straight at Roger's telex, immediately cabling back "frightfully interested in working with enfant terrible." Asking Caras to contact his agent, he wrote, "What makes Kubrick think I'm a recluse?"

Caras's first message was followed by an airmail letter delivered just under a week later, but evidently written the same day as the cable. "I was talking with Stanley Kubrick today, and he indicated that he would like to be in touch with you in the not too distant future," he wrote. "I have taken the liberty of giving him your address, and you will probably be hearing from him." Under Columbia Pictures letterhead and in the practiced patter of a PR flack, Caras continued:

> Since you insist on escaping the rigors of civilization by resid-
> ing in your palm-fringed paradise, it is conceivable you're not
> fully conversant with Mr. Kubrick's latest phenomenal accom-
> plishment. I'm enclosing for your edification a set of reviews
> of Mr. Kubrick's latest film—a true masterpiece with the ex-
> tremely unlikely title of *Dr. Strangelove or: How I Learned to Stop*
> *Worrying and Love the Bomb.*

Mike Wilson's local Sinhalese-language productions aside, Clarke had
been trying to break into film for years, and was very aware that sci-fi
writer Robert Heinlein had done so more than a decade before with
his script for the 1950 feature *Destination Moon*, which had been followed by a novella of the same name. Another friend, Ray Bradbury,
had also done film work, such as writing the script of *Moby Dick* for
John Huston—though he'd had a terrible time of it, because the director had proven to be such an abusive, mean-spirited egomaniac.

Just recently, Clarke had written his writer friend Sam Youd (better known by his pen name, John Christopher), "I remain the most
successful writer in the world who's never had a movie made." True,
his novel *Childhood's End*—which had covered the arrival in giant
starships of an extraterrestrial master race who imposed peace on
warring humanity—had been optioned by neophyte producer Arthur
Lyons in 1958. Lyons had hired screenwriter Howard Koch, an Oscar
winner for *Casablanca*, and Koch had done a creditable job, but the
project had gone nowhere—though allegedly MGM was now looking at it.

As for Kubrick, Clarke had seen *Lolita* at the Regal, one of Colombo's colonial-era movie palaces, and been duly impressed. And
as a daily listener to the BBC World Service, he was well aware
that *Dr. Strangelove* was making waves internationally. In fact, just
days before Caras's letter, he'd received one from his rocket engineer
friend Val Cleaver—designer of the Blue Streak missile that was supposed to serve as Britain's nuclear deterrent but was now suffering a
slow death due to budget cuts. Cleaver had just seen Kubrick's new
film.

Dr. Strangelove is, in a word, a masterpiece. I'd never have believed it possible, and went into the cinema quite hostile and
convinced it must be in the worst possible taste; merely curious
because of the universal critical praise . . . Well, it comes off.
You *must* see it. I guess maybe it was a stroke of genius, and that's
the *only* way to make a film about this subject.

He responded to Roger's letter as well, mentioning that he'd seen *Lolita* and looked forward to seeing *Dr. Strangelove* as well. "Kubrick is
obviously an astonishing man," he observed, before proudly filling in
Caras on Mike's second feature, which was "passed by the censors with
acclamation yesterday, and it's released in twenty cinemas next week.
It's in black and white this time and is a social satire, with an exciting
speedboat race ending, right out of *Ben-Hur*."

Since then, however, not much had happened. It was a bit frustrating waiting for further word, and Clarke had spent some of the
time working on revisions to his book for the Time-Life imprint *Life
Science Library* and filling in his New York lawyer, Bob Rubinger, on
his rather serious case of "matrimonial trouble." Before getting into
how he hoped to end the currently intolerable situation—in which his
literary agent, Scott Meredith, was required by law to put all his earnings into escrow, and his New York account had been frozen pending
a hearing—Clarke explained the background of the case. "To put you
into the picture, I was married in New York in 1953," he wrote. "The
marriage was fraudulently contracted, as my wife did not inform me
that she had undergone a hysterectonomy [*sic*] as a result of her first
marriage: she did not inform me of this until several days after we were
married."

He didn't mention, of course, that there had been another level of
fraudulence at work: namely, that Clarke was gay. In short, a complete
fiasco. In any case, he'd agreed to a separation agreement soon after the
marriage. But he'd done so without legal advice, and had been paying
her a monthly allowance ever since, or, at least, until he had become
"completely incapacitated by polio (or a spinal injury—the specialists
disagree) in March 1962. As my earnings ability was then zero, and I

discovered soon after that my wife had left England and was no longer in a position to harass my family, I stopped payment."

This had led to the current situation, which Clarke was hoping to resolve with a final settlement and, if possible, a divorce. In fact, he wrote Rubinger, there may well be other good reasons to settle.

> There is also a possibility that a really big deal may come up at any time: several of my books are under discussion with film companies, and only last week I heard that Stanley Kubrick (whose *Dr. Strangelove* seems to be breaking all critical records) was anxious to get in touch with me. So I am prepared to consider a reasonable settlement.

Meanwhile, the weeks had passed, and by now he'd almost given up— he knew from experience how tenuous such film-world approaches usually were—when Pauline entered, nodded without a word, and left the latest mail in a neat pile.

Clarke seized it and riffled through, grimly noting one from Rubinger, but then arriving at what he'd been hoping to see for weeks: another New York return address, this one from Polaris Productions Inc., 120 East Fifty-Sixth Street. Could it be Kubrick? He reached for the small Ceylonese dagger he used as a letter opener, slit the envelope from corner to corner with practiced precision, and deftly extracted the paper inside.

· · ·

Years after collaborating with Mike Wilson on three films in the early 1960s, Sri Lankan director Tissa Liyanasuriya retained a vivid memory of Arthur C. Clarke at work in the house he shared with Wilson on Gregory's Road. Liyanasuriya had been assistant director on Mike's first feature, *Ran Muthu Duwa* ("The Island of Treasures"). Packed with action, catchy songs, and underwater scenes, it had been a huge success, with more than a million people seeing it in 1962 and 1963—approximately one-tenth of the country's population. Its songs are still popular today. However, no prints appear to have survived.

Tissa had then worked even more closely with Mike on a second feature, *Getawarayo*. After directing its village scenes himself and also being called upon to take the director's chair by the exasperated producer, Sheha Palihakkara, on the frequent occasions when Wilson left the set early "to relax," Mike had generously insisted that Liyanasuriya receive codirector credit. In part due to this boost, Tissa went on to a long and distinguished directing career. *Getawarayo*, which ended with a prodigious boat race across the glittering green surface of Bolgoda Lake, was released in February 1964. It was also a success, though a bit less so than *Ran Muthu Duwa*. It, too, has vanished.

Liyanasuriya had many opportunities to observe Clarke, who'd provided most of the funding for both films, because the office of Clarke, Wilson, and Palihakkara's production company, Serendib, was adjacent to the writer's study on the ground floor of the house, and when things were reasonably calm, Clarke would leave his door partially open. More than a half century later, Tissa still clearly remembered the *tik-tik-tik* of the typewriter keys issuing forth through that door, with a muted *ding* signaling the end of each line—that long-gone sound of authorial industry. Curious to observe the great man at work, he positioned himself discreetly at the correct angle to peek inside.

"I saw he starts writing like this," said Liyanasuriya, hunching over an invisible typewriter to demonstrate. "And then all of a sudden, he stops writing. He takes the specs, cleans the specs a little, puts them back, and then starts writing again. At the desk. At the typewriter. He was *working* on it. And then what he does is, all of a sudden, he gets up. And then he walks into the garden."

Tall, balding, and with an earnest manner leavened by his quick wit and sense of humor, Clarke had adopted the Ceylonese male habit of wearing a brightly colored sarong with no shirt in the daily tropical swelter. Only he hadn't really gotten the hang of tying it at the waist, instead tucking the cloth over the elastic band of his briefs. As a result, his sarong usually started falling off as he loped to the garden door, and he was forced to grab it and shove it back under the elastic. "He was not used to wearing the sarong," Tissa said, tittering gleefully at the memory.

"And so he goes into the garden, where he has this armchair. He sits on the armchair like this." Liyanasuriya imitated Clarke leaning back, legs akimbo, looking up, lost in thought. "And he's looking at the sky. Thinking. *Thinking*. Looking at the sky for some time; maybe five or ten minutes. And then he gets up, comes running into the room, and starts writing. So I liked this very much. He was very lovable; he was a very lovable person."

. . .

Clarke unfolded the two-page letter, which was dated March 31, and saw that it was indeed from Kubrick. Fairly brief, quite to the point, it seemingly had two clear agendas. One was picking his brain about a possible telescope purchase (the director mentioned a Questar telescope in the first and last sentences). The other was his desire to discuss "the possibility of doing the proverbial 'really good' science fiction movie." This line—the second after the Questar bit—would become well known, and certainly served as the initial aim of the nascent project Kubrick was proposing.

"My main interest lies along these broad areas, naturally assuming great plot and character," Kubrick wrote. "1. The reasons for believing in the existence of intelligent extraterrestrial life. 2. The impact (and perhaps even lack of impact in some quarters) such discovery would have on Earth in the near future. 3. A space probe with a landing and exploration of the Moon and Mars." Caras, Kubrick continued, had made him aware that Clarke was intending to come to New York soon, and he wondered if he might perhaps be able to come a bit early "with a view to a meeting, the purpose of which would be to determine whether an idea might exist or arise which would sufficiently interest both of us enough to want to collaborate on a screenplay." Should that "most agreeable event" occur, he was "reasonably certain" that an understanding concerning Clarke's services could then be reached. And he concluded with his second Questar reference—asking what model medium-sized scope the writer might recommend.

Clarke loved being in the position of an expert, and to be consulted about telescopes by an internationally recognized director—a subject he could discuss all day—was an unexpected bonus to the main thrust of the letter. He didn't necessarily agree that good science fiction movies hadn't been made already but certainly was aware that most were utter bilge, and Clarke was prepared to concede that a *really* good one hadn't been made yet. More to the point, he had been chafing for the chance to break into movies for a very long time. If not now, when—and who better than Kubrick?

Having concluded this rumination, Clarke put a blank sheet of letterhead—"Arthur C. Clarke; Clarke-Wilson Associates," cable address: "Undersea, Colombo"—into his typewriter carriage. After some prefatory sentences covering Roger Caras, his interest in seeing *Dr. Strangelove*, and his having already seen *Lolita*, he continued with information about his arrival in only ten days in New York, and suggested that his work at the Time-Life book department shouldn't interfere with getting together to discuss the proposed collaboration.

He also informed Kubrick that he would have to return to Ceylon "pretty promptly; that is, probably around the middle of June, as I have a big organization here with lots of problems." In fact, he had spent nearly every rupee, pound, and dollar he possessed on Mike's film. "Also, as I owe the local tax department incomputable thousands of rupees, they have only let me out of the country on condition I am back in two months. To ensure this, they have injected me with a mysterious Oriental drug, unknown to Western science, which will cause me to die in convulsions on June 15, unless I am back at the Tax Department (with cheque), to receive the antidote." Having transmitted something of his financial situation and devastating sense of humor, he circled back to the subject at hand.

> As to the main point of your letter, I also feel, as you obviously do, that the "really good" science fiction movie is a great many years overdue. The only ones that came anywhere near qualifying for this were *The Day the Earth Stood Still*, *The Forbidden Planet*, and, of course, those classic documentaries, *Destination Moon* and *Things to Come*. *The War of the Worlds* and *When Worlds Collide* also had their moments, catastrophe-wise.

There: despite the caveats, he'd gone and done it—he couldn't help himself. He'd listed not one, or two, but *six* science fiction films he approved of. (Why he'd identified two of them as documentaries is a mystery; all are features. Clarke's predecessor, the influential British science fiction writer H. G. Wells, had written both *Things to Come* and *The War of the Worlds*.) He continued by citing *Childhood's End*— "which everybody agrees is my best book, and . . . deals with the impact of a superior race on humanity"—repeated that he would really have to "hurry back to Ceylon," and extended an invitation. In the event that they did "get a worthwhile idea cooked up in New York," he hoped Kubrick could come work further on it with him in Ceylon: "We have our own organization here, having produced two films in the last year, including the first Technicolor Sinhalese movie ever made. I think I could promise you a very interesting time."

He closed by advising Kubrick that the Questar was indeed the best small scope, and told him he was bringing his own to New York for reconditioning and would happily show him how to use it. Off the letter went that afternoon, and there it might have remained until their meeting in New York, but that night Clarke fell to thinking. The sun always sets around six in Ceylon, and though he typically went to bed early, that still left a few hours of consciousness, during which he considered the question of making "the proverbial 'really good' science fiction movie" and his potential involvement. Which of his ideas could best serve as basis?

As if on cue, the Moon rose over the palms fringing their property—on its side as always in the tropics, its poles not vertical but horizontal. Clarke could hear a faint Buddhist chanting coming from an open window in a neighboring lot, issuing from some hidden speaker system somewhere—an oddly soothing *sing-song-sing-sing-song-sang-sang*. Practicing Buddhists left this chanting on all night, some of them, a live feed from the temples of Kandy, and there were radio stations that played nothing but. Even if he couldn't share the faith behind it, it didn't bother him. In fact, it humanized the evening.

Anyway, Clarke reflected, at least Buddhism and Hinduism, of all the Earth's religions, had somehow managed to intuit an approximate sense of the vast scale of space and time that science had demonstrated—the infinity on either side of that thin flash of light constituting an individual lifetime. Mike and his wife, Liz, had gone out. Having seen to Arthur's needs, surveyed the scuba tanks, oiled the air compressor in the garage, and wheeled out his trishaw, Clarke's body man Hector Ekanayake had motored off to the gym and was doubtless now pounding on a sandbag with gloved hands. Arthur's loyal shepherd, Laika—named after the dog the Soviets sent into Earth orbit in 1957, never to return—had settled down comfortably on the grass near his feet, with only her ears pinioning occasionally toward some distant car backfire. The house was otherwise silent.

It should really be "The Sentinel," Clarke realized.

. . .

The next day he rose early, made a pot of tea, seated himself back at the typewriter, and inserted another blank sheet into the carriage roller. It was a time of day he valued above all others. The sun hadn't yet risen, but the fox-like Ceylonese fruit bats had all returned to their trees and were hanging upside down, snoozing peacefully like nightmarish Christmas ornaments, their immense leather wings folded.

Kubrick had mentioned the Moon and clearly was interested in extraterrestrial intelligence as the core of his film. His short story "The Sentinel" killed two birds already. Maybe they could add Mars later. Or not. He began to type:

Opening: Screen full of stars. A completely black disc slowly drifts in from the left until it fills the center of the screen, eclipsing the stars. Its right-hand edge lightens, the glow of the corona appears, and the sun rises. As it does so, we realize that we are looking down on the night side of the Moon and are racing around, as if on a close satellite, to the daylight side.

He stopped and meditated for a minute: not too bad. The black disc was a nicely graphic touch. This was a movie, after all.

The crater-filled crescent expands steadily from a thin bow to a half moon. As it does so, voices come up on the soundtrack. They are American, Russian, British—the various lunar bases and exploring parties, talking to each other, asking for supplies, exchanging information, joking, grumbling . . .

Also okay. International competition. Let's keep the Brits in the game.

We hear the end of a countdown, and there is a brilliant, moving glare against the darkened face of the Moon as an outward-bound ship takes off for Mars.

He wanted Mars. I'll give him Mars.

The soundtrack singles out one exploring party, driving a tractor across the Mare Crisium. The camera moves down onto it, and the other voices fade out. We can tell from the rising excitement and incoherent phrases that the survey team has found *something*— something which, even on the Moon, is quite out of the ordinary.

Okay, maybe that was just about enough—a kind of teaser. But what would Kubrick make of it, without any further explanation? Typing a line of dots below the short prose sketch, Clarke added another paragraph, steering the director to "The Sentinel" and describing briefly the essence of the story, in which a survey team discovers a diamond-hard crystal pyramid of alien origin, which has clearly been on the lunar surface for millions of years. When broken open after great effort, its signal to the stars ceases. "Some of the scientists in the resulting debate decide, correctly, that it can only be a monitor—the equivalent of a celestial fire alarm."

Then he wrote a cover letter: "I have thought of a nice opening for a space movie . . . It might lead into a great number of situations, not only the one described in *Childhood's End*."

And he sent that off as well; two letters in two days.

CHAPTER THREE

THE DIRECTOR

You seldom get what you pay for, but you
never get what you don't pay for.

—STANLEY KUBRICK

Ambiguous indications, odd clues, and cryptic hints had been issuing forth from the office of Polaris Productions for weeks. Occasionally they were from Kubrick himself, but more usually they were from his associates. Like tea leaves or magnetic filings, they didn't say much individually. But when collated together, they produced a heuristic pattern—a kind of tracery outlining what was on the director's mind.

On March 10, for example, Kubrick's assistant Ray Lovejoy had written Sky Publishing in Cambridge, Massachusetts, expressing interest in acquiring back issues of *Sky & Telescope* magazine. And on March 19, Lovejoy (whose name was a source of fascination for the director, so much so that he'd toyed with it, early in 1963, finally arriving at an intriguing new compound: "Strangelove") wrote Pocket Books on West Thirty-Ninth Street, asking the company to please send a copy of the Arthur C. Clarke novel *The Sands of Mars* to Polaris.

In between were many phone calls and bookstore visits. Then almost a month later, Lovejoy, who'd served as assistant film editor on *Dr. Strangelove*, reached out once more. This time it was a letter to Cinerama, a company formed in response to television's encroachment into the American psyche—and consequently, into the country's revenue stream, causing Hollywood to writhe in voluptuous distress—resulting in a perceived necessity to up filmmaking's game by increasing resolution, quality, color, and overall splendor. The Cinerama process accomplished this by using larger 65-millimeter negatives and thus bigger 70-millimeter prints, which were threaded through multiple projectors firing away simultaneously at immense curving screens, all of which produced an enveloping, panoramic cinematic experience.

So far, the best example of Cinerama had been MGM's *How the West Was Won*, a production so epic that it had deployed three film directors to follow four generations of settlers as they moved westward in screenings requiring three projectors. Released in 1962, the picture's grand sweep had been a big financial success and had managed to put television in its place—at least for a time. In any case, on April 13 Lovejoy wrote the company asking for tickets "to any preview you may be considering" for an experimental Cinerama short, *To the Moon and Beyond*, at the 1964–65 New York World's Fair. This one, as he understood it, required only one projector with a fish-eye lens, pointing directly upward at a planetarium-style dome.

All these feelers, of course, indicated a thematic thrust—namely, upward, toward the empyrean. Occasionally, however, they were accompanied by seemingly off-topic communications from Kubrick himself. In mid-April, for example, the director responded to a query from a Dutch magazine.

> In answer to your question as to why I produce my own films, I can say straight off that it is much simpler that way. There is an enormous difference between having to convince someone else and being in the situation where someone has to convince you. Even if you have a wonderful producer who understands you and has taste, you are still obligated to spend a great deal of time convincing him of things which may not be immediately apparent to him, when you first bring them up, and also fighting against ideas which he may have or may like and in which presentation you do not share his opinion.

During this period, he usually came into the production office in the late morning or early afternoons, if at all. Despite getting even less sleep than usual, he was in a reliably expansive mood. Under normal circumstances Kubrick slept only four or five hours, with the rest of his time spent engaging restlessly with the world—usually via words, be they printed, over the phone, or even in person. But these days, things weren't normal, because his daughter Vivian, only four, had a seemingly perennial case of croup. This inflammation of the larynx and trachea can be life-threatening in young children, and was exacerbated by the city air, so he and his wife, Christiane, had been taking turns sitting up next to her bed and monitoring her breathing as she slept. You can have the most exclusive address in New York—and the Kubricks had recently moved into a penthouse apartment on Lexington Avenue and Eighty-Fourth Street—but you can't remove yourself from the city air.

Still, Kubrick was in an excellent frame of mind. His film *Dr. Strangelove*, which had been out since late January, was doing exceedingly well, both commercially and critically. On February 5 *Variety*'s front-page headline had been an all-caps affair: "Flash! Stanley Kubrick's Dr. Strangelove Breaks Every Opening-Week Record in History of Victoria Theatre (New York), Baronet Theatre (New York), Columbia Theatre (London)."

Riding that wave, he was casting about for a new project, writing letters, and had even surprised Christiane by being unusually gregarious socially. After arriving at Polaris, having a brief conference with Lovejoy, surveying the latest box office reports, and riffling through his mail, he would typically pull a chair up to the typewriter parked on his banged-up metal desk and fire off brief, pointed messages. Kubrick admired precise language and tried to keep even crucial letters down to a laconic single page. "Any longer, and they'll think you've got nothing better to do," he observed, citing a World War II transmission from a US Navy patrol plane over the North Atlantic as a perfect model: "Saw ship sank same."

On March 31, a Tuesday, he wrote eleven letters, with only the one to Clarke being more than a page. To Michael Vosse in Los Angeles: "I haven't the faintest idea what film I'm going to make next, so it is a bit difficult to talk about casting." To Shael C. Harris of Canada Life Assurance, in Winnipeg: "The name Bat Guano refers to the manure droppings found in bat caves and used as fertilizer. I should say you win your bet." (USAF Colonel "Bat" Guano, played by Keenan Wynn, was a character in *Dr. Strangelove*.)

A week later, on April 6, he wrote *seventeen* letters—possibly a Kubrick record. To his old producer and sidekick Alexander Singer, by then a director himself: "Sue Lyon is supposed to be great in *Night of the Iguana*. You should see Ingmar Bergman's new film *The Silence*. Enough of that. I'm working on a couple of ideas for a film, though I haven't really decided what to do next." To Gilbert Seldes, founding dean of the Annenberg School for Communication at the University of Pennsylvania, responding to an invitation to speak: "I never make speeches or write articles. I like to think that I do this out of humility, but it is probably a form of the most supreme egotism. Seriously, I always feel that there is something not quite right about filmmakers or writers who decide to become critics or lecturers."

Despite the latter, Kubrick was keenly involved in the promotion of his films—just not by putting himself front and center. After moving back to New York in the fall of 1963, he focused on ensuring that Columbia Pictures devote as much time and money to promoting *Dr. Strangelove* as it had to their 1961 release *The Guns of Navarone*—the second-highest-grossing film that year. In doing so, he'd acquired the reputation within the studio's publicity department of being a bit of a control freak. Still, he had time on his hands, and he and Christiane became part of a social circle that included jazz great Artie Shaw, who'd sworn off performing a decade previous and was deep into film distribution and writing fiction; Shaw's wife, the actress Evelyn Keyes; writer Terry Southern and his wife, Carol; and UK director Bryan Forbes, who passed through New York frequently.

One day in the winter of 1963–64, Forbes was visiting, and the subject of a next step after *Strangelove* came up. Kubrick had enjoyed Forbes's praise of his work in the UK papers over the last few years, and he'd appreciated Forbes's films, which included the critically acclaimed *Whistle Down the Wind* and *Séance on a Wet Afternoon*. Though they'd become friends of a sort, Kubrick kept finding the man's judgmental attitudes getting in the way. It was a cold winter, and in his typical state of oblivion over things sartorial, Kubrick had distractedly grabbed a cheap fake-fur nylon cap in a drugstore and parked it on his head—the kind of move the stylish Christiane cringed at and tried to control, usually unsuccessfully.

Despite some well-publicized attempts to render the genre respectable over the previous decade or so, in the early 1960s, science fiction was only a step or two above pornography on the social acceptability scale. It was "little green men stuff," as Christiane recalled. One day Forbes and Kubrick could be seen walking along in a roar of traffic, puffs of condensation dispersing above their heads in the subzero air as they chatted. They may have been discussing the relative merits of filming in London versus Los Angeles—the UK government had put significant financial incentives in place, designed to lure Hollywood productions to Britain, and the country had superb studio facilities. In any case, Forbes, who'd started off as an actor, was speaking in a plummy upper-class diction derived from his attendance at the Royal Academy of Dramatic Art, and Kubrick in his semisardonic outerborough New Yorkese, when Forbes asked what kind of project he was thinking of doing next. Kubrick responded that he was looking into the possibility of doing a science fiction film.

"Oh, Stanley, for God's sake!" Forbes exclaimed, turning. "*Science fiction?* You've got to be kidding." Kubrick observed him dispassionately. No, he was not, he replied. Actually, he was looking at several ideas. By now they were facing each other on the frozen pavement. Forbes, seeing that his counterpart was indeed serious, now scrutinized Kubrick's hat with distaste. "You know, Stanley, you really can't walk around like that," he said.

Kubrick stared at him for a long moment. "You're . . . *unhappy* about my clothes?" he asked incredulously. "You sound like my mother."

On returning home, he described the incident to Christiane. He'd wanted to like Forbes, but he couldn't quite believe what he'd heard. He tried to ascribe it to cultural differences. "I guess it's an English thing," he said.

"No, Stanley," said Christiane. "This is an asshole."

Later, she remembered the falling-out with Forbes as the beginning of a decisive chapter in Stanley's life—the one in which he made *2001: A Space Odyssey*.

.   .   .

During the previous few years, while working on *Lolita* and *Strangelove*—both shot at UK studios—Kubrick had listened to the BBC radio a lot on weekends. In November and December 1961, he'd caught a new science fiction radio drama, *Shadow on the Sun*, and it had captured his attention. Written by Gavin Blakeney and with a lead role played by American expat actor William Sylvester, the plot involved a succession of mysterious events after a large meteorite fell to Earth, something occurring alongside a mysterious darkening of the sun. As temperatures dropped due to the obstructed sunlight, Sylvester and another character discovered that an alien virus had arrived along with the meteor. It caused humans to be impervious to the cold, but it also gradually made them lose all sexual inhibitions.

Although a seemingly hackneyed plot, one of Kubrick's maxims was that good books make bad films and vice versa, and he saw real potential in the story. He was contemplating optioning it, but first wanted a writer to look at it. He was intrigued by its portrayal of a global crisis triggered by the need to stay warm, fused with a sexual frenzy as the characters inexorably spread the alien virus. In the midsixties, the strictures of the Motion Picture Production Code were gradually being loosened, and having been limited in what he could get away with on *Lolita*, his 1962 film based on Vladimir Nabokov's controversial novel, Kubrick was interested in exploring original ways of being sexually explicit on-screen. *Shadow on the Sun* could potentially allow that, while simultaneously exploring a science fiction theme.

In fact, *Dr. Strangelove*'s script had contained a sci-fi framing device until shortly before shooting commenced in 1963. The film's opening credits were supposed to start with "a weird, hydra-headed, furry creature" snarling at the camera under the opening title "A Macro-Galaxy-Meteor Picture." After an effects shot in which the camera moved through stars, planets, and moons, a narrator evidently of alien origin was to have explained that the "ancient comedy" the audience was about to see had been "discovered at the bottom of a deep crevice in the Great Northern Desert by members of our Earth probe, Nimbus-II." At the end, following the film's crescendo of exploding hydrogen bombs, scrolling titles were supposed to conclude by noting that "this quaint comedy of Galaxy pre-History" was "another in our series, *The Dead Worlds of Antiquity*."

Like Stanley, Christiane had been intrigued by science fiction since her primary school days in Germany. If anything, Bryan Forbes's knee-jerk rejection of the genre only increased both of their interest in it, even as they quietly dumped him. Stanley had played jazz drums since high school and still practiced regularly. He'd developed a rapport with Shaw, one of the greatest jazz clarinetists, and Christiane and Stanley would periodically visit Shaw and his wife, Evelyn. Apart from jazz, Stan and Art shared a love of guns—Shaw was nationally recognized for marksmanship—and Kubrick, who had his own gun collection locked away at home, would cast an appreciative eye across his host's instruments and firearms both.

"I was disgusted," Christiane recalled years later, referring to the latter. "I didn't like his face when he played with those things and cleaned them."

During one social call in the winter of 1963–64, it emerged that Shaw was also a science fiction fan. Intrigued, Kubrick told him that he was looking seriously at the genre for material. "I want to make the first science fiction film that isn't considered trash," he said. Describing *Shadow on the Sun*, he told the musician that he was looking for a top writer to adapt the radio play. At this, Shaw suggested he read Arthur C. Clarke—and, specifically, his novel *Childhood's End*. The Kubricks wasted no time getting a copy, which they read beside Vivian's bed while monitoring her breathing.

Absorbing for the first time Clarke's vision of all-powerful extra-terrestrials arriving to intervene in human affairs, Kubrick grew increasingly excited. "You gotta read this," he kept saying, peeling chapters out of their binding and passing them over to Christiane as soon as he finished them. "We had to take turns staying awake," she recalled, "and so we were forever totally overtired, and we were reading these things, and Arthur, we thought, was the ultimate." With a typical paperback price being only twenty-five cents, this literal form of book consumption was common practice in those days. Down finally to the last tattered fragment—in which the human race was reborn as a new species, and the Earth evaporated into vapor—Kubrick read the bio on the back, which located the writer in Colombo, Ceylon.

The next day, he asked Lovejoy to determine the rights status of *Childhood's End*, but it soon emerged that the book had been optioned in the 1950s. There was no way Kubrick was getting the property— not without a lot of money down—and Clarke himself was evidently off in the tropics somewhere, doing God knows what. Distracted by the final stages of *Dr. Strangelove*'s publicity campaign, the director put the matter out of his mind—though he continued plowing diligently through science fiction novels in the evenings.

On February 17, just a couple weeks after *Strangelove*'s premiere, he met Roger Caras for lunch at one of his favorite hangouts, Trader Vic's in the Savoy-Plaza Hotel. A tall Falstaffian presence with a booming, exuberant voice, Caras had been a senior Columbia Pictures publicist for a decade. Given the restaurant's tacky Polynesian motifs—its ersatz grass island house entrance, looming tiki totem poles, and giant clam lighting fixtures—it wasn't the classiest environment, but "at heart, Stanley was a peasant," Caras observed. "He was without pretense. He just happened to be a genius. It was a funny thing that happened along the way."

After chatting about *Dr. Strangelove*'s successful rollout, he asked what the director was thinking of doing next. Kubrick surveyed him with his olive eyes. "You'd laugh," he said, remembering Forbes's reaction. "I don't think so," said Caras—who, recalling the conversation three decades later, commented, "I would never laugh derisively at Stanley Kubrick. Even then, it was clear what he was . . . It revealed itself over the years, but clearly even back then, it showed. And he said,

Stanley Kubrick and Roger Caras.

'Well,' and he had a funny little sparkle in his eyes and watched my facial reactions and determined if I knew what he was talking about. And the expression 'ET' was not as commonly held or used as it is today, because of Spielberg. And he said, 'I want to do a film about ETs.'"

At this, Caras immediately responded with, "Fantastic. As a matter of fact, I'm doing a radio show on extraterrestrials tonight. On Long John Nebel, starting at twelve." Kubrick sometimes listened to Nebel, a popular fixture of late-night radio on WOR-AM New York for decades. The host focused on unexplained phenomena, including UFOs, witchcraft, and parapsychology. Caras explained that he appeared regularly on the show, and that Nebel frequently showed up for only the first hour or so, leaving it to his guests to hold it down for the rest of the night.

Kubrick mentioned that he'd asked his assistant to get him a lot of books, and that he'd been reading everything by everybody. When he started listing authors, the publicist interrupted. "Why are you going through all that? Just hire the best and get going," he said.

"Who's the best?" Kubrick asked.

"I said Arthur C. Clarke," Caras recalled. "And he said, 'Yeah, but I understand he's some kind of a nut; a recluse that lives in a tree in India.'" Years later, he laughed at the memory—and, no doubt, in 1964 also. And he quoted his response: " 'Not really, he lives in Ceylon.' It was then Ceylon, not Sri Lanka. 'He lives in Ceylon, he's not a nut, and has a very nice house there and servants, and a good lifestyle, a driver, and everything else.' And he said, 'Do you know him?' I said, 'Extremely well. Arthur and I have been friends for a number of years.' "

"Jesus, get in touch with him, will you?" Kubrick said.

·   ·   ·

Clarke had experienced even more than his usual difficulty extricating himself from Ceylon. Despite the best efforts of his barrister, the tax authorities had demanded a substantial deposit before he left the country—but with his US accounts frozen and all his cash invested in Wilson's film, he had few funds left. On April 9 he sent an urgent cable to his US agent, Scott Meredith, who immediately intervened with Time-Life, asking for the balance of the book advance. This had been duly wired and deposited—but then the Air Ceylon Comet had experienced mechanical trouble, and the flight had been delayed not by hours but by two days as they waited for jet parts to be flown in from London.

Finally en route to New York, Clarke stopped in London, where he stayed as usual for a few days with his brother Fred and family, and went to see *Dr. Strangelove* at the Columbia Theatre—a modernist pile on Shaftesbury Avenue. Apart from marveling at Peter Sellers's dominance of three roles—including the eponymous scientist, clearly modeled on Clarke's friend Wernher von Braun—he was impressed by the way Kubrick had handled the film's technology. On the evidence, Kubrick put a premium on realism, particularly the interior of the nuclear-armed B-52 bomber. The exterior views, which had been shot in front of back-projected aerial footage of cloudy peaks, weren't bad, either. "Its impressive technical virtuosity certainly augured well for still more ambitious projects," the author wrote a few years later.

Arriving in New York on April 18, Clarke checked in to the Chelsea Hotel. He always stayed at this shabby red-brick structure on West Twenty-Third Street, where he mingled with other regular visitors such as Arthur Miller, William S. Burroughs, Allen Ginsberg, and

Gore Vidal. The Muhlenberg Library was directly across the street, which was useful, and breakfast of a sort could be had at the Automat on the corner of Seventh Avenue. True, the elevators reeked intermittently of pot smoke, and the lobby occasionally resembled a freak show. But nobody paid much heed to alternative lifestyles at the Chelsea, which, among other things, was a gay hangout.

Clarke's meeting with Kubrick was at Trader Vic's the following week, and so he had a few days to shake off the jet lag, buy a Smith Corona electric portable, and see some friends. On Monday he established himself in a "lovely office on the thirty-second floor" of the Time-Life building in Midtown Manhattan and began working with editors there on preparing his book *Man and Space* for print. "It was strange, being back in New York after several years of living in the tropical paradise of Ceylon," he wrote. "Commuting—even if only for 3 stations on the IRT—was an exotic novelty, after my hum-drum existence among elephants, coral reefs, monsoons, and sunken treasure ships. The strange cries, cheerful smiling faces, and unfailingly courteous manners of the Manhattanites as they went about their affairs were a continued source of fascination."

On Wednesday he descended from his glass-walled aerie and walked to the old Savoy-Plaza Hotel on the east side of Fifth Avenue. It was April 22—the day the World's Fair opened. Fresh from his elephants and monsoons, he surveyed the restaurant's faux-tropical décor with some amusement and made his way to the bar. He was early.

Kubrick arrived on time and threaded his way between crowded tables to the author, whom he recognized from his book jacket photo. As they found their table and sat down, Clarke noted "a rather quiet, average-height New Yorker (to be specific, Bronxian) with none of the idiosyncrasies one associates with major Hollywood movie directors, largely as a result of Hollywood movies . . . He had a night-person pallor." The Kubrick of spring 1964 was clean shaven, personable, deadpan humorous, and with "the somewhat bohemian look of a riverboat gambler or a Rumanian poet," as Jeremy Bernstein, the *New Yorker* writer and physicist, would put it a couple years later. After ordering, they began an intensive talk marathon that would last for the better part of the next four years.

One characteristic that struck him almost immediately was "pure intelligence," Clarke wrote. "Kubrick grasps new ideas, however complex, almost instantly. He also appears to be interested in practically everything." Their first meeting, which lasted eight hours, covered such topics as science fiction, politics, flying saucers, the space program, and *Strangelove*. When Clarke told the director that he'd seen the film in London, and that he counted German rocket engineer Wernher von Braun among his friends, Kubrick said, "Please tell Wernher that I wasn't getting at *him*." Later, Clarke would comment, "I never did, because (a) I didn't believe it, (b) even if Stanley wasn't, Peter Sellers certainly was."

The key concepts behind what would become *2001: A Space Odyssey* were born in 1964 from discussions that took place between Kubrick and Clarke with no third person present. They were "in camera," as Caras put it—a Latin legal term meaning "in private" (literally "in chambers") but with the same root as Kubrick's main axe, if we add the word "Panavision" and clamp a Zeiss lens on the front. We know, however, that in some ways the first meeting of Kubrick and Clarke was a kind of perfect match of content meeting ability, and vice versa.

Clarke was then forty-seven and had spent most of his life absorbing everything there was to know about space, cosmology, rocketry, astronomy, futurism, and science fiction. Apart from his own fictional output, which was considerable, he had campaigned vociferously for the expansion of humanity into the solar system in a series of influential works of nonfiction. He was articulate, witty, egocentric—something he managed to go about in a remarkably inoffensive way—and happy to be considered a world authority. Although used to getting his own way when writing, he knew very well that film was a collaborative medium and that the director was boss.

Kubrick, thirty-six, was more than a decade younger and was hitting his peak as a creative force. He was patient, soft-spoken, polite, savvy, trenchant, utterly implacable, and capable of keeping multiple intellectual balls in the air at any given moment. When it came to his absorbing content necessary to create visually compelling, intellectually provocative films, the metaphor almost all his collaborators used was that of a sponge. Still, a sponge is a largely passive object. His wife described a more active process.

"Stanley had a great ability to concentrate, and if somebody knew something that he wanted to know, he would sort of officially suck it out of him!" Christiane said with a laugh. "He was a hungry student of anybody that knew something he didn't know and wanted to know. So it was quite fun to teach him, too, because nobody could pay better attention."

Over a heavy meal of barbecued beef—both were largely carnivorous—they discussed the writer's suggested opening sequence and the possible use of "The Sentinel," which had been reprinted in one of the anthologies Scott Meredith had sent to Polaris a few weeks before. Clarke was disappointed to discover that Kubrick seemed intent on adapting a BBC radio drama for the screen, and wanted him to look at it. After diplomatically hearing the director out as he summarized *Shadow on the Sun*, Clarke expressed the preference that they work on an original story based either on his own concepts, or perhaps ideas they might develop in collaboration.

Three years later, Clarke described their conversation in a draft piece for *Life* titled "Son of Dr. Strangelove: Or How I Learned to Stop Worrying and Love Stanley Kubrick": "Somewhat to my chagrin, I found that Stanley had already acquired an interest in a conventional 'Invasion of Earth' script, and I made it clear that I was *not* interested in working with anyone else's ideas." According to his own notes, however, *Shadow on the Sun* remained in the running until May 2—more than a week and two meetings later—so he may have been less emphatic than he made it sound. In any case, by 1967, Clarke had been persuaded to submit anything he wrote about *2001* to the director for comments, and a typescript of his draft piece with Kubrick's marginalia has that sentence crossed out, with the comment "This is a bit ungenerous and makes me seem a bit of a jerk." (The article was never published because the editors found it too hagiographic.)

After tabling the BBC drama for the moment, they discussed in general terms what the director wanted to accomplish. "Even from the beginning, he had a very clear idea of his ultimate goal, and was searching for the best way to achieve it," Clarke wrote. "He wanted to make a movie about Man's relation to the universe—something which had never been attempted, much less achieved, in the history of motion pictures." Kubrick, Clarke wrote, "was determined to create a work of art which would arouse the emotions of wonder, awe—even, if appropriate, terror."

He was also keenly interested in what Clarke had to say on the topic of UFOs. "When I met Stanley for the first time," the author commented, "he had already absorbed an immense amount of science fact and science fiction, and was in some danger of believing in flying saucers; I felt I had just arrived in time to save him from this gruesome fate."

He surprised Clarke by asking him not to contact Caras for the moment, because he was afraid the publicist would distract them from their work. Somewhat taken aback, Clarke agreed, wondering if this kind of stage management would be an ongoing feature of their collaboration. (He hadn't seen anything yet.) After eight hours of talking, during which they agreed to visit the World's Fair together, they exited the hotel onto a darkened Fifth Avenue. Each was pleased. Kubrick because he'd encountered a knowledgeable, congenial, potentially very useful collaborator, and Clarke because Kubrick's self-evident intellectual capacity and absence of pretension had been something of a revelation.

Many years later, he would describe the director as "perhaps the most intelligent person I've ever met."

■　■　■

That Friday, they convened at the Kubrick penthouse, where Clarke met Christiane and the couple's daughters, Katharina (from Christiane's first marriage), Anya, and Vivian, for the first time. Low ceilinged but huge, their labyrinth of rooms had originally been two apartments before several walls were knocked down. It was entirely surrounded by a patio, above which an incinerator chimney on the building's roof produced a low, choking rumble and continuous rain of fine ash flakes. Inside, protected from the soot, Christiane's vivid oil paintings hung on the walls. The living room had the usual collateral damage from an ongoing churn of children.

Kubrick's study was jammed with recording equipment, amps, speakers, and the like, with a boxy silver-fronted Zenith Transoceanic shortwave radio that he'd been using in an intermittent attempt to gauge Moscow's reaction to the emerging American escalation in Vietnam. "The preoccupation with hi-fidelity equipment and portable tape recorders for dictation purposes is part of my general interest in all things that save time," he commented to the *New Yorker*'s Jeremy Bernstein, whose profile on the director was published in November 1966.

As promised, Clarke had brought his Questar and they mounted it on a tripod, shrugged into their coats, and set it up outside, aiming it at the Moon. It was only four days shy of full, just shaded enough in the west for a clear view of the mountains rimming crescent-shaped Oceanus Procellarum, their shadows spiking dramatically across the vast Grimaldi and Riccioli craters. The sixth test flight of von Braun's Saturn-1 rocket was scheduled to launch in just a few weeks, and though it was only going into Earth orbit, they may have speculated about where NASA would eventually land. (The Saturn boosters would eventually take astronauts to the Moon, but in the spring of 1964 NASA was also still conducting unmanned launches of their two-man Gemini capsules. The Earth-orbiting Gemini program was designed to demonstrate spaceflight techniques necessary for Apollo to reach the Moon and return safely.)

What followed was an intensive month or so of regular meetings during which a great deal of information was exchanged, mostly in one direction: from Clarke's amiably voluble intellect to Kubrick's keenly hungry one. "Every time I get through a session with Stanley, I have to go lie down," he would later comment. No agreement had been made and no deal signed, though things were clearly heading in that direction. Asked years later if Kubrick had dominated their conversations, Clarke recalled them as evenhanded. Pressed on the point by his biographer Neil McAleer, who observed that Kubrick always controlled all aspects of his productions, Clarke said, "I don't control easily."

It sounded good. But that wasn't quite how things transpired.

With the author still busy finishing his book at the Time–Life building, on April 30 Kubrick went alone to Flushing Meadows, Queens, to see a preview screening of *To the Moon and Beyond* at the Transportation and Travel Pavilion of the New York World's Fair. The pavilion featured a ninety-six-foot Moon Dome with a stylized topographical landscape of craters and mountains on its curving roof. The eighteen-minute experimental production had been made using a special Cinerama 360° process involving a fish-eye lens and vertically mounted projector, permitting projection of the 70-millimeter print on the inside of the dome.

As promised, the film featured views of a rugged lunar surface (with the Earth low on the horizon) and a further voyage into the solar system by a large interplanetary spacecraft with a rotating crew compartment. This was followed by animations made by Los Angeles visual effects pioneer John Whitney, including a depiction of the Big Bang, with rings of material expanding into the newly created universe and multicolored nebulae blooming. A spiral galaxy was also seen, gradually forming from a collapsing cloud of hydrogen.

Although these sequences were made with animation techniques and couldn't be confused with photorealistic depictions of the kind he had in mind for his project, Kubrick was impressed with the integrity of the whole production. Examining his comp ticket, he saw the film had been written and directed by Con Pederson and produced by an LA-based outfit, Graphic Films. It didn't say so on the ticket, but the rotating spiral galaxy had been painted by a young employee of theirs named Doug Trumbull.

The following Saturday, May 2, Clarke dropped by the Kubricks again and gave Stanley a primer on positional astronomy—an ancient technique for locating objects in the sky. "Arthur was really like talking to the ultimate granduncle, who will absolutely tell you everything there is to know about science fiction, about science," Christiane remembered. "They went on the roof, and Arthur showed him how

Stanley Kubrick and Arthur C. Clarke with a Celestron
telescope on the patio of the Kubrick penthouse on
Lexington Avenue, New York City, 1964.

to look up certain planets and stars. It's a hard thing to actually focus
on one, and very cold, and we learned a lot from Arthur. We were just
like children."

After retreating from the chill, they resumed discussing their putative project. Clarke repeated his strongly held view that "The Sentinel"
would be an excellent basis for the film, arguing that *Shadow on the
Sun* wasn't a story designed to feature space sequences but was closer
to a *War of the Worlds*–type drama—one entirely set on Earth. Kubrick
finally agreed to scrap the radio play and focus on expanding Clarke's
short story to feature length. One early question was whether the discovery of the alien artifact on the Moon would come at the climax or
merely be one aspect of the plot. If not positioned at the conclusion,
how would they end their film?

"I was working at Time-Life during the day and moonlighting with
Stanley in the evenings, and as the Time-Life job phased out, Stan-

ley phased in," Clarke told Jeremy Bernstein in 1969. "We talked for weeks and weeks—sometimes for ten hours at a time—and wandered all over New York." They also screened many films, including *Destination Moon*, *The Day the Earth Stood Still*, and *Forbidden Planet*. Kubrick soon discovered that Clarke was far too forgiving of material he found embarrassingly shallow and badly made. When the author insisted he screen the 1936 British science fiction classic *Things to Come*, written by H. G. Wells and based on a number of his stories, Kubrick "exclaimed in anguish, 'What are you trying to do to me? I'll never see anything you recommend again!'" Clarke remembered in 1972.

Although he hadn't been feeling well—he had a cough and felt a general lack of energy—Clarke went to Washington, DC, from May 11 to May 13, where he dined with most of NASA's top brass. Noting without rancor that the space agency's top speechwriter had been "busily cribbing" from his 1962 book *Profiles of the Future* to the benefit of NASA administrator James Webb, he was especially tickled by Apollo project director George Mueller's soliciting his ideas on what NASA should do after landing on the Moon. Back in New York, he wrote Mike Wilson, "Still spending every spare minute with Stanley K., trying to get basic story worked out. Think we have it nearly O.K. now, but there's still no definite commitment. Keep your fingers crossed . . ." He also reminded Wilson that the overdraft on their joint account was due and asked his partner if he was in a position to put some money in it.

Around this time, Kubrick grabbed a tape recorder and hopped a cab over to Joseph Heller's apartment at 390 West End Avenue, near the American Museum of Natural History. He was a confirmed admirer of Heller's novel *Catch-22*, which, apart from its nonlinear structure—the book's various story lines were woven skillfully out of sequence—possessed the kind of "nightmare comedy" effect he believed he'd achieved in *Dr. Strangelove*. A film that Heller, in turn, had much appreciated. While their purpose in taping their conversation that day is unknown, it was a revealing one.

Sitting in the very foyer where Heller, then forty-one, had written

Catch-22, Kubrick said, "A very good plot is a minor miracle; it's like a hit tune in music."

"That's very good," Heller responded.

"In *Aspects of the Novel*," Kubrick continued, "E. M. Forster talked about how regrettable it is that you have to have a plot, but how necessary."

> How the first cavemen sat around a fire, and if the storyteller didn't hold their interest, they went to sleep or hit him with a rock. But you pay a terrific price for a good plot, because the minute everybody's sitting there wondering what's going to happen next, there isn't much room for them to care about *how* it's going to happen or *why* it happened. One of the neatest tricks is not to have a good plot and yet to sustain interest either by dealing with something incredible and making it realistic— which is where the surrealism and fantasy and dreamlike quality of your book take over—or getting so close to the heart of a fact or character that they're held quietly even though their pulses aren't pounding.

Heller agreed. "You really succeed as a creator when you win the audience on your own terms," he observed.

Kubrick continued: "Because of the film form and its ability to generate a lot of emotion, there's what you might call the no-plot story or the antiplot film. Once they get under your skin, you're vibrated by the subtlest kind of vibrations. It works especially in films, but it works in books, too."

. . .

During their walks around New York—which extended across Central Park, to the Guggenheim Museum, and on to the East River, spanning "restaurants and Automats, movie houses and art galleries"—Kubrick described the Cinerama format to Clarke. So high resolution that it was capable of taking the audience on a kind of voyage, the name fused the words "cinema" and "panorama." Cinerama films were released first in the major cities as "road-show" releases, with reserved seating, a printed program, and an intermission, as in the theater. People even dressed up to go to road-show screenings.

Having discussed the matter at some length with his cinematographer friend Robert Gaffney, Kubrick wanted their film to be a Cinerama production—but made using the newer process, which required only one 65-millimeter camera and a single projector. He mentioned *How the West Was Won*, a film in which multiple generations of characters are seen expanding westward. At almost three hours long, it was the last of MGM's big-budget epics to succeed commercially, and Kubrick thought it was well worth evaluating as a template, particularly given the emergent scope of their own enterprise. With five major sections and an epilogue, and without a single dominant character, it was more a docudrama—albeit a spectacular one—than a drama in the conventional sense.

Clarke agreed that the struggle of pioneers to establish themselves on other worlds would have a kind of futuristic, space-age echo of the opening of the American West. But in his view, interplanetary travel was the only form of conquest still compatible with civilization. In the previous two weeks, they'd occasionally touched on the problem of what to name their space epic. Now they decided that their private title would be *How the Solar System Was Won*. "What we had in mind was a kind of semidocumentary about the first pioneering days of the new frontier; though we soon left that concept far behind, it still seems quite a good idea," Clarke wrote in 1972.

As they continued their discussions, the author undoubtedly referred to an idea he'd already presented in print: that the true parallel between humanity's leap into space and something from history actually went back much further than the opening of the American West, or Christopher Columbus, or even Odysseus. Space travel, Clarke contended, would provide an evolutionary jump as significant as that of life moving from the sea to the land—only by contrast, expansion into space was a conscious move by a sentient species. "We seldom stop to think that we're still creatures of the sea, able to leave it only because from birth to death we wear the water-filled space suits of our skins," he had observed. "Only the creatures that dared to move from the sea to the hostile, alien land were able to develop intelligence. Now that this intelligence is about to face a still greater challenge, it may be that the Earth is no more than a brief resting place between the sea of salt and the sea of stars."

Another point Clarke made in their wanderings across Manhattan was that the first tool users weren't humans but prehuman primates—and their tool use doomed them. Because even the most primitive tools forced their users to develop manual dexterity, and even change their postures—for example, by walking upright. "The old idea that man invented tools is misleading, it's a half-truth. It's more accurate to say *tools invented man*," Clarke wrote in 1962. "They were very primitive tools, in the hands of creatures who were little more than apes. Yet they led to us—and the eventual extinction of the ape-men who first used them."

■　■　■

After four extended meetings and a dozen or so phone calls, the shape of the movie was emerging "from the fog of words," as Clarke put it. On Sunday, May 17, he returned to the Kubrick apartment, and the director said he was prepared to make an offer for work on a film that he estimated would take about two years to complete. Polaris Productions would option six of Clarke's stories, which would be linked together to serve as the movie's backbone. The film he wanted to make would encompass the next half century or so—a span comparable to *How the West Was Won*—covering the first chapter of the space age and culminating in the first contact with an extraterrestrial intelligence.

Clarke would have to defer his dream of returning to Ceylon anytime soon, and Kubrick's close associate, the Los Angeles lawyer Louis Blau, was prepared to talk terms with Scott Meredith. Polaris would offer a weekly salary while Clarke worked on the script. Blau had the specifics, Kubrick said, but he hoped Clarke would be satisfied. He envisioned that the writing would take from fourteen to twenty weeks, with a minimum payment guarantee for fourteen in the unlikely event they were done before.

Clarke, who sorely missed Mike Wilson, his young friend Hector Ekanayake, and his life in Ceylon, agreed reluctantly to remain in New York for much of the rest of the year. They discussed the additional stories they'd contemplated as constituent elements of a film based on an expanded "The Sentinel." "Before Eden" was a vignette concerning the discovery of a primitive, ground-hugging form of life on Venus by a scouting expedition from Earth. When the Earthmen departed, they left a buried bag of refuse behind. The Venusian organism proceeded to extend tendrils into the garbage, absorbing "a whole microcosm of living creatures—the bacteria and viruses which, upon an older planet, had evolved into a thousand deadly strains." The story concluded, "Beneath the clouds of Venus, the story of Creation was ended."

"Into the Comet" involved the penetration of a comet by a spaceship that got stuck within its fuming shell of gas and dust due to a computer malfunction. Using homemade abacuses, the crew calculated their way out of danger. "Breaking Strain" concerned the effects of a meteor strike that had punctured the oxygen tanks of a nuclear-powered interplanetary spacecraft, leaving only enough breathable air for one of the two men on board to survive until their destination. The ship was described as having a dumbbell shape, with one end being a large sphere containing living quarters, connected via a long cylinder to the propulsion system, with the latter kept at a distance from the former due to the radiation. "Who's There?" concerned an apprehensive astronaut in a space pod—a kind of one-person spacecraft—growing increasingly anxious as faint stirrings made him start to suspect that the pod, previously used by a deceased crew member, was haunted. After a warm, fuzzy sensation on the back of his neck produced the shock of his life, he discovered that what he'd taken to be a ghost was actually one of several squirming, weightless kittens that the ship's cat had deposited inside.

Neither Kubrick nor Clarke was satisfied with their working title, but they weren't yet sure enough of their film's identity to come up with anything better. Clarke's fifth story, "Out of the Cradle, Endlessly Orbiting," was first published in 1959 in the *Dude* magazine: a midcentury rival to *Playboy* with literary pretensions and plenty of cleavage. It began with a curmudgeonly protagonist meditating on dates from his position in a lunar base.

> Before we start, I'd like to point out something that a good many people seem to have overlooked. The 21st Century does not begin tomorrow; it begins a year later, on January 1, 2001. Even though the calendar reads 2000 from midnight, the old century still has twelve months to run. Every hundred years we astronomers have to explain this all over again. But it makes no difference, the celebrations start just as soon as the two zeroes go up.

It ended with the narrator shedding his sour tone to describe "the most awe-inspiring sound I've ever heard in my life. It was the thin cry of a newborn baby, the first child in all the history of mankind to be brought forth on another world than Earth." The story's title referenced the visionary Russian rocket scientist Tsiolkovsky's pronouncement, "Earth is the cradle of the mind, but humanity can't remain in its cradle forever."

And finally, of course, there was "The Sentinel," which they'd already agreed would serve as a major building block of their film. Referring to its eponymous alien artifact, the story concluded ominously.

> Now its signals have ceased, and those whose duty it is will be turning their minds upon Earth. Perhaps they wish to help our infant civilization. But they must be very, very old, and the old are often insanely jealous of the young.
>
> I can never look now at the Milky Way without wondering from which of those banked clouds of stars the emissaries are coming. If you will pardon so commonplace a simile, we have set off the fire alarm and have nothing to do but wait.
>
> I do not think we will have to wait for long.

One story neither of them thought to include in their discussions that May was "Encounter in the Dawn." First published in the magazine

Amazing Stories in 1953, the tale tracked an alien survey expedition as it arrived at prehistoric Earth, where it discovered a primitive hominid tribe in some respects not unlike themselves, only at a far earlier stage of evolution—"savage cousins waiting at the dawn of history." One of the three-member crew, Bertrond, befriended a hunter named Yaan, described as "wearing the skin of an animal and . . . carrying a flint-tipped spear." The galactic empire that had sent them on their mission appeared to be in trouble, however, and their ship was recalled before their interstellar anthropology was complete.

"I'd hoped that with our knowledge we could have brought you out of barbarism in a dozen generations," said an agitated Bertrond to an uncomprehending Yaan, "but now you will have to fight your way up from the jungle alone, and it may take you a million years to do so." Leaving him with some tools, including a knife—"it will be ages before your world can make its like"—Bertrond and his friends withdrew into their vessel and rose "no more swiftly than smoke drifts upward from a fire." Watching the alien ship turn into "a long line of light slanting upward into the stars," Yaan understood dimly that "the gods were gone and would never come again." Behind him, a river could be seen winding through fertile plains "on which, more than a thousand centuries ahead, Yaan's descendants would build the great city they were to call Babylon."

. . .

After shaking hands on their deal, or at least their intention to negotiate one, Clarke and Kubrick repaired to the patio. They had established a real rapport over the past month, and any guardedness had long since dropped. Both were excited and didn't mind showing it. It had been a beautiful late-spring day, with temperatures reaching 75 degrees, and was now a perfectly mild evening, with a crescent Moon hanging in a slight haze several degrees above the southeastern horizon. Thankfully the building's heating system had been switched off weeks before, and the ash-spewing chimney was now silent. To the south, all of Midtown Manhattan was spread out before them, its lights winking.

Suddenly they noticed a brilliantly bright, unwavering point of white light rising above the horizon in the southwest. Radiant as a navigational beacon, it climbed steadily in the night sky. Clarke had seen the *Echo 1* satellite from Ceylon many times, but this object seemed far brighter. After about five minutes, it had climbed to the absolute zenith of the sky—and there it appeared to stop. A sense of awe and exhilaration filled both of them. "It's impossible," Clarke sputtered. "At its closest point, an artificial satellite has to be moving at its maximum apparent speed!" A thought flashed through his mind: "This is altogether too much of a coincidence. *They* are out to stop us from making this movie."

Belatedly, they hastened indoors, grabbed Stanley's new Questar, and hauled it up a set of metal stairs to a higher perch on the building's roof. The object still appeared to be almost vertically overhead. Fumbling with knobs, the tripod planted on roof tiles, Clarke managed to get it into the telescope's field of view. It remained, however, simply a brilliant point of white light, with no visible dimensions. They took turns viewing it as it gradually descended toward the northeast, passing through Ursa Major—the Great Bear—before finally winking out in the horizon's haze. The whole episode had lasted no longer than ten minutes. "That's the most spectacular of the dozen UFOs I've observed in the last twenty years," Clarke said, his voice wavering.

Descending excitedly back into the living room, they located the *New York Times* and paged back to the Visible Satellites table the paper had started printing a few years before. In the early 1960s, NASA had launched a pair of gigantic inflatable satellites meant to serve as passive reflectors of microwave and communications signals. *Echo 2*, launched into a polar orbit on January 24, 1964, was 135 feet in diameter and surfaced in bright reflective aluminum. But no transit was given for it at nine o'clock at night—though the slightly smaller *Echo 1* was listed as passing overhead at eleven and then again at one in the morning.

Although Clarke had tried to persuade Kubrick that most UFOs had a rational explanation and shouldn't be taken as evidence of extraterrestrial intelligence, Kubrick had reserved judgment. Now that he'd actually seen one, however, he felt vindicated. As for Clarke, he was genuinely shaken. The most confounding thing, he observed, was the object's apparent motionlessness at the zenith. This defied all logic. It simply wasn't how any satellite should or could behave.

At eleven they went back on the roof and had an excellent view of *Echo 1*, which rose exactly as predicted, looking almost identical to what they'd seen previously. It didn't stop at the zenith, however, but just kept motoring along, inscribing a seemingly perfect, geometrically straight line across the heavens. "So Stanley had seen his first artificial satellite," Clarke wrote, "and was duly impressed. The nine o'clock apparition, however, still remained a complete mystery." While he was sure there must be an explanation, he couldn't think of one.

They discussed filing a sighting report with the US Air Force, but Kubrick was reluctant. He believed the USAF must still be irritated by its satirical portrayal in *Dr. Strangelove* and would regard the whole thing as a publicity stunt. Eventually they did send in a report, however, and meanwhile, Clarke asked his contacts at the Hayden Planetarium to look at its satellite transit time database. In the end, both the air force and the Hayden reported back that *Echo 1* had indeed transited over New York at nine, with subsequent passages at eleven and then one o'clock. The "satelloon" had an orbital period of only 118 minutes—almost exactly two hours—which conformed perfectly to what they'd seen.

"Why this wasn't listed in the *Times* was the only real mystery involved," Clarke concluded. "The illusion that it had hovered for a while at the zenith resulted from astronomical, meteorological, and— let's admit it—psychological factors." Later, he cited the absence of reference points in "the brilliantly moonlit sky" as one of them—though the Moon was less than half full at the time.

Whatever the explanation, it was an appropriately strange, and strangely well-timed, incident to mark the formal start of what would become a landmark alliance.

. . .

Over the next few days, they conducted a contract negotiation that in many ways would set the tenor of their work together. As with everything else in his professional life, Kubrick was scrupulous about such matters and unwilling to give anything away that he didn't absolutely have to. In Louis Blau, he was working with an accomplished entertainment attorney, whose other clients included such figures as Lana Turner, Walter Matthau, and François Truffaut. Although he trusted Blau, and had made him president of Polaris Productions, Kubrick usually closely monitored and even micromanaged what the lawyer was doing.

By contrast, Clarke didn't like to get too involved in such negotiations, preferring to leave them to Scott Meredith. Unfortunately for the author, despite Meredith's nearly twenty years of experience—not to mention his having already roped together Peter George and his novel *Red Alert*, the basis for *Dr. Strangelove*, with Kubrick via Blau— the agent found himself outmatched and outmaneuvered.

It was in Kubrick's interest to keep Clarke effectively subject to his will throughout their work together. One way to do that was to persuade him to in effect become a salaried employee of his production company. While the contract they eventually signed included several lump-sum payments—including a $10,000 flat fee to option Clarke's six stories—they came due either upon commencement of principal photography or on October 1, 1965, whichever came first—realistically, more than a year later, either way. Having agreed to defer his return home in order to work with Kubrick, the only money that Clarke—a renowned author with some twenty books to his credit—would be receiving for his services was $1,000 per week for the duration of his writing efforts on a screenplay, with a fourteen-week minimum guarantee. There was no mention of a novel based on their screenplay, or vice versa.

While the contract promised that Polaris would try to get whatever studio backed the film to hire Clarke as a technical consultant for the duration of the production, there were no guarantees. The total sum he would receive depended on a number of imponderables, including how many weeks they took to write the script, but the minimum guaranteed figure was $38,000. Through Blau, Kubrick argued that the figure and the deal structure were justified because so far, all the money had come from his own personal funds, and he wouldn't see any return on his investment until the script was in shape to sell to a studio. With the additional money in consultancy fees, the Kubrick side contended, Clarke should make something like $100,000.

The most critical element missing from the offer, however—and certainly its Achilles' heel from Meredith's and Clarke's standpoint—was that Polaris offered no percentage whatsoever to the writer. Absolutely no points, be they of net or gross, and therefore no stake. Whatever may be said about Meredith's negotiating skills, or, for that matter, Clarke's tendency toward acquiescence, it's hard to look at this as anything other than a failure of omission on Kubrick's part—even a moral deficit. In the early sixties, major studios such as MGM had inflexible ground rules disallowing the sharing of a film's profits with their writers. Kubrick and Blau may even have used this as justification during the negotiation. But no such rules existed with the independent producers that the studios were increasingly relying on to make films for them. And Clarke was hardly a studio rewrite man. He was a leader in the science fiction field, and his ideas would be foundational to the project. That much was already clear.

In a letter to Wilson on May 22, Clarke tried to put a gloss on it. "The deal is a complicated one, settled after a certain amount of haggling," he wrote. "I can see Stan's point of view, as at this stage all the money put up is his, and there will be nothing to show for it until we are well into the script. However, I can't think of anyone who will have less difficulty finding a distributor—he'll have to beat them off at the door." Clarke's ability to see Kubrick's point of view would be a recurrent feature of the next four years, even if it would repeatedly be tested to the breaking point.

He described the director's mood. "Stan, who is a ball of fire, and mad keen on the project, wants to shoot for release at Christmas '66. So we have to get cracking on the script immediately." When the first draft was done, he wrote, he would try to come back for a visit. He also wanted "you and Hector to come over at separate times during the production, so that I won't feel too homesick." He would make an effort to get Mike a position as script advisor and "do my damnedest to get you a credit line somewhere." But he instructed Wilson not to let such possibilities interfere with his own filmmaking activities and also asked him not to talk about the agreement, because his attorney was on the verge of making a settlement offer to his wife, Marilyn. "If she finds out about this deal, she'll skin me."

The longest paragraph ruminated on the implications for Clarke's career.

> You realize just as well as I do (perhaps better) what all this means. It's simply impossible to imagine a more effective way of promoting my name. And on top of the *Time* book . . . I've now been commissioned by *Life* to do a major piece on Comsats [communications satellites] in a special Space double issue . . . What with *Time*, *Life*, and Kubrick, I seem to have reached the top of three fields simultaneously; of course, I still have to deliver for the last, and there are many possible snags, but with him as a collaborator, I feel absolutely happy. His technical skill and artistic sense are both uncanny.

And yet, Kubrick's technical skill extended to producing his own work—and producing is about using whatever instruments are available to control all aspects of a film's creation. Although Clarke seemed largely unaware of it, his letter's conclusion reflected the deftness with which this had been brought about in his case: "Will try to get some money across soon—but oddly enough, this deal will force me far into the red, since I won't get most of the cash for many months, and will have to borrow heavily at once to make a settlement with M. (Assuming she is prepared to be reasonable.)"

Even as he proceeded to sign what would become by far the most important deal of his professional career, Clarke's expensive personal entanglements and high-stakes negotiation miscalculations were putting him into debt. Oddly enough.

CHAPTER FOUR

PREPRODUCTION: NEW YORK

SPRING 1964–SUMMER 1965

> *You can never have enough information, and*
> *you can never ask enough questions.*
>
> —STANLEY KUBRICK

For the rest of the spring and into summer and fall, Clarke and Kubrick continued meeting so regularly that their days and discussions effectively blurred into one extended brainstorm that seemed to span the seasons, because it did. "Stan's a fascinating character—I've been living with his family, very nearly, ever since reaching New York," Clarke confided to his writer friend Sam Youd on June 19. When the constant parade of kids and dogs tromping through their deliberations got to be too much, they attempted a strategic retreat to the Polaris office, which had moved to Central Park West at Eighty-Fourth Street—a fairly short latitudinal traverse across the park.

It didn't take. Stanley grew restless away from his kids. "When your children are young, there's always one that has a horrible cold," Christiane recalled. "There's another one that has *this* and another one that has *that* . . . and he would constantly phone up, 'Is she all right? How's she been? I don't like this doctor, get another one.' That family involvement for the outsider is hugely boring and interfering and really just not on. And Stanley didn't play that game; he just said, 'Sorry,' you know, 'I have to do it.' And Arthur, he's a practical person, he said, 'Let's go to your flat, and then you can see if she's throwing up again, or not!'"

The schedule they'd drawn up imagined three and a half months for the script, a brisk two weeks of consultations, a couple months to complete script revisions and find studio backing, just a month to ready the visual identity of the film—a particularly naïve detail, in retrospect—and then twenty weeks of shooting. Following that, another twenty weeks of editing was envisioned, and about three months to prepare for the picture's release, with a premiere to take place around Christmas 1966—or two years in total. It was, as Clarke commented later, "hilariously optimistic."

With his book heading for the printer, Clarke was more available, and Kubrick attempted to install him, at least, in Polaris. With the director himself rising late and working at home much of the time, the office had been left largely to Lovejoy and a secretary and was underutilized. On the way there on the late morning of May 26, Clarke steered the director to the center of Central Park's Great Lawn, scanned the sky intently, and then pointed out a fleck of white light nearly lost in the pale blue. It took a minute, but Kubrick too caught a glimpse of Venus—faint but discernible. The planet can be seen even in broad daylight if you know where to look. Suddenly the universe was there, beyond a thin skin of air.

After being shown his desk, Clarke joined Kubrick and Artie Shaw for lunch. The musician was thrilled to meet one of his favorite authors, and Clarke discovered that Shaw had "put Stan onto me." Clarke lasted only a day on the Upper West Side, however, before retreating to the Chelsea, where he proceeded to generate about a thousand words a day. They emerged as prose, of course—his natural medium—and not in the rather arcane, cryptic format of the shooting script, something Clarke had never written before.

Kubrick intuited that this might be a problem and proposed that they let the film's structure emerge in this way first. "We will not sit down and write a screenplay, we will sit down and write a novel," he announced one day. "We'll get much more depth." They hadn't really settled on a firm story yet anyway, and it would allow their imaginations freer rein. They could turn it into a shooting script later. Clarke was relieved to accept, later quoting British novelist John Fowles on the subject: "Writing a novel is like swimming in the sea; writing a film script is like thrashing through treacle." A couple years later, Kubrick detailed his reasons for proceeding in this way.

> I decided trying to do an original story in screenplay form was, in a sense, putting the cart before the horse. In order to write a good screenplay, you have to not only know what you want to say, know what your ideas are, but you've got to figure out how to dramatize the idea. In other words, how to make the idea implicit in the action or the emotional structure of the story. While you are still trying to work out the ideas, it is rather paralyzing to, at the same time, have to work out how you will dramatize them. You are drawn to scenes which suggest themselves and seem interesting, but which may not really be accurate as far as what you are trying to say. On the other hand, if you work in the novel form, it is not necessary to be concerned with the method of dramatizing the idea. Then, when the story is written and everything is there, you can start all over again, to synthesize the dramatic structure from the much looser and freer and longer form of the novel. This may be why so few good original screenplays are ever written.

Of course, having to "start all over again" isn't necessarily the most efficient use of a director's time. In many ways, though, it would become the modus operandi of Kubrick's and Clarke's incipient project.

One of their earliest conundrums was how to portray the aliens that they thought would inevitably have to make an appearance at the film's climax. On May 23 Clarke wrote a friend, the brilliant geneticist and biologist J. B. S. Haldane: "It is extremely difficult to represent any alien on the screen without either scaring, or amusing, an audience. But unless you show *something*, people will feel cheated. Of course, a really advanced ET might be wholly inorganic, which helps solve the problem." The comment would prove prescient.

Late May and early June were spent exploring some of the implications of inorganic versus organic extraterrestrials. On May 28 Clarke suggested they "might be machines who regard organic life as a hideous disease. Stanley thinks this is cute and feels we've got something." The idea was soon dropped. In discussion three days later, they came up with a "hilarious idea we won't use. Seventeen aliens—featureless black pyramids—riding in an open car down Fifth Avenue, surrounded by Irish cops." Although the *form* of what would eventually become *2001*'s monolith hadn't yet been established—and wouldn't for more than a year after much shape-shifting—its color (black) and quality (featureless) had already emerged. Two important elements retained from a brief, seemingly throwaway word sketch.

As for *organic* aliens, Kubrick and Clarke soon found themselves locked in a good-natured but contentious dispute. In his 1953 story "Encounter in the Dawn," Clarke had envisioned that both the superior alien race and the "savage" terrestrial one were essentially human in form: "As she must do often in eternity, Nature had repeated one of her basic patterns." But his views had since evolved, and he now thought exactly the opposite: that aliens were far more likely to be extremely different from us, and perhaps unimaginably so. Kubrick, however, was essentially of Clarke's original opinion, and was particularly intrigued by Swiss sculptor Alberto Giacometti's weirdly spindly, elongated human forms, which he was contemplating using as a kind of alien template. And, of course, bipedal humanoid extraterrestrials would be much easier to depict in a film.

Clarke, who'd been looking for an excuse to meet Carl Sagan anyway, proposed that they invite the astronomer down from Harvard to adjudicate their disagreement. On Clarke's recommendation, Kubrick had read Sagan's paper on extraterrestrial contact, and he immediately agreed. A dinner invitation was extended, and the meeting took place on Friday, June 5, at the Kubricks' penthouse.

It's a nice tableau to contemplate: the three of them, each already a major player in his respective world or soon to be one, seen in the amber light of a vanished summer evening. They're gathered around the table, the dishes cleared away, the ideas flowing. Manhattan glints through the windows, and the kids have been put to bed. But actually, it wasn't a meeting of the minds, or even an exchange of views, that anyone apart from Sagan remembered later with any satisfaction.

Sagan's take on the dinner, written a decade after, came across as both self-regarding and self-aggrandizing. After hearing out the two sides, he came down in favor of Clarke's position: "I argued that the number of individually unlikely events in the evolutionary history of man was so great that nothing like us is ever likely to evolve again anywhere else in the universe." He proceeded to take credit for much of the film's tantalizing opacity: "I suggested that any explicit representation of an advanced extraterrestrial being was bound to have at least an element of falseness about it, and that the best solution would be to suggest, rather than explicitly to display, the extraterrestrials. The film . . . opened three years later. At the premiere, I was pleased to see that I had been of some help."

In fact, it opened four years later, and a number of other details cited in Sagan's account are demonstrably untrue. For example, his riff on an early title, *Journey Beyond the Stars*: "The film's title, by the way, seemed a little strange to me. As far as I knew, there is no place beyond the stars. A film about such a place would have to be two hours of blank screen—a possible plot only for Andy Warhol. I was sure that was not what Kubrick and Clarke had in mind." That rested on shaky ground, because at the time of their meeting, it hadn't been thought of yet. When it eventually was, almost a year later, neither Kubrick nor Clarke was satisfied with it, viewing it as just a stopgap conceived for an MGM press release rather than the film's actual title.

Sagan also wrote that the movie was "nearing studio production in England," and yet "this fairly important plot line—the ending!—had not yet been worked out by the two authors." The ending would indeed remain unresolved for years, which is probably why he remembered things that way. But as they sat around Kubrick's table early in the summer of 1964, they were a good year and a half away from production, however optimistic the director may have been about it. He hadn't even made a deal with a studio yet. Sagan's rhetorical disbelief was off base. He had, after all, been invited to a working meeting at an early developmental stage, the purpose of which was to solicit his ideas regarding exactly the final scene that left him so incredulous.

Throughout the meal, Kubrick had been solicitous of Sagan, asking for his views and listening politely. He even agreed to the suggestion that they reconvene the following day to resume the discussion. But actually, he'd been irritated by what he saw as the young astronomer's supercilious, patronizing manner. After seeing his guests to the door, he waited an hour and then called Clarke at the Chelsea. "Get rid of him," he said. "Make any excuse, take him anywhere you like. I don't want to see him again."

Kubrick proceeded to ignore Sagan's opinion by engaging in multiple attempts over the next four years to depict extraterrestrials in the film, just as Clarke wrote thousands of words describing them. If they ultimately settled on ambiguity—on something both black and featureless—it was a decision arrived at over time and through their own efforts, and can't be ascribed to Sagan's views, however correct they may have been at the time.

·   ·   ·

In his May 23 letter, Clarke had written Haldane, "Kubrick is absolutely brilliant . . . we see eye to eye on everything." Spurred on by his conversations with the director, his output was prodigious during the second half of 1964. Every other day, they would meet at various city locations and discuss, then he would return to his electric typewriter on the tenth floor of the Chelsea and write. In this, he was fueled by a diet of liver paté on crackers, an artery-clogging repast sometimes served by a questionable new interest of his: an Irish merchant seaman with a room down the hall named Peter Arthurs.

On June 12 Clarke scribbled the following observation in his journal: "nb Stan's enthusiasm for material later discarded—part of his technique with authors/actors?" ("Nb," for the Latin nota bene, means "note well.") He'd become aware of a practice familiar to most of Kubrick's collaborators—one sometimes capable of driving them to the brink of nervous collapse, but also of goosing them to the absolute limits of their creative capabilities. On June 20 Clarke reported, "Finished the opening chapter, 'View from the Year 2000,' and started on the robot sequence." Both were later dropped, but each contained elements that would surface in the film. The initial opener contained a riff on how residents of millennial Earth would know, on looking at the Moon, that people were up there looking back: "And they would remember that there were some whom Earth could never reclaim, as it had gathered back all their ancestors since the beginning of time. These were the voyagers who had failed to reach their goals, but had won instead the immortality of space, and were beyond change or decay." Within a few years, Kubrick would realize the passage with an unforgettable shot of astronaut Frank Poole spinning off into the silent black immensity of interplanetary space, forever entombed in a canary-yellow space suit with a severed oxygen line.

As for Clarke's "robot sequence," it proved to be a missing link between the concept of artificial intelligence best embodied by Isaac Asimov's midcentury "I, Robot" stories (1940 to 1950)—as well as *Forbidden Planet*'s clunky, whirring Robby the Robot (1956)—and the gleaming, elegantly disembodied HAL-9000 mainframe of the final film. In the chapter draft, HAL's precursor is named Socrates, is "roughly the size and shape of a man," and walks on legs comprised of "intricate assemblies of sliding shock-absorbers, universal joints, and tensioning springs, held in a light framework of metal bars. They flexed and yielded at each step with a fascinating rhythm, as if they possessed a life of their own." Socrates is "no more intelligent than a bright monkey," but when switched to "independent mode," he transforms into an autonomous individual. He also speaks, "generating the words himself." This first-draft AI would go through several more name changes, upping its IQ each time.

In early July Clarke was averaging two thousand words a day. Reading the first five chapters, Kubrick pronounced, "We've got a bestseller here." On July 9 Clarke spent the better part of the afternoon showing him how to use a slide rule: "He's fascinated." On the twelfth, Clarke jotted down the following dialectical observation: "Now have everything—except the plot." On July 26, Stanley's birthday, Clarke found a card in Greenwich Village depicting the Earth "coming apart

Director and writer in Kubrick's office, New York, 1964.

at the seams and bearing the inscription 'How can you have a Happy Birthday when the whole world may blow up any minute?'"

In late July Kubrick's idea of generating a story in prose, and then converting it to a script, and then going to a studio for backing, underwent a further change. Now he proposed that the novel itself be used as the basis of a film deal. Only later would it be converted into a shooting script. As for the book, they could publish it before the film's release, he said. Considering their joint authorship, they would have to make another agreement, with more money involved. But that could wait.

Resonances with *The Odyssey* emerged gradually. In another handwritten note dated August 17, Clarke wrote, "Chose hero's name— D.B." They'd settled on the name Dave Bowman—though it took them some time to realize they'd arrived at a reference to Homer's itinerant protagonist. "It was literally months later that it occurred to us that the bow is a symbol of Odysseus," Clarke recalled in 1968. "This was entirely unconscious. I can't believe it was a coincidence."

Investing their astronaut with a measure of Ithaca's wily king didn't yet extend to the film's name, however. "The Odyssean parallel was clear in our minds from the beginning, long before the title was chosen," Clarke remembered in 1972. Having already reduced their working title from *How the Universe Was Won* to simply *Universe*, on August 21 they expanded it again: it became *Tunnel to the Stars*.

In early chapter drafts, the character who would become David Bowman is named Bruno. He rides his computer-guided Rolls along the "auto-highway" bisecting "the great Washington–New York complex" with his son Jimmy and the family dog, whose grey hairs give evidence of advanced years. Their destination is the launch complex from which Bruno will vault into orbit to join an exploratory mission to Jupiter.

> Suddenly, out of the half-forgotten reading of his childhood, Bruno remembered how Ulysses's old hound had recognized his returning master at the end of his long wandering, had wagged his tail, and died. The memory of that brief episode from the greatest of all the epics made the tears start in his eyes, and he turned to a side window so that Jimmy would not notice. (And what, he wondered, would Homer have thought, of the Odyssey on which he was embarking now?)

By midsummer, Kubrick had already processed a staggering amount of information, including books by astrophysicist Alastair G. W. Cameron, computer pioneer and artificial intelligence theorist Irving J. Good, astronomer Harlow Shapley, *New York Times* science writer Walter Sullivan, and many others. On July 28 he fixed Clarke with his penetrating eyes and announced, "What we want is a smashing theme of mythic grandeur." Clearly their project was already undergoing mission creep, from "the first science fiction film that isn't considered trash" to something bolder and potentially more profound. Clarke recalled this phase of their collaboration in the first draft of his unpublished 1967 piece for *Life*.

[D]uring this period, we went down endless blind alleys and threw away tens of thousands of words. The scope of the story steadily expanded in both time and space. To our surprise, we found ourselves involved with nothing less than the origin and destiny of Man, and what had started out primarily as a story of exploration began to have complex philosophical overtones.

Granted the right of revision, Kubrick crossed out most of the last sentence, commenting, "This sounds pompous, and I believe statements like this invite later deflation by critics." Although in retrospect his observation was both entirely accurate and to the point, Clarke dutifully killed the sentence. He continued:

There were times when I became a little scared of what we were letting ourselves in for; on such occasions, Stanley would say reassuringly: "If you can describe it, I can film it." Though I managed to disprove this dictum, I must also admit that Stanley later filmed things I couldn't possibly describe.

Most of this, too, had been crossed out by the director—but not before he modified the quote to "If it can be described, it can be filmed," writing, "Again, implication you do the creative writing." It was one of several such comments.

Kubrick's prickliness concerning authorial credit, which reared its head in a very public slapdown of Terry Southern in the fall of 1964, when he perceived that Southern was receiving too much acclaim for *Dr. Strangelove*, occasionally surfaced in his work with Clarke as well. In a 1966 profile in the *New York Times*, he was quoted as saying "We spent the better part of a year on the novel. We'd each do chapters and kick them back and forth. It seemed to me a better kind of attack." Actually, there's no evidence that Kubrick wrote a word of the novel—though without question he contributed significantly to its contents and was certainly the chief author of the screenplay, which would undergo nearly daily revisions throughout the actual shooting of the film.

After reading Clarke's latest output on September 7, Kubrick was ebullient. "We're in fantastic shape," he exulted, and proceeded to write "a 100-item questionnaire about our astronauts, e.g. do they sleep in their pajamas, what do they eat for breakfast, etc." Two days later, Clarke went to bed with an upset stomach and dreamed "I was a robot, being rebuilt." After breakfasting at the Seventh Avenue Automat, he returned to his room, experienced a wave of energy, and revised and retyped two chapters. That evening he brought them uptown to a pleased Kubrick, who fired up a frying pan and cooked Clarke "a fine steak," commenting, "Joe Levine doesn't do this for his writers"—a reference to a well-known film producer. Despite this evidence of directorial equanimity, on September 29 Clarke slept fitfully again, dreaming "that shooting had started. Lots of actors standing around, but I still didn't know the story line."

Throughout the year, they continued screening countless movies and reading many books, a number of which decisively influenced their emerging concept. One was a superb Oscar-nominated black-and-white short directed by Colin Low and produced by the National Film Board of Canada. Titled *Universe*—the source of their own short-lived working title—the half-hour film used innovative techniques to represent planets, star clusters, nebulae, and galaxies. Low's visual effects collaborator Wally Gentleman had filled tanks of clear paint thinners with suspended inks and oil paints, filming them under bright lighting at high frame rates, a technique that when projected at regular speed seemed to convey with a previously unknown realism the floating majesty of a cosmos illuminated by flaring stars and incandescent, ionized hydrogen gas. For Kubrick, who'd suffered through endless hours of cheesy animations and badly done matte work, the Canadian film was a revelation. He ran *Universe* again and again, studying it closely, and wrote down Low's and Gentleman's names—though he'd not yet noted the film's narrator: Toronto-based actor Douglas Rain.

Clarke's shrewd observations concerning human origins were clearly having their effect, and playwright–science writer Robert Ardrey's foray into paleoanthropology, *African Genesis*, soon became another major influence. Published in 1961, the book elaborated on a theory held by anthropologist Raymond Dart, best known for his 1924 discovery of the fossil bones of *Australopithecus africanus*: the earliest human progenitor thought to walk upright. By the mid-1950s, Dart was convinced, based on blunt force injuries evident in the fossil record, that civilization was rooted in an ancient propensity for violence. The survival of our apelike ancestors rested on the development of lethal weapons, he argued, titling one paper "The Predatory Transition from Ape to Man." Ardrey's book was founded solidly on Dart's ideas and became a hugely influential international bestseller. It contained potentially useful resonances with Clarke's story "Encounter in the Dawn."

When he finished reading *African Genesis* on October 2, Clarke noted in his journal, "Came across a striking paragraph which might even provide a title for the movie: 'Why did not the human line become extinct in the depths of the Pliocene? . . . we know that but for a gift from the stars, but for the accidental collision of ray and gene, intelligence would have perished on some forgotten African field.' True, Ardrey is talking about cosmic-ray mutations, but the phrase 'A gift from the stars' is strikingly applicable to our present plot line." And they changed their working title yet again, to *Gift from the Stars*.

Kubrick had his own favorite quote from *African Genesis*:

We were born of risen apes, not fallen angels, and the apes were armed killers besides. And so what shall we wonder at? Our murders and massacres and missiles and irreconcilable regiments? Or our treaties, whatever they may be worth; our symphonies, however seldom they may be played; our peaceful acres, however frequently they may be converted into battlefields; our dreams, however rarely they may be accomplished. The miracle of man is not how far he has sunk but how magnificently he has risen.

On September 26 the director had handed his collaborator another book to read: Joseph Campbell's *The Hero with a Thousand Faces*, an extraordinarily panoramic look at the commonalities between human mythologies. Kubrick had already quoted Campbell's core thesis to

Clarke earlier in their collaboration: *"A hero ventures forth from the world of common day into a region of supernatural wonder: fabulous forces are there encountered and a decisive victory is won: the hero comes back from this mysterious adventure with the power to bestow boons on his fellow man."* (Italics in the original.) Campbell referenced, among other examples, Prometheus ascending to heaven, stealing fire from the gods, and descending, and Jason sailing through the "Clashing Rocks into a sea of marvels," circumventing the dragon guarding the Golden Fleece, and returning "with the fleece and the power to wrest his rightful throne from a usurper."

By late fall, Clarke's writing had described a staggering number of scenes and situations. Many were discarded, but all were somehow necessary to the process of understanding what their story was going to be—much as actors find it useful to conceptualize their character's prior life stories before playing a role. The film's outlines had definitely emerged, and its temporal boundaries had expanded radically from their initial conception—namely, upstream into prehistory, to encompass an opening sequence based on "Encounter in the Dawn."

In Clarke's original story, humans had already settled in villages and learned to use spears and other rudimentary tools. Under the influence of *African Genesis*, however, the new version was set further back in time, when our ancestors were barely distinguishable from apes. "Moonwatcher," a particularly intelligent australopithecine, replaced the hunter Yaan of the original story. Rather than simply being handed a knife by an extraterrestrial anthropologist, in the new opening chapters, a mysterious "crystal slab" accompanied by a "pulsing aura of light and sound" appears one night in the African savanna. After an extended display of visual pyrotechnics during which Moonwatcher feels "inquisitive tendrils creeping down the unused byways of his brain," the slab plants the idea of tool use in the hominid's mind—specifically, the deployment of a sharp, pointed stone weapon.

Meanwhile under a steady fusillade of keystrokes from Clarke's typewriter, a futuristic equivalent of Jason's "sea of marvels" was gradually taking form in the film's last quarter. The discovery and excavation on the Moon of an ancient artifact buried there by an alien

intelligence, originally conceived as the climax of the film, had gradually been moved up to a position directly after the man-ape prelude.
There followed an expedition by a team of astronauts to Jupiter—the
destination of a focused beam of energy fired off by the alien object
when hit by sunlight for the first time in millions of years. The Sentinel
itself was variously described as a tetrahedron, a crystal, a crystal block,
and a perfect cube.

On arrival at Jupiter in their giant nuclear-powered spacecraft, the
astronauts discover a Star Gate cut into one of the planet's moons—
apparently a wormhole-like shortcut to another part of the galaxy that,
at least in some versions, functioned as a kind of galactic Grand Central
Terminal. Different sequences of events followed. In one, most of the
crew has survived, and they proceed into the Star Gate together. In
another, only Bowman makes the trip, in a small auxiliary capsule: a
"space pod" not unlike the one in Clarke's story "Who's There?"

It's in describing the journey through and beyond this mysterious
portal that Clarke's concentrated powers of imagination translated
into some of the most vividly compelling, almost tactile prose he ever
achieved. Hunched over his humming electric portable, a wastebasket
filled with discarded paper parked by his desk in the dim light of his
north-facing room, Clarke described extraordinary vistas of star clusters, red suns, and ambiguously fog-shrouded alien worlds. He wrote
of Bowman gripped by a "sense of wonder . . . as powerful as any force
that had held him during the journey" as he emerges "into a sky as wild
as the hallucinations of some mad painter." His Odysseus-astronaut
passes "the charred corpse of a world as large as Earth. Here and there
in the slagged lava of its surface—on which even the mountains had
been melted down—the faint ground-plan of erased cities could still
be dimly seen."

Gliding past this victim of a star that had gone "nova—as must
all suns, during some stage of their evolution—and had murdered its
orbiting children," Bowman witnesses the "glorious apparition" of a
globular star cluster spanning "almost a quarter of the heavens . . . a
perfectly spherical swarm of stars, becoming more and more closely
packed toward the center until its heart is a continuous glow of light."

He transits a vast sun that he can look at directly without discomfort—a red giant whose face appears "no hotter than a glowing coal. Here and there, set into the somber red, were rivers of bright yellow—incandescent Amazons, meandering for thousands of miles."

Other surviving pages included descriptions that would find almost literal realization three years later when Kubrick's visual effects team took Clarke's words and transmuted them into filmic language: "Now the turning wheels of light merged together, and their spokes fused into luminous bars that slowly receded into the distance. They split into pairs, and the resulting set of lines started to oscillate across each other, continually changing their angle of intersection. Fantastic, fleeting geometrical patterns flickered in and out of existence as the glowing grids meshed and unmeshed; and the hominid watched from his metal cave—wide-eyed, slack jawed, and wholly receptive." It was an echo of a paralyzed Moonwatcher being programmed by an alien intelligence in prehistoric Earth.

Even nominally rejected sections featured glinting shards of content later reflected and refracted to various degrees in the film. Bowman eventually arrives at a planet, where he flies over "a most peculiar ocean—straw-yellow in some areas, ruby-red over what Bowman assumed were the great deeps." In one attempted ending, he's grateful to find shelter in a building because it "offered mental security, for it would shut out the view of that impossible sky." In another, he experiences "something that could not possibly be real. He was no longer inside the pod . . . He was standing outside it . . . looking through the window at his own frozen image at the controls." Several of these descriptions would appear, virtually shot for shot, late in the film.

As Clarke commented later, "Stanley Kubrick and I were still groping toward the ending which we felt must exist—just as a sculptor, it is said, chips down through the stone toward the figure concealed within." In view of the prose on offer, Kubrick had reason to feel euphoric, and cook his collaborator steaks.

Not that Clarke was the only idea man. His own notes amply attest to the director's vital role in what was clearly a joint effort. On October 17 he scribbled: "Stan's idea 'camp' robots who create Victorian environment to put heroes at ease." (Later revised to "Stanley has invented the wild idea of slightly fag robots" doing the same.) As with their extraterrestrial pyramids on Fifth Avenue, it was dropped, but the idea became the seed of two important aspects of the final film: a supercomputer with a studiously neutered male persona, and a Louis XIV hotel room that serves as a kind of holding cell for Bowman after his epic journey—one providing "mental security, for it would shut out the view of that impossible sky."

Regarding the "slightly fag" thing, as Clarke became fonder of Kubrick throughout 1964, an anxiety grew within him. What would this man he'd grown to like and admire think if he discovered his collaborator's sexual orientation? There was no telling, and it gnawed at him. Finally, he decided to confront the issue head-on. During one of their meetings, he chose his moment and then announced abruptly, "Stan, I want you to know that I'm a very well-adjusted homosexual."

"Yeah, I know," responded Kubrick without missing a beat, and continued discussing the topic at hand.

As an utterly blasé "Next," it could hardly be equaled, and it brought a relieved smile to Clarke's face. Describing the scene to Christiane later, Kubrick said that Clarke had sounded "like a schoolteacher."

"He was very pleased that I don't care, and he doesn't know how much I don't care," the director concluded.

■　■　■

Meanwhile, both Kubrick's sprawling, low-ceilinged apartment and the Polaris Productions office were filled inexorably with books, charts, images, 16- and 35-millimeter film cans, and the like. The question of how to depict cosmic immensity preoccupied Kubrick, and he devoured the space art illustrations of Chesley Bonestell and Czech artist Luděk Pešek—which arrived at Central Park West in the form of lavishly illustrated large-format books—as well as hard-core technical reports of the kind produced by the US Air Force–affiliated RAND Corporation think tank, NASA, and the scientific journals.

Having learned to use a slide rule and absorbed the basics of the celestial coordinate system from his tutor, Kubrick turned his attention to the nature of infinity. At one point in July, he and Clarke suspended their discussion of plot development to engage in a lengthy exegesis of Cantor's paradox, which is based on the idea that the number of infinite sizes can itself be infinite. This, in turn, raises the paradoxical prospect that if there are infinitely many infinities, then the former all-encompassing one must be larger than any of the individual latter infinities. "Stanley tries to refute the 'part equals the whole' paradox by arguing that a perfect square is not necessarily identical with the integer of the same value," Clarke wrote. "I decide he is a latent mathematical genius."

Kubrick himself was aware of his own intellectual and creative capacities, and while he may have quietly reveled in them, he didn't let them go to his head. In Clarke, he had found an ideal sparring partner and foil. Clarke always gave as good as he got, responding with relevant insights and knowledge, and frequently disrupting Kubrick's assumptions in productive directions. Each was self-absorbed in a different way. Kubrick was single-mindedly devoted to achieving the goal of producing a significant work of art, a project inevitably encompassing a complex interplay of finances, logistics, ideas drawn from all fields, filmmaking and acting techniques, time management strategies, dramatic structures, and the like.

Filmmaking is a form of Gesamtkunstwerk—a "total art form"—and Kubrick afforded himself no real outside interests, because he drew everything—all the diverse materials of the project currently preoccupying him—within the capacious realm of his work. Almost by definition, there was no room for anything else, including most extrinsic human relationships apart from his family. While he engaged in intense friendships, they were largely centered on the achievement of his vision. Fishing for ideas everywhere, he sifted out convenient assumptions—"the passports of the lazy"—from the genuinely useful concepts. "Keep asking the question until you get the answer you want" was one of his maxims. Despite Kubrick's abilities and considerable achievements, he wasn't the least bit arrogant—though he could get impatient with fuzzy thinking or dissembling. Nor did he harbor any internalized notion that he was a significant artist. "He *hoped* he was good," said Christiane. This, in turn, colored his way of interacting with the world. Many who encountered him for the first time were surprised by his humility.

Clarke also reveled in his abilities and achievements but was less cautious in letting people know about them and in transmitting a sense of self-regard. And yet his self-absorption was suffused by an almost childlike sense of wonder, both at the fact of his own prominence and at the universe in which it was set like a particularly entrancing bauble. In part because of his essentially optimistic view of life, there was a disarming quality to his egotism—an invitation to share in the wealth, as it were. Like many writers, Clarke was in some ways a lonely figure, and his collaboration with Kubrick also provided him with an enjoyably liberating sense of release from professional solitude.

Still, he was homesick for Ceylon, and in particular missed Hector Ekanayake, his young Ceylonese friend, who had once been the country's flyweight boxing champion. Ekanayake, who'd acted in Wilson's boat racing film *Getawarayo*, had narrowly avoided decapitation the year before when he lost control of his speedboat and veered directly under a low pier—ending up in the hospital with a fractured skull, alive thanks only to his helmet and lightning-fast turtle reflex. On top of that, he'd been partially cross-eyed from birth, and Clarke had promised to fund an operation in the United States to fix his vision. In late November the author went to JFK Airport to meet Ekanayake, who'd never left Ceylon before. The surgery would take place in California. Clarke had also promised to pay for Hector to take diving training.

On December 21, Ekanayake tagged along when Clarke went up to Lexington Avenue in the early afternoon. Christiane and the kids had purchased a Christmas tree the previous day and set it on a stand in the living room, but they hadn't had time to decorate it. Although Kubrick did his best to give Clarke, who was trying to finish the novel before Christmas, his undivided attention, he was distracted by his Academy Award campaign for *Dr. Strangelove*, and their discussion kept on being interrupted by phone calls. After providing Hector with a cup of tea, Christiane announced that she was going out for a few hours with the girls to do some Christmas shopping. At this, Hector, who'd been left at loose ends, offered to decorate the tree while they were gone. Christiane had been planning on ornamenting it with the kids, but on seeing that Arthur's friend had nothing to do, she kindly agreed.

With Clarke and Kubrick in their stop-and-go discussion down the hall, Hector attacked the tree with vigor, flattening it out by tying branches together, creating an evenly shaped, temple-like silhouette. He proceeded to hang the ornaments in concentric lines vaguely suggestive of a pagoda. With its columnar flattened central form, elaborately tied-together limbs, and neatly horizontal lines of tinsel without the slightest sag, the tree had become a symmetrical evocation of Buddhist Asia—more sacred Bodhi than Tannenbaum, suggestive of incense, chanting, and meditation rather than shepherds, kings, and the infant Jesus. Returning some hours later, Christiane swiftly converted her astonishment into a simulation of pleasure for Hector's benefit. "Mommy, Mommy!" cried Anya, tugging her mother out of earshot down the hall. "That guy made a *very strange tree!*" she whispered urgently.

Returning to the Chelsea that evening, Clarke found a note from Allen Ginsberg and William S. Burroughs under his door, inviting him to join them at the bar. With all the interruptions, it had been a frustrating afternoon, and he gratefully went off to find them "in search of inspiration." After completing the last chapters of the book on December 24—a draft of about fifty thousand words—he took what he fondly regarded as a completed manuscript back uptown to Stanley as a Christmas present. Reading the new material with a rapt intensity, Kubrick finally set the pages aside with satisfaction. "We've extended the range of science fiction," he said, smiling.

Were they done? "We were, indeed, under that delusion—at least, *I* was," Clarke wrote eight years later. "In reality, all that we had was merely a rough draft of two-thirds of the book, stopping at the most exciting point. We had managed to get Bowman into the Star Gate, but didn't know what would happen next, except in the most general way."

Still, as of Christmas 1964, it seemed all Kubrick had to do was develop a screenplay based on the novel, achieve studio backing, hire some actors and production staff, and start filming. Accordingly, on December 31, Clarke received a copy of a letter from Kubrick to Scott Meredith informing the agent that as per their agreement, Clarke's services had been terminated.

Happy New Year.

. . .

In his "Son of Dr. Strangelove" notes, Clarke wrote, for December 24, "Delivered complete copy—sacked!" It's hard not to sense a world of hurt in those words. We don't know what happened next, exactly. In his draft article, written as always with an eye to posterity—not to mention the knowledge that Kubrick would read every word—Clarke wrote, "The first version of the novel was handed to Stanley on 24th December 1964, and he promptly fired me. True, I began work next day under a new contract, but I like to claim I was sacked on Christmas Eve." (Predictably enough, Kubrick crossed this out with the comment "Confusing. No one will know what you mean.")

Most likely, after some serious pushback from his erstwhile collaborator and a bit of pragmatic reconsideration, Kubrick rethought his cost-cutting measure, and by the time that happened in early January, Clarke's new contract was tweaked retroactively to start directly after the previous one. He wouldn't have done so if he hadn't realized that they truly weren't finished. Neither suspected that under the inexorable drive of the director's uncompromising perfectionism, the film's final narrative form would require another three years of unrelenting effort.

And apart from realizing that the story was really only partway there, Kubrick also knew that even without his services as a writer, Clarke was well connected in the aerospace community and was valuable simply as a consultant. In any case, Polaris would soon need to shift the burden of its financial commitments to a studio, because other significant expenses were looming as well. If the intention was to start shooting later that year, Kubrick would soon have to start hiring production staff. Studio support had become an urgent necessity.

But first he wanted to have some sequences to show. After studying *Universe* for much of 1964, early in the new year Kubrick decided to replicate the film's techniques, now on 65-millimeter color stock. In January he flew in a camera from Los Angeles, contracted with a small film visual effects outfit called Effects-U-All, and rented an abandoned brassiere factory on Seventy-Second Street and Broadway. There he and his collaborators set up tanks of black ink and a particularly noxious World War II–era paint thinner called banana oil (isoamyl acetate), surrounded these with high-intensity film lights, and shot the first frames of what would become *2001: A Space Odyssey*. Conducted under high security, they called it the Manhattan Project—a riff on the American nuclear weapons program of World War II.

Powerful lights permitted high camera speeds, crucial in capturing the high-speed alchemy of surface tension, color change, and chemical reaction that they were after. The overcranked camera shooting at seventy-two frames per second produced a smoothly nuanced "galactic" slow motion as they used toothpicks to drip blobs of white paint into the ink-thinner mixture. Reacting to the banana oil, the paint sent ersatz star flows and galactic tendrils streaming into cosmic space. A macro lens made an area the size of a playing card look like a nebula light years across. Parts of what would become the film's trippy Star Gate sequence were conjured in this way on the Upper West Side in early 1965, with Kubrick himself manning the camera.

Wally Gentleman, the visual effects pioneer behind *Universe*, had conceived of the technique in the late 1950s, realizing that "many of the truly cataclysmic things that can occur in nature really occur on a very tiny scale, so providing you can get the right flux of elements moving and photograph it, you can have something that looks gigantic. By dropping paint onto oil or other paints, you can get these explosive effects and changes in color . . . And the combinations are really quite endless."

Christiane Kubrick remembers the brassiere factory scene vividly. Big, low tables supported shallow square-sided metal tanks and cans of paints and chemicals. A stink of thinner, ink, and lacquer "rotting" under the hot film lights filled the air. The materials Stanley was working with fostered bacteria growth and became "unspeakably disgusting." Life arose in Kubrick's ephemeral micro-universe, replicating exponentially within the expanding star clusters and morphing nebulae even as they were captured on film at high frame rates.

The tenacity that the director would display throughout the production was already evident. Returning from the factory in the early hours of the morning with eyes red and swollen from the fumes, he ignored the foul reek for weeks on end, scrupulously writing down what percentages, temperatures, and densities of which liquids required what heights to drop from to create a given effect. "The difference between a lot of us and Stanley is that he hung in there long after any of us would have lost patience, to get it right," Christiane recalled. "And it becomes an enormously boring filing of each particular effect so you can repeat it, and repeat it with another combination, and another combination that doesn't look like ink spreading but looks like the universe. And that is the madness that artists should have."

· · ·

A bank of darkening grey clouds sat low over Manhattan in the chilly afternoon of Friday, January 22, when Clarke's phone rang in the Chelsea. It was an old acquaintance, Fred Ordway—until recently an employee of Wernher von Braun's at NASA's Marshall Space Flight Center in Huntsville, Alabama. Ordway was in town with a colleague of his, another von Braun associate, German graphic designer Harry Lange. Would Arthur have time to meet? Although they both had dinner engagements, they arranged to convene a few hours later at the Harvard Club, where Ordway always stayed when in New York.

Seated by one of the fireplaces in the club's grand oak-paneled central room, the three chatted about von Braun's Apollo booster test flights and the American lunar program. A thick snow was falling outside. To Clarke, Ceylon felt a million miles away. Ordway, the Harvard-educated offspring of a prosperous New York family, was in his element. An athletic, preppie type in his late thirties, he was a snappy dresser. He had published numerous articles and books on astronautics, rocketry, and astronomy, and had been on the staff of the Army Ballistic Missile Agency. He'd known Clarke since meeting him at an international astronomical conference in Paris in 1950.

Harry Lange had fled East Germany after the Second World War, studied design in the western half of the divided country, and immigrated to the United States in the 1950s. Also immaculately dressed and in his midthirties, he had an unruffled demeanor and intelligent eyes. Below a prominent widow's peak, a can opener of a nose dominated a somewhat equine face. After he spent a detested year in advertising in New York, the US Army drafted him during the Korean War. Foreign nationals weren't sent to fight, however, and Lange was given a job illustrating technical manuals in Alabama. Following his discharge, he got a job at the Redstone Arsenal, where von Braun was already designing missiles with a posse of ex-Nazi engineers, most of them graduates of Adolf Hitler's V-2 rocket program. He ended up working with Ordway, then a technical writer under von Braun, and when NASA was formed in 1958, they became employees of the space agency. Together they had produced a series of illustrated books with the rocket scientist, among other projects. More recently, they'd quit NASA to focus more attention on their own company, General Astronautics Research Corporation.

When Ordway asked Clarke why he was in New York, he reported that he was working on a film with Kubrick involving the first contact with an advanced alien intelligence. At this, Lange's eyes widened. "It happens we've just published a children's book for Dutton on life in other solar systems," he said in his mild *ostdeutsch* accent. "And we're working on a big Prentice-Hall book, a professional work to be called *Intelligence in the Universe*. We're delivering galleys to them and a lot of my artwork." He excused himself, went upstairs to retrieve a folder, and for the next half hour, he and Ordway explained their work. "How extraordinary," Clarke said, shaking his head. "Intelligence in the universe . . . I'm working with Stanley Kubrick on that same general subject."

Checking their watches, they rose to say their goodbyes. "Arthur went off and, apparently, went down to the next corner and called Kubrick," Ordway recalled. "Harry and I went up and got our overcoats, and just as we were leaving the front door, one of the employees said, 'Mr. Ordway, there's a call for you.' I thought it was from the host and hostess that evening, wondering if we were going to have trouble getting up there, 'cause it was snowing hard . . . And I heard, 'This is Stanley Kubrick.' 'Ah,' I said, 'Well, how do you do?' We talked about his film and that Arthur had just called him, and would I be available to meet with him. Well, we were somewhat flexible."

The next day, they reconvened in Kubrick's apartment for a "mentally exhilarating" afternoon. "He had immersed himself in the field," said Ordway, describing an office jammed with science fiction magazines, books, and trade journals. "It was exciting to hear what he was doing. He sort of sketched out what he had in mind, and they gave me a copy then and there of *Journey Beyond the Stars*"—their latest title, the one Sagan had taken exception to—"which was called a film novel, or film story, by Kubrick and Clarke."

During their conversation, Lange explained that until recently, he'd been head of the Future Projects Graphics group at NASA, where he had supervised a staff of ten illustrators working to visualize upcoming space technologies. Selling spaceflight to congressional appropriations committees and the American public were high priorities for von Braun, who understood the central role that public relations played in drumming up support for his extraordinary ambitions. At one point, Lange recalled that Walt Disney had arrived in Huntsville, and the rocket scientist had convened a meeting of his top engineers. In front of Disney, he said, von Braun had commented, "Harry, your work makes money. Everyone else wants to spend it."

Kubrick, who'd been listening intently, waited for him to finish. "Well, I can get better illustrators than you; they're a dime a dozen in Greenwich Village," he said bluntly. "They're all running around starving. But they don't have your background. That's what I need. You've been around rockets of all sizes and purposes; you know what they look like, in detail as well."

"Yes, yes, I've seen enough," said Lange, who was not offended.

Kubrick asked them if they would like to work as technical consultants on his film. He explained they were looking for scientific accuracy and a thoroughly researched believability. The film had to look utterly real, with none of the cheesy falseness of prior science fiction efforts. The audience had to feel that they'd been transported into the early years of the next century. The story would be set in a period about three decades in the future—far enough away to ensure that its spacecraft and computers wouldn't be undermined by reality, and close enough that they had a realistic chance of being roughly congruent with what might actually happen.

Ordway and Lange looked at each other. Their answer was never in doubt. They'd taken the risky move of quitting NASA and starting their own company precisely for such opportunities. For much of the rest of the year, they divided their time between Huntsville and New York, with Ordway researching aerospace technologies and his partner designing and illustrating them. Although hired as a consultant, Lange's contribution would expand significantly, and he would end up earning a production designer credit and a career in the film business.

With a little help from Clarke, Kubrick had found a pair of collaborators who couldn't have been better suited for his purposes.

. . .

Despite the epic fail of MGM's 1962 film *Mutiny on the Bounty*, the studio's new president, Robert O'Brien, was committed to continuing its otherwise successful "road-show" distribution of big-budget Cinerama features. He'd already approved the production of David Lean's three-hour wide-screen *Doctor Zhivago*, which was set for release in December 1965, and had stood by the director through significant delays and cost overruns. He had high hopes it would refill fiscal reserves drained dry by *Bounty*.

With a ten-minute show reel of Kubrick's Manhattan Project cosmological footage and Clarke's draft novel in hand, the director's LA consigliere Louis Blau presented *Journey Beyond the Stars* to MGM in early February. "They had a two- or three-day deadline during which they had to come back to me, and that's how I made the deal," Blau recalled. Although the studio was used to receiving completed scripts—not unpublished novels with awkwardly truncated endings, albeit accompanied by beautiful but highly experimental filmic abstractions—Kubrick was the hot director of the moment, and O'Brien had little doubt he could emulate the success of *How the West Was Won* with a similar frontier drama, this one set in the future.

Accordingly, the studio chief agreed to Blau's terms within the required three days. The deadline had been calculated to cover for any narrative deficiencies by making it unlikely anyone would have time to actually read the book. O'Brien green-lighted a budget of $5 million, with release planned by late 1966 or early the following year. With the financial hit of *Bounty* still stinging, studio stockholders and executives criticized his decision immediately, a backbiting chorus that would only grow as Kubrick's project spiked well above its original budget and repeatedly overshot its release dates. This pushback reportedly came in part from some of the more conservative members of MGM's board, who'd found *Dr. Strangelove* manifestly unpatriotic and offensive.

Kubrick had chosen his studio boss wisely, however. In the years to follow, O'Brien never wavered in his support. His decision to back the project was certainly due to a complex nexus of factors. Although by 1965 the science fiction genre had produced few films that were truly respected, it had nonetheless established itself rather rapidly in the previous decade and a half. Figures tell a tale of exponential growth. Only three science fiction feature films were released in 1950, but by the middle of the decade, the genre was averaging around twenty-five per year, and by its end over 150 had been released—an unprecedentedly fast expansion for a new genre, even if most of the productions were schlocky B-grade material. What it needed was an unqualified upmarket success.

Added to that, of course, was the actual race to the Moon, now in full swing and occupying a lot of media bandwidth. Popular attention was focused on the exploits of American astronauts, and at the time MGM announced Kubrick's space epic on February 23, 1965, NASA's economy-sized Mercury capsules had been supplanted by Project Gemini's midrange ones, which were soon to begin transporting two-man crews to orbit. Given all the attention being paid to American space efforts, O'Brien sensed considerable untapped commercial potential.

Plus, the director was as close as it came to a safe bet. He'd already handled a film of approximately this size and budget with *Spartacus*, and had just shown himself capable of achieving significant critical acclaim and commercial success with *Dr. Strangelove*.

For Kubrick and Clarke, the MGM press release hailing *Journey Beyond the Stars* crystallized continuing doubts about the film's short-lived series of titles. *New Yorker* writer Jeremy Bernstein had first met Clarke a year or so before, and in the spring of 1965 Clarke brought him up to Lexington to see the director. Bernstein, who in his day job was an astrophysicist, recalled the scene years later: "When I first saw Kubrick and the apartment, I said to myself: 'He is one of ours.' What I meant was that he looked and acted like almost every eccentric physicist I had ever known. The apartment was in chaos. Children and dogs were running all over the place. Papers hid most of the furniture. He said that he and Clarke were doing a science fiction film, an odyssey, a space odyssey. It didn't have a title."

While Clarke would later state that the final title had been entirely Kubrick's idea, he'd evidently forgotten the meditation on millennial dates in "Out of the Cradle, Endlessly Orbiting," the short story with a narrator grousing about how the twenty-first century actually begins in 2001, not 2000. A surviving cover of the *Journey Beyond the Stars* prose treatment contains a record of Kubrick's evolving thoughts. In what amounts to a freeze-frame of a day in late April 1965, a set of titles appears in red ink. They're in the director's hand:

Earth Escape
The Star Gate
A Space Odessy [*sic*]
Jupiter Window
Farewell to Earth

And at the top left, for the first time:

2001: A Space Odyssey

According to Clarke's bifurcated notes for "Son of Dr. Strange-love," the title was decided on either April 29 or 30, 1965.

The new title stuck.

. . .

Having signed a Memorandum of Understanding sufficient to announce their collaboration on January 14, and then having announced the film on February 23, Kubrick (via Blau, representing Polaris) and O'Brien (via in-house lawyers) proceeded to negotiate a seventeen-page, thirty-five-clause contract for "a motion picture now tentatively entitled *2001: A Space Odyssey*." Read carefully, the surviving draft, dated May 22, 1965, is a fascinating cornucopia of Kubrickiana. It undermines some myths, reinforces others, and reveals both Clarke's significance to the project and the extent of his exclusion from the film's financial prospects. (While the actual signed contract may still exist somewhere, it's not readily available to mere mortals.)

Kubrick approached such contracts with the same level of fastidious concentration that he applied to all other aspects of directing, not least because they impacted all aspects. "He made his deals very carefully, and he was very much a chess player in these things—really careful," remembered Christiane. "He said, 'Don't relax too soon. That's when you make mistakes.' He was very attentive to all the details . . . He did get very good contracts that nobody else got. He didn't trust. When he finally got through, and he got what he wanted and the right film, the right time, the right money, the right everything, he'd say, 'I think I've done it.' He would never say, 'I've done it!' Just 'I think I've done it. I think I have.'"

The contract, which noted that Polaris had optioned "The Sentinel" for $5,000 and also paid Clarke $30,000 to write the treatment, stated that MGM would reimburse all of Kubrick's preproduction expenses, including all future commitments to Clarke (meaning the $15,000 he was entitled to at the start of principal photography and a further $15,000 on completion). The total figure of the reimbursement as of May 1965 was $53,429, most of which was payments to Clarke. The film's budget was specified as $5 million—not the $6 million sometimes reported. That figure included $350,000 for acquisition of the work, which at the time constituted Clarke's draft novel, Kubrick's Manhattan Project footage, and nothing else.

Thus to call a spade a spatulous device for abrading the surface of the soil, according to the contract draft, Clarke made something like $35,000 for having "The Sentinel" optioned, and writing the novel, and eventually stood to make a further $30,000. Adjusted for inflation, that's about $500,000 in today's dollars. Kubrick got more than three times that for selling the property to MGM, with no participation by Clarke. Meanwhile, according to a separate agreement, Kubrick would get 40 percent of the novel to Clarke's 60 when it was eventually decreed ready for print. Who would decide that? Why, the director—who'd carefully seen to it that he had the right of approval. (Apart from its being the basis of the film, MGM had no further claims on the book.)

On top of that, *2001*'s budget included $200,000 for Kubrick's services as producer-director, and $50,000 for his efforts in crafting a screenplay. When adjusted for inflation, that's just under $2 million more accruing to Kubrick—making a total of more than $4.5 million going to the director in today's dollars. None of this included Polaris's share of the profits, of course, which were specified as 25 percent of the net, with the net starting after MGM had recouped 2.2 times its expenses. If Kubrick went over budget—and he would more than double it—that net figure would start as high as 2.7 times MGM's total expenses. Still, Kubrick stood to earn a significant figure if the film was a success. And, again, Clarke had no skin in the game.

The contract contained other items of considerable interest as well. Over the years, a legend has accrued that following *Spartacus*, Kubrick never made a deal in which he didn't have absolute control of his work. His contract with MGM for *2001: A Space Odyssey* makes clear this wasn't always the case. The draft contained a provision specifying that the "then-president" of MGM would have the right to request changes. If the director disagreed, two versions would be previewed and measured by an audience reaction test: one with the requested changes and one without. If the test favored the changes, he was obligated to make them. (Kubrick's marginalia here suggested only slight tweaks to the language, and it seems reasonable to suppose he accepted the clause after these small modifications.* While unsigned, the existing draft was clearly the result of intensive negotiations.)

Other interesting details include a short list of possible directors, only one of them named Stanley Kubrick. The others were Alfred Hitchcock, David Lean, and Billy Wilder. (Try to imagine *2001: A Space Odyssey* directed by Wilder, the man who made *Some Like It Hot*.) The film would be shot in MGM's Borehamwood Studios, north of London—one of Europe's top production facilities. And the contract included the following definition of MGM's rights: "Distributor's Territory shall also include any and all space vehicles, lunar shuttles, space stations, service and orbital life support systems pertaining, but not limited to, the planets, planetoids, and moons in all of the galaxies of the Universe (said territories are herein referred to as the space territories)."

The boilerplate had coincided with the destination.

Finally, MGM retained a say over the choice of actors, and the contract contained an annex with casting suggestions from Kubrick that MGM had approved. A good four months before Keir Dullea became aware of the director's interest, the young actor was approved for David Bowman, the lead role. Dullea, then twenty-nine, was best known for playing a coldly remote, emotionally damaged psychiatric patient in Frank Perry's film *David and Lisa*. Other actors that MGM greenlighted included, for the role of high-flying presidential science advisor Heywood Floyd: Robert Montgomery, Joseph Cotten, Robert Ryan, Henry Fonda, Jason Robards, and George C. Scott. For Moonwatcher, the lead man-ape, Kubrick had listed, and MGM approved, Robert Shaw, Albert Finney, Gary Lockwood, and Jean-Paul Belmondo—the last being the star of Jean-Luc Godard's 1960 film, *Breathless*. French accents aren't a problem for nonspeaking parts.

Gary Lockwood's presence on the list is interesting, and Kubrick

---

* For example, he wanted to add "a certified" before the phrase "audience test."

would return to him later for another role. But it seems clear who he initially favored as Moonwatcher. In February he'd written Shaw, enclosing a graphic depiction of a broad-faced, hairy Neanderthal type and commenting, "Without wanting to seem unappreciative of your rugged and handsome countenance, I must observe there appears to be an incredible resemblance." Shaw, a glowering British actor then best known for playing military roles and bad guys in such films as *The Dam Busters* (1955) and *From Russia With Love* (1963), would later indelibly embody shark hunter Quint in *Jaws* (1975). While his response to Kubrick's note, if any, is lost to history, the problem of how to present convincing prehuman primates would prove almost intractable and would be solved only late in the production, without the benefit of Robert Shaw's semi-simian visage.

. . .

By the spring of 1965, the formerly sleepy Polaris office on Central Park West had been jolted into hyperactivity. Lange's drawings were pinned onto bulletin boards, along with work mailed in from Graphic Films, the LA outfit that had made *To the Moon and Beyond* for the New York World's Fair, and which Kubrick had contracted to produce spacecraft and Moon base designs. Storyboards wallpapered another room, as well as detailed early color renderings of projected scenes. These included a Frisbee of a space station being approached by a switchblade space shuttle. Phones rang ceaselessly, and cigarette smoke swirled through the air as people came and went, portfolios of drawings and sheaves of documents in their hands. On April 19, Clarke visited.

> Went up to the office with about three thousand words Stanley hasn't read. The place is really humming now—about ten people working there, including two production staff from England. The walls are getting covered with impressive pictures, and I already feel quite a minor cog in the works. Some psychotic who insists that Stanley *must* hire him has been sitting on a park bench outside the office for a couple of weeks, and occasionally

comes to the building. In self-defense, Stan has secreted a large hunting knife in his briefcase.

Ordway remembered the park bench episode a bit less charitably. "He was an odd person in New York," he said—not of the benched guy but Kubrick. "I always remember one time he said, 'Fred, you see that guy sitting over there on the bench across the street?' I said, 'On the park bench?' He said, 'Yeah, he was there yesterday and the day before . . . what do you think he's doing there?' You know, he was watching things like that. When we'd go out to eat dinner . . . he'd always ask for a table and sit and face the wall, because he had that obsession."

The go-to filmmakers for NASA and the US Air Force, Graphic was a niche visual effects house that had been founded in 1941 by former Disney animator Lester Novros. Apart from *To the Moon and Beyond*, it had recently completed other short films projecting future space developments, including details such as animations of small one-man space pods with mechanical arms. In May Novros and the director of *To the Moon and Beyond*, Con Pederson, flew in from Los Angles to consult. Their first impression of Kubrick was of an earnest type, quiet but extremely direct.

"We had a very good time," recalled Novros in 1984. "He said a lot of the script was undefined, especially the ending. He wanted to introduce men from some extraterrestrial environment. He thought of them as being about twenty feet high, very thin, and attenuated. He asked if we knew the sculptures of Giacometti, and we said 'yes,' and he said there was a show of his at the Museum of Modern Art. So we went down there, and he said, 'I have sort of an idea that our spacemen should look like this. So at the end of the film, they could reach down and grab this little guy from Earth by their hands and walk into the sunset.' It was that banal."

Kubrick told his visitors that, in his view, there were three factors to consider in every film: Was it interesting? Was it believable? And, was it beautiful or aesthetically superior? At least two of the three had to be in every shot of the film. He was pleased to discover that, like Ordway and Lange, Pederson had worked with Wernher von Braun when he was in the US Army in the 1950s, and had also worked on the Walt Disney Company's TV shows. Three of these airing on ABC from 1955 to 1957 had featured guest appearances by the German rocket engineer and animations of his space exploration concepts, which included Earth-orbiting shuttles and wheel-shaped stations—the same technological vocabulary that *2001: A Space Odyssey* would soon appropriate.

Kubrick arranged a screening of his newly shot Manhattan Project cosmological sequences for his visitors. "He was very coy about it," Novros recalled. "He wouldn't tell anybody how he had achieved them. He was a very creative man and a very egocentric man, and so that was it." Pederson, at least, was suitably impressed, and Kubrick signed a nonexclusive one-year contract with Graphic to provide visualizations and storyboards, on the understanding that it would later be asked to take the lead in providing visual effects for the film. Eventually he would decide that working with a Los Angeles company from distant London would be too unwieldy, however, and he ended the connection—but much to Novros's irritation, he followed this decision by hiring away some of Graphic's best people. Despite Novros's strong effort to hold on to him, Pederson would move to Britain and join Kubrick's team by the end of the year.

Meanwhile, the question of whom to entrust with the film's all-important production design preoccupied the director, who'd been playing footsie with Ken Adam—the mastermind of *Dr. Strangelove*'s sets, including its stunning war room—since the previous summer. Still shell-shocked from Kubrick's relentless perfectionism, Adam was ambivalent about working with him again, and in any case, he'd signed on to do the fourth James Bond film, *Thunderball*. He politely declined.

This left an important position unfilled, and, at the recommendation of *Dr. Strangelove*'s associate producer, Victor Lyndon, Kubrick asked production designer Tony Masters to fly in from England for an interview. A tall, lanky type with bushy eyebrows and a perpetually good-humored gleam in his eye, Masters, then forty-six, wasn't yet considered a front-rank film designer, but Lyndon knew he was a superb draftsman and was thoroughly familiar with organizing studio-based films. Masters had worked his way up through the ranks of the

British film industry from draftsman, to assistant art director, to set decorator. His first art director credit had been in 1956, and since then, he'd designed dozens of feature films. Among other things, he'd worked closely with production designer John Box on David Lean's 1962 wide-screen epic *Lawrence of Arabia*—a significant credit.

Masters had a slight speech impediment, but he'd largely overcome it through sheer force of will. Although the director would be taking a chance with a man he didn't know, he trusted Lyndon's instincts, and Masters passed his initial interview beautifully. The first thing Kubrick wanted to know was if the designer was interested in science fiction. "And, well, I am," Masters recalled in 1977. "I like it, and I like that kind of design. I like designing for the future. A lot of people don't."

He would soon prove himself more than equal to even the Ken Adam standard. Masters knew how to manage large film design departments, and his training as an architectural draftsman would prove an indispensable asset. He'd soon be supervising a staff of forty as they labored over *2001*'s sets—but first Masters and set decorator Bob Cartwright, who'd also flown in from London, spent three months with Kubrick in New York, "just sitting in an office, talking about the script and the story, and just figuring out how the hell we were going to make this kind of crazy picture," he remembered. "I mean, he has a very good sense of humor, and often the days would end up with both of us laughing too much to work anymore—it just seemed ridiculous; what we were trying to do was just *too* much. When you do a science fiction film, it's very difficult to discipline yourself to stay within reason. There are no limits, and you can go berserk." Luckily, Lange and Ordway were there to provide reality checks, however, and Masters remembered Lange in particular as playing the role of "anchorman"— a stabilizing influence protecting them from themselves.

In many ways, the Huntsville spaceflight geeks and the British film designer were a perfect combination. Ordway was already sending a blizzard of letters to his many contacts in aerospace technology, soliciting ideas and concepts. Lange was well versed in converting them into the two-dimensional space of the graphic image, but he had no experience with the highly specialized realm of film set design. Although the

Tony Masters, Fred Ordway, and Harry Lange.

story continued morphing relentlessly, they managed to establish the basic design for a set of spacecraft types. These included a streamlined, futuristic-looking space shuttle called *Orion*; the giant double-wheeled Space Station 5, which had evolved from the flattened disc of early renderings; a semispherical lunar landing spacecraft with four legs, the *Aries*; and the nuclear-powered Jupiter-bound interplanetary spacecraft *Discovery*. Apart from that, there were numerous small-fry vehicles, including one-man space pods, an oblong rectangular Moon Bus, and a small multinational squadron of orbiting nuclear bombs. It being a good three decades before computer-generated imagery, detailed models would have to be constructed for all their exteriors.

The first and most ambitious design the three of them collaborated on in New York was *Discovery*'s internal centrifuge, a thirty-eight-foot-diameter, ten-foot-wide rotating carousel that was supposed to provide artificial gravity for the crew. Because they knew they needed to mount a camera dolly on a thin blade of steel projecting through the floor, they conceived of two matching discs fashioned to meet perfectly at a narrow gap between the wheel's two sides. Done right, it would permit the blade to lock the dolly in place, allowing the set to turn around the camera. When joined, the two sides would become *Discovery*'s central crew quarters, containing the main supercomputer console, the coffin-like hibernaculum modules with their entombed sleeping astronauts, a narrow dining table, and the like. The exterior framework for a set of this scale and level of precision wasn't something a normal film design department could be expected to construct, however, and they decided to farm it out to a British aviation firm.

As for the centrifuge interior, it became a kind of design template for the rest of the film. "We decided to use a lot of white material," Masters recalled. "Everything was white—and if it wasn't, black. Black and white and blue. And the white gave a nice feeling; it gave a sort of suspended look to everything, because there were no shadows anywhere. That worked quite well. Originally we did things in colors, but once again, Stanley was saying, 'Well, what color would it *be*?' 'I don't know what color it would be, Stanley. I mean, what's a nice color? Maybe we'll try it blue—IBM blue.' He says, 'No, no, no. Blue is too "today"—you see blue everywhere.' So, in the end, we said, 'Well, let's not use any color—I mean, we can't go wrong there.'"

This left the question of who would supervise the film's visual effects: the "trick" photography, the construction and filming of the models, the production of content for the film's innumerable flat-screen displays, and so forth. Graphic Films was working on visualizations and schematics, but it was largely an animation house and not particularly well versed in achieving the photorealistic quality Kubrick had in mind. Plus, its inconvenient distance would become less tenable when they moved from New York to London. For some time, Kubrick had been attempting to recruit Colin Low and Wally Gentleman, the team behind *Universe*, but Low, who was manifestly uninterested in working with the director, warned his colleague that Kubrick "would walk over you with hobnail boots on." Finally, Gentleman agreed warily to take charge of the film's effects—but he wouldn't sign a contract. After Low's warning, a bit of disconnect existed between Kubrick and Gentleman from the beginning.

Among the first things they discussed in the spring and early summer of 1965 was enforcing the film's believability by keeping all the effects shots on the original negative—not compositing and duping shots in an optical printer, a clumsy predigital technique that inevitably added additional generations of film grain each time. This meant instigating a complex, potentially error-prone production chain in which the same negative stock that had been used to film, for example, live-action footage of astronauts visiting an excavated alien object on the Moon, would be stored, sometimes for many months, as a "held take," and then run through an animation camera to add rugged lunar terrain in surrounding areas of the picture left dark for that purpose during the live-action shoot. Sometimes a third or fourth pass would be necessary as well, to add the Earth, the stars, and other elements such as people moving around in spacecraft windows. The method was potentially fraught with hazard, but calculated to produce a flawlessly pristine, first-generation believability.

"I was engaged as the director of special effects for the entire movie," Gentleman recalled in 1979, "but my main purpose at the outset was to act as Kubrick's general instructor in basic effects techniques. He had a commanding knowledge of still photography, but he was lacking precise knowledge of special effects operation. I must say he was the most able student I've ever taught, because he could soak up information like a sponge." (There's that simile again.) Gentleman's assumption of a pedagogical role would begin to grate, however, as would his reluctance to work late at night and then field calls from Kubrick at all hours of the morning. His tenure on *2001* would only last a year.

Two years later, the Oscar for Best Visual Effects would go not to Gentleman but his erstwhile pupil, Stanley Kubrick.

.   .   .

If Ordway, Lange, Masters, and Gentleman were collectively in charge of producing a realistic and plausible technological environment for the journey to the Moon and Jupiter, all of it, in turn, was supposed to provide a launch pad for Dave Bowman's quasi-surreal voyage through the Star Gate and beyond. As tools used to tame and control nature, the film's technologies were intended in part to lull the audience into a state of suspended disbelief, thus permitting their open-minded receptivity to the film's kaleidoscopic, kinetic, transcendental final chapter.

Most of the letters sent out to various companies, institutions, and individuals by Fred Ordway in the summer of 1965 had to do with hardware of various kinds. But one of them—a particularly intriguing communication written on June 7—dealt with "wetware": the intricate neurochemical wiring of the human mind. His letter that day to Dr. Walter Pahnke of the Massachusetts Mental Health Center made no reference to a film project, and though it contained Polaris's street address, it failed to mention the production company at all. Rather, it was supposedly written in his capacity as an editor of aerospace technical journals.

Ordway asked for more information concerning a brief published account of an experiment Pahnke had conducted as part of his PhD thesis at Harvard, in which he gave Boston divinity grad students doses of a potent hallucinogen, psilocybin, within a supervised, controlled environment. His letter was almost certainly return-addressed anonymously because Kubrick wanted to disguise his own interest in the study. "We were particularly intrigued by the fact that the subjects experienced a heightened insight into philosophical matters as a direct result of receiving the drug," Ordway wrote. "We have examined all conceivable possibilities for reducing the awareness of time of future astronauts including hibernation, hypnotism, and, of course, drugs."

Kubrick had caught wind of Pahnke's so-called Marsh Chapel Experiment, conducted in 1962 under the supervision of his thesis advisors, Timothy Leary and Richard Alpert. The study was designed to see if hallucinogenic drugs, when taken in a religious setting such as a Boston University chapel, could induce religious experiences comparable to those recorded by the great mystics of history. Nine of the ten theology students participating reported that they had indeed experienced sensations indistinguishable from religious revelation. Some described vaulting beyond past, present, and future into a realm where three-dimensional space-time was merely one among infinite possibilities.

One participant, forty-six-year-old religious scholar Huston Smith, experienced "the most powerful cosmic homecoming I have ever experienced." Another, Mike Young, had the sensation of being "awash in a sea of color . . . Sometimes it would resolve into patterns with meaning, and other times it would just be this beautiful swirl of color. It was by turns threatening and awe-inspiring. It was a radial design, like a mandala, with the colors in the center leading out to the sides, each one a different color or pattern."

You might say that Pahnke's subjects had passed through a kind of Star Gate. While Ordway's letter was perfectly in keeping with Kubrick's and Clarke's goal of keeping their project accurate scientifically by tying it to current research of all kinds, the subject of Pahnke's experiment—conducted under the supervision of two men who would soon become high priests of sixties psychedelia—provides a direct link between *2001: A Space Odyssey* and the quasi-metaphysical explorations of that decade's drug culture. (Both Leary and Alpert were dismissed from Harvard in 1963. After overtly endorsing the recreational use of LSD, Leary would coin such influential generational catchphrases as "Turn on, tune in, drop out." Alpert soon changed his name to Ram Dass and became a kind of guru. His best-selling 1971 book, *Be Here Now*, is still in print.)

Following his dose of psilocybin, test subject Mike Young soon found himself paralyzed, and experienced the terrifying sensation of his own death, something he later interpreted as the expiration of his ego—something necessary for him to "live in freedom . . . I had to die to become who I could be." The Marsh Chapel Experiment is remembered today as the last formal trial evaluation of the potential spiritual and psychological benefits of hallucinogens before the War on Drugs put a stop to such investigations in the 1970s.

While no record remains of whether Walter Pahnke ever responded to Ordway's request, it's striking how closely the experiences of his subjects—including Young's voyage of terror and rebirth—corresponded to Kubrick's depiction of Dave Bowman's trip "beyond the infinite" in the last quarter of *2001: A Space Odyssey*.

. . .

By June 1, Victor Lyndon, who'd agreed to be Kubrick's associate producer for a second film, had flown in from London and set about compiling a 115-page set of production notes. Jammed with rigorously delineated detail, it left Clarke "completely overwhelmed." In many places, it provides intriguing glimpses of paths not taken and colors not quite used.

An "important personality" in a midsixties setting and a plain business suit was supposed to open the film on a pedagogical note, expounding with the aid of animations on reasons to believe in an "abundance of extraterrestrial civilizations." A prehistoric sequence immediately following would be shot "mainly on location, most probably in South West Africa," with a group of trained animals, including a "leopard or lioness" and warthogs. Other, "nontrusting" warthogs were to be killed in a hunting scene. Apart from its various technological devices, the briefcase of a VIP space traveler, Dr. Heywood Floyd, should contain "a beautifully made and finished pill case, with segmented areas inside containing smooth, sugar-coated tablets of various colors and sizes."

Arriving at an Earth-orbiting space station, Floyd was to replace his old suit with a new one expelled from an automatic dispenser in a four-inch square container, which, when unfolded, "should appear creaseless and in good shape." The old one should be disposed of by "rumpling it up into a small ball" before tossing it into a waste receptacle. Checking into a hotel room in the station, Floyd was supposed to consult a menu on a wall display screen offering such delicacies as "Venusian Mock Turtle Soup" and "Boeuf à la Gagarin," named after the first man in space, Soviet cosmonaut Yuri Gagarin. When delivered "very quickly by the automatic dispenser, but . . . beautifully and elegantly cooked and served," his meal was to be accompanied by fruit "grown in a zero-gravity environment and . . . very, very large—perhaps an eight-inch strawberry or a bunch of giant grapes."

Apart from that, 2001's production notes contain a number of startlingly prescient glimpses of the world we live in today. As of mid-1965, approximately the same time that the US Department of Defense was conceiving of the internet's direct predecessor, ARPANET (Advanced Research Projects Agency Network), Kubrick's intrepid band of futurists had seemingly already visualized important aspects of the new technology's implications. One document sent from Tony Masters to Roger Caras on June 29 listed matter-of-factly—under a letterhead replete with the roaring MGM lion—nine props that he asked Caras to help him with. Number one was "2001 newspaper to be read on some kind of television screen. Should be designed television screen shape; i.e., wider than it is high."

Within a week or so, Caras had indeed made an agreement with the New York Times permitting use of its logotype in a mocked-up electronic edition of its front page. If it had made it into the film, it would have been read by an astronaut on the iPad-type tablet computers seen on board Discovery. And had Kubrick followed through and actually presented the newspaper in this way, there's no doubt that 2001: A Space Odyssey would be remembered today as an important harbinger of the internet. Instead, the director elected to use only a TV transmission, seemingly from the BBC, on the portable tablets.

As a logical spin-off of the tablet computer concept, a sheaf of production notes written in December 1965—immediately prior to the start of live-action shooting—contained an offhanded description of a sight so common today that it's hard even to remember when it wasn't ubiquitous in the world. "A rig should be made for the newspad, so that if one is looking straight at it over a man's shoulder when he is reading, we can illuminate a fixed transparency of one page of a book," Ordway wrote. "On the reverse shot, we will have to place a small light on the hidden face side of the newspad, so that a little light can be shining on his face."

The description is so innocuous, so seemingly blasé, it's easily passed over. Ordway was simply proposing a design enabling a reverse-angle shot of a person's face, illuminated from below by a screen. At this very moment, of course, a sizable percentage of the planet's population is lit in exactly this way. But inevitably this technology, and its resulting lighting geometry, was described for a first time and in a first place. Probably here, decades before the sight became so omnipresent as to be unworthy of comment, as item thirty-two of "Production Meeting No. 6" for *2001: A Space Odyssey*, where it was articulated in the offhandedly dry language of filmmaking, such that the effects team could produce the effect without ambiguity or confusion for "Scene C8: Centrifuge."

. . .

As they continued hashing out their ideas, one by one, the six stories that Kubrick optioned in 1964 dropped away, leaving only "The Sentinel" as the kernel of their narrative—an acorn that would grow into an oak, as Clarke would put it. "Encounter in the Dawn," his tale of an alien survey expedition reconnoitering prehistoric Earth, hadn't made the list, though it would, in fact, serve as the template for the film's Dawn of Man prelude. At one stage that summer, Clarke woke up to the fact that Kubrick still owned the rights to some of his best material, and told him he wanted his five unused stories back.

He clearly wasn't prepared for the response. In his "Son of Dr. Strangelove" draft, Clarke tried to tell the tale in a pithy sentence, writing, "Later, I had the quaint experience of buying back my unused stories from Stanley (a very shrewd businessman) at many times the figure any editor had ever paid for them." This triggered an emphatic response from Kubrick, who crossed it out in bold horizontal pen strokes. "This makes it sound like I swindled you," he complained in the margins. "You paid *less* than I paid you for them."

In the fall of 1964, the author finally received permission to contact his old friend Roger Caras. "I said, 'You bastard! You've been here six months, and you didn't call me,'" Caras remembered decades later. Actually, he was neither offended nor surprised by Kubrick's moratorium on his company. "That's typical Stanley," he observed. "If Stanley wants to work on something intensely, he would stay away from me. He would cut himself off from the world." Invited back into the fold, Caras proceeded to join Clarke and Kubrick for a series of lunches and dinners, and was quickly brought up to speed on the state of the production.

By now, Caras had been with Columbia Pictures for ten years, rising to become head of merchandising promotion for the United States and Canada—a significant job, with thirty-six people working for him. "Then one night, it was a Sunday night, it was eleven thirty, and the phone rang," he said. "And there was Stanley. He said, 'What are you doing tomorrow?' I said, 'I'm going to work, Stanley, like any other slob in the motion picture industry. I'm going to work. That's what I do.' He said, 'Well, quit . . . Quit and join Arthur and me. We're going to move to England and make a film. It'll be called *2001: A Space Odyssey*, but that's just a working title . . . we'll have a lot of fun together.'

"And I said, 'What do I get, Stanley? I got a wife and two children.' He said, 'Obviously, all the expenses and everything else, moving and all that.' And I forgot what the salary was, but it was better than what I was making at Columbia. In London, you need it. 'And you can be vice president of my two companies.' And it really was a decent salary. So I yelled over to Jill, my wife, who was sitting on the couch, and I said, 'Do you want to move to England?' And I'll never forget, without batting an eye, she said, 'Sure.'"

Kubrick's contract with MGM gave him substantial say over the film's promotion. Apart from taking the lead on *2001*'s publicity for Polaris and Kubrick's UK company, Hawk Films, Caras would mastermind outreach to a broad range of US firms. During the first phase of the film's production, he would arrange corporate support in exchange for product placement—at a time when these things were not nearly as commonplace in the film industry as they are today.

By late June, Kubrick's production staff was furiously packing and shipping materials and decamping for the MGM studios at Borehamwood. Ordway and Lange had agreed to extend their consultancy contracts and move to England as well. The Kubrick family

alone accounted for forty-eight steamer trunks—but this didn't include a large library occupying a half dozen floor-to-ceiling book-
shelves in the Polaris office on Central Park West. It contained many
books precious to the director, including ones he'd had since childhood. Now officially Kubrick's VP, Caras came in one day to help
him decide which to ship via transatlantic freight service and which
to get rid of.

"People who deal with Stanley and don't know him perhaps don't
understand that he was an alien," Caras recalled in late 1999. "He
wasn't like us. As friendly as he was—God knows he was friendly—he
just lived on a different plane, operating by himself."

He could be endlessly thoughtful and endlessly thoughtless. I
remember I gave him, when we first got to know each other,
I gave him a copy of my first book, which I, as young authors tend to do, inscribed in flowery terms.* And later, he
was getting rid of his apartment on Central Park West. And
I was up there, and we were deciding what's gonna go where
and who's gonna get what and so forth. And he went into the
library—which was huge, a billion books—began taking books
down, and he stacked some to the side to go into the pack and
take 'em with him, and others he would just throw out. And he
pulled my first book off the shelf, looked at it, looked *in* it, and
he said, "Do you mind?" And he threw it in the rubbish! And
I said, "No, I don't mind." Which was a downright lie. I was
insulted as hell. He would do things like that. But he would be
horrified if told later that he hurt someone's feelings. And a way
you could absolutely throw him into a fit of despair was to say,
"You really hurt her feelings, or his feelings." That would be
very cruel. Jesus, he'd want to go jump off a bridge; he hated
that thought. But he did it, he did it.

---

* *Antarctica: Land of Frozen Time* (Philadelphia: Chilton Books, 1962). Following
2001, Caras became a leading wildlife preservationist and TV personality. He
would ultimately write seventy books, mostly on animals and ecological issues.

In mid-July 1965 Clarke flew to Ceylon for the first time in more than a year for a much-needed break. Not long after, Kubrick and family boarded an ocean liner bound for Southampton.

Within a month or so, it became clear that all of Kubrick's most precious books had been lost in transit.

CHAPTER FIVE

BOREHAMWOOD

*Perhaps our role on this planet is not to
worship God, but to create Him.*

—ARTHUR C. CLARKE

Tony Frewin was vexed. "I don't want to get involved in the film industry," he'd insisted yet again over eggs that morning. He loved his father, of course, but, like most teenagers, simultaneously found him exasperating. The old man had been harping about the same subject for weeks, insisting that he go in for an interview with some big-shot Yank at MGM British Studios in Borehamwood—Boreham Wood, actually; the frauds at the local council had conflated the two in some kind of ludicrous rebranding gimmick—and no matter how much he protested that he didn't give a toss for British films, or American films, and was interested only in European ones—you know, the Nouvelle Vague, for God's sake, or Bergman—Eddie Frewin wouldn't take no for an answer. He kept insisting that Tony give it a shot, that he can't spend all his time getting his arse thrown in the nick—three nights behind bars last time for want of a miserable few pounds, at the age of fifteen. Dad hadn't been amused by that one, and nuclear disarmament be damned. Look, said Eddie, Tony needed to make himself useful, and stood to make a few quid besides, and, anyway, the American— goes by the name of Stanley Kubrick—had just made a film everyone thought was brilliant. It was called *Dr. Strangelove*. Had he seen it? Shot at Shepperton Studios.

This caught Tony's attention. It was the first film by an American director that any of his friends had ever bothered with. In fact, practically the entire UK Campaign for Nuclear Disarmament had gone to see it, coast to coast, and they'd come away doubly determined to fight the global epidemic of insanity best boiled down to the acronym MAD—for mutually assured destruction, appropriately enough.

Actually, *Strangelove* had done more than that. It had shattered Tony's picture of America. Everyone knew the country was insipid, stupid, silly, and complacent. So how could it have produced a film like this?

"Right," Tony said to himself. "The only way I'm ever going to get Dad off my back is to go down for an interview." Eddie bought Tony a pair of new shoes, drove him to the studio—it's what he did: he drove to MGM, he drove *from* MGM; he was a studio driver—and went off looking for Kubrick. Leaving Tony here, now barely seventeen, on a September Sunday in an office in Building 53, a room that managed to be both cramped due to its low ceiling and big enough for a large central conference table with green baize on top—vaguely suggestive of card sharpery or gambling of some dubious sort—big enough also to contain upward of a couple hundred books, including large, full-color art books. Several of which Tony had taken down from the shelves lining the walls, since nobody had shown up to interview him. In fact, nobody had come in at all, apart from weekend cleaning staff in the form of one average-height, dark-haired guy in his thirties wearing baggy blue trousers, an open-necked shirt, and a rumpled blazer bearing evident signs of food stains and ash impacts. He'd entered, smiled in Tony's general direction, and gone off again as Frewin sat there, quietly engrossed in the books.

And what books! They seemed to span every subject—though if you put your mind to it, a discernible pattern began to emerge. There was surrealism, futurism, cosmology, UFOs. There was a well-thumbed copy of a US Department of Defense *Troop Leader's Guide*. There were dozens of science fiction novels. And there were stacks of magazines on every conceivable subject, from nuclear physics to computer science to psychology. "I wouldn't mind working here just so I can get my hands on these books," Frewin thought, turning the pages of Patrick Waldberg's study of German surrealist painter Max Ernst.

The cleaner came back in again, and Tony thought to himself, "Shouldn't he be off with his duster somewhere?"

"Listen, I'm sorry; you Eddie's son?" he asked.

"Yes," said Tony, startled.

"I'm Stanley. Stanley Kubrick."

Jolted into action, Tony rose, and they shook hands. "Gee, you have a wonderful collection of books here," he offered, gathering his wits. "Why've you got so many about surrealism, Dadaism?"

"Well, one of the problems I've got on this film is to come up with convincing extraterrestrial landscapes," said Kubrick, sitting down. "As for Max Ernst, there are all sorts of ideas here. It might suggest something. For example, have you seen *Europe After the Rain*?"

"Yeah," said Tony, regaining his seat, "it's a dazzling painting." It *was* pretty extraterrestrial, come to think of it. "Well, I've always been interested in that sort of thing," he continued, like a man with more years under his belt.

"Really?" said Kubrick, extracting a small notebook and a pen from his jacket. "Are there any other artists I should be looking at?" He scrutinized Frewin inquisitively.

Nobody had ever asked Tony's opinion about anything before. "Well, yeah," he said. Apart from Ernst, the director might take a look at works by Giorgio de Chirico and Jean Arp. And he provided some other names as well. He'd first learned about surrealism four years previously upon reading George Orwell's essay "Benefit of Clergy: Some Notes on Salvador Dali." Tony was from a working-class family, but he'd caught the wave and had been reading ever since.

Kubrick wrote down the names. He explained that the film he was working on concerned the discovery of intelligent alien life, and he asked if Tony had read much science fiction. "Yeah, I've read the novels." Kubrick said that he had, too, and had also absorbed lots of pulp magazines when he was a kid, but he'd always thought that while the ideas could be wonderful, the characterizations were frequently shallow. "Most of the people are treated like robots," he observed. As for the movies, "Cinema has let science fiction down." Most sci-fi films, said Kubrick, were embarrassingly shallow and unconvincing, despite the fact that all kinds of interesting special effects techniques were now available.

Concerning extraterrestrials, the director was confident they were out there somewhere. The problem was the immense distances between the stars. These were so great that entire civilizations might rise and fall over millions of years without other civilizations ever knowing they existed.

And their conversation meandered on in this way for quite a while, covering evolution, the origins of consciousness, the future of human civilization, and other middlebrow topics. Soon two hours had passed, but still Kubrick didn't seem to have anything better to do. In fact, he looked like he was enjoying himself.

Nobody had ever really spoken to Tony like a human being before, let alone bringing him directly into his innermost thoughts like that. In fact, thought Tony to himself, he'd been treated like shit, by idiots, ever since he'd left school. This was rather different.

Finally, Kubrick stood up. "I need a runner," he said.

"When do you want me to start?" Frewin asked.

"How about seven o'clock tomorrow morning?"

"You've got a deal."

Tony Frewin. Photograph by Stanley Kubrick.

And he never went back to his job as a temporary filing clerk at the baker's union on Guilford Street.

. . .

MGM British Studios in Borehamwood, twelve miles north of Charing Cross station, was the most modern British film studio. Sometimes referred to loosely as Elstree, that being the name of the civil parish within which it was once located, it covered 114 acres, making it the largest studio complex in the United Kingdom. It boasted ten soundstages of various sizes, two dubbing theaters, five preview theaters, twenty-two cutting rooms, and processing facilities for all gauges of film stock between 16 and 65 millimeters. Green fields surrounded the complex, sometimes dotted with fluffy white flecks, as though a fine-point brush dipped in titanium white had touched down in the greenery here and there. Sheep.

Usually just referred to simply as MGM by its denizens, the studio's expansive back lot was filled with large standing sets such as Asian and Mediterranean rural villages, a complete Boeing 707 fuselage with cockpit, a Battle of Britain–era Spitfire fighter plane that seemed in flight even while at rest, and a rusting, torpedo-like Japanese kamikaze flying bomb. When *The Dirty Dozen* set was constructed in spring 1966, a large French château rose to dominate a far corner of the complex, and kept on catching fire and exploding in the evenings for the benefit of that World War II drama, with no discernible ill effects visible the next day. Another corner was defined by a fifteen-thousand-square-foot water tank equipped with wave and wind machines, backed by a large billboard ready for whatever kind of nautical sky might be required, from tropical blue to gunmetal grey.

From the front gate, however, it all looked innocuous—a large white-collar industrial park of some sort, pleasantly accessorized in the English style with rosebushes and greenery, and featuring ample parking. Its most distinctive feature was a white two-story art deco central administration building with a square central clock tower about twice as high. Directly above the clock face were three words: Metro, Goldwyn, and Mayer.

At 5,437 miles from Hollywood, it wasn't really far enough for Kubrick. But it would have to do.

It is difficult to exaggerate the extent to which *2001: A Space Odyssey* would dominate MGM's UK studios over the next two and a half years. Of the complex's ten soundstages, nine would be used by the production, with a plurality reserved concurrently. Although the facility had been producing between ten and twelve films per year, it would manage less than half that for the duration of Kubrick's production. Furthermore, MGM chief O'Brien had agreed that the complex's considerable overhead costs would not accrue to the film's budget. A big gamble, particularly given that Borehamwood was already considered something of a financial liability.

. . .

Building 53, MGM Borehamwood. Exterior. Late Summer, 1965.

A low, flat-roofed single-story prefab structure with twelve evenly spaced windows just to the right of the MGM administration building, Hawk Films HQ stands in front of the giant red-brick shoe box housing Stages 6 and 7. A sure sign that the doctor is in, Stanley Kubrick's shiny new Mercedes 220 is parked directly beside his offices at the west end, in front of the Theater 3 entrance—a conveniently situated screening room where yesterday's rushes will soon be scrutinized by the director every morning with something like the attention a brain surgeon pays to the next incision.

For all its capabilities, MGM didn't have the largest stage in the country. That would be Stage H at Shepperton, another major British studio compound, south of Heathrow Airport. At 250 by 120 feet, Stage H had 30,000 square feet to play with and had been reserved by associate producer Victor Lyndon for use from December through early January. It's where *2001*'s immense TMA-1 set would be constructed— for Tycho Magnetic Anomaly 1, an alien artifact the size, shape, texture, and material of which was even now under intense discussion in Kubrick's office.

The director wanted the alien object made of an absolutely clear material, a transparent tetrahedron that would materialize in Africa—

because plans still called for the man-ape prelude to be shot there—and then be excavated on the Moon four million years later, meaning late December at Shepperton. Let's make it in Plexiglas, Kubrick said, which you Brits call Perspex. Go forth and seek the best Perspex people in London, Masters, seek and ye shall find. And Masters said, Right, if that's the decision, he'd look into it right away.

And fortuitously, a Perspex trade fair materialized in London that early fall. All the best manufacturers were there, and so was Masters—for an hour or so, anyway—evaluating. Finally, he approached the gent hosting the most impressive of Perspex sizes and forms. "I'd like you to make a large piece of Perspex for me, please," he said.

"Oh yes, sure," said the gentleman. "How big would you like it?"

"I'd like a sort of pyramid made in Perspex," said Masters.

"Oh yeah, fine," said the gent.

"I want it to be about twelve feet high," continued Masters.

"My God," replied the gent, somewhat put off. "That *is* a big piece of Perspex. May I ask what you intend to do with it?"

"Yeah, I'm going to put it on top of a mountain in Africa," said Masters.

"Ah, well," said the Perspex man, examining his interlocutor covertly to ascertain if anything in his expression provided grounds for the sudden supposition that his leg was being pulled. "Ask a silly question . . ." At this, Masters smiled cryptically. Kubrick demanded absolute discretion, and the designer wasn't natively a blabbermouth anyway. "What's the biggest piece you can make?" Masters inquired.

"Well, I've never made anything that size—I mean, *nobody* has," said the gentleman, frowning thoughtfully. "But we'd like to do it. For various reasons, we can make it best in the shape of a pack of cigarettes, like a big slab."

With this information still in his ear, Masters returned as swiftly as traffic allowed to Building 53, where he found Kubrick behind his desk. "Well, okay," said the director, rapidly reevaluating his position. "Let's make it that shape." After several days of further discussions concerning height, dimensions, aspect ratio, and the like, off went Masters down the motorway again.

"It'll take quite a long time to do the pouring and everything," said the man at the fair. He was pleased, however, to see Masters again. This was what in the trade was known as a rather substantial commission. "And then it takes a month to cool, because it has to cool very slowly; otherwise it'll shatter." It would then have to be polished, also for quite some time—several weeks at least—in order that it be perfect.

A perfect piece of Perspex, the largest ever made.

And Masters brought it unto Kubrick. Before the director saw it, however, a crew set it up on a soundstage, directed lights upon it, and gave it a final polish. It looked magnificent, but it looked like a piece of Plexiglas. Masters and his deputy Ernie Archer went to inform the director that it had arrived.

"Oh, right, let's go and have a look at it," Kubrick said, rising from his desk. He accompanied them down glass-roofed pathways, up metal staircases, through soundproofed doors, and onto the stage area.

Three hominids approached the gleaming, brightly lit monolith. "Oh God," said Kubrick. "I can *see* it. It's sort of greenish. It looks like a piece of glass."

"Yes, yes," said Masters, "afraid it does—it looks like a piece of Plexiglas." He'd unconsciously adopted the American term.

"Oh God," Kubrick repeated. "I imagined it would be completely clear."

"Well, it *is* nearly two feet thick, you know," said Masters, frowning slightly. They contemplated the gleaming slab of greenish, refractive, reflective polymethyl methacrylate—over two tons of it. Several workers in blue overalls stood a slight distance away. Plexi. Lucite. Perspex. Whatever one chose to call it, it was not the magically ultraclear, almost invisible alien artifact of Kubrick's imagination.

"Ah," he sighed regretfully. "File it."

"Do *what?*" Masters asked incredulously.

"File it," Kubrick repeated.

"Oh," said Masters. "Okay." He turned to the workers. "Take it away, boys."

As to the expense, one of Stanley's young assistants would later esti-

mate that it cost more than a fair-sized home within Greater London.*
Masters and Archer accompanied the director back to his office.

"I don't believe it," Kubrick said ruefully. "It looks like a piece of glass."

"Well, I'm afraid it does, yes, you're right there," said Masters, who had a reputation for thinking nimbly on his feet and devising alternate solutions with great rapidity. "So, let's just make a black one, because then we won't know *what* that is."

"Okay, make a black one," said Kubrick.

The size, shape, and color of *2001*'s monolith had been established.

■ ■ ■

Clarke's homecoming in Colombo lasted only a month or so, long enough to cover the usual bounced checks, see to it that Mike Wilson had adequate financial resources to start shooting his next film—a James Bond parody officially titled *Sorungeth Soru* but often referred to simply as *Jamis Bandu*, featuring the adventures of the eponymous Sinhalese secret agent—catch up on his mail, and hop a jet bound for London.

Arriving at Elstree on August 20, he discovered that Kubrick had been so worried when NASA's unmanned *Mariner 4* spacecraft had flown by Mars a few weeks previously that he'd contacted Lloyd's of London and asked the insurance company to draw up a policy to compensate him if their plot was demolished due to the discovery of extraterrestrial life. "How the underwriters managed to compute the premium I can't imagine," Clarke wrote wonderingly, "but the figure they quoted was slightly astronomical, and the project was dropped. Stanley decided to take his chances with the universe."

Throughout the year and well into 1966 and actual production, Kubrick and Clarke remained enmeshed in seemingly perpetual story development. Despite their best efforts, the film's ending was repeatedly deemed unsatisfactory. As of August 1965, an "unbelievably graceful and beautiful humanoid" was supposed to approach Bowman, the crew's sole survivor, and lead him into "infinite darkness." How

---

* Fred Ordway later pegged it at about $50,000, or a bit under $400,000 today.

to achieve such grace and beauty had been left indeterminate. In any case, it wasn't just inadequate, it flirted with risibility. Kubrick didn't do risible.

Earlier in the year, they'd decided that astronaut Frank Poole, Bowman's sidekick, would need to die in an accident. This would spice up the journey to Jupiter, but the exact nature of the accident was in flux, and even the finality of his death was sometimes in question; one Clarke journal entry had him "fighting hard to stop Stan from bringing Dr. Poole back from the dead. I'm afraid his obsession with immortality has overcome his artistic instincts." * In May, Clarke had devised a scene where Poole's space pod smashes into *Discovery*'s main antenna, breaking their link to Earth and sending both antenna and Poole spinning off into space. What had caused the accident wasn't yet clear, however. The spacecraft's computer, then named Athena, after the goddess who'd helped Odysseus out of many a scrape, wasn't yet fully developed as a character.

Soon after Clarke's return from Ceylon, Kubrick told him he'd devised another plot twist in which Poole and Bowman, the only two Jupiter-bound astronauts not in suspended animation, had been kept in the dark about their mission's true purpose. According to his new idea, only the "sleeping beauties" locked away in *Discovery*'s sarcophagus-like hibernaculums had been informed that they were seeking contact with intelligent extraterrestrial life—and they weren't supposed to be revived until arrival at Jupiter.

While Clarke wasn't particularly pleased by Kubrick's revision, he also wasn't happy that he hadn't been able to come up with a mutually satisfactory ending. On the twenty-fourth, he produced a two-page, nine-point memo. "I've put my finger on a flaw that worried me," he wrote Kubrick. "It's simply insulting to men of this caliber to assume they can't keep a secret that hundreds of others must know. Also it

---

* Kubrick had read Robert Ettinger's book *The Prospect of Immortality* (Garden City, NY: Doubleday, 1964), which introduced cryonics: the freezing of human bodies immediately after death in hopes that scientific advances would one day revive them.

introduces an unnecessary risk." In response, Kubrick had scribbled, "You can construct just as logical a reason for it if you try." Clarke: "This element of suspense rather artificial and improbable." Kubrick: "Don't agree. Only if you fail to try to make it work."

He wasn't giving an inch.

A day later, Clarke conceived of yet another new ending to the film in a second message, written from his brother's house on Nightingale Road.

> It's amazing how long it takes to see the obvious. The weakness of our ending was that we never explained what happened to Bowman, but left it entirely to the imagination. Well, we can't explain it—but we can symbolize it perfectly in a way that will push all sorts of subconscious and even Freudian buttons . . . Remember the beautiful little spaceship Bowman sees on landing? We used it merely to draw a contrast with Earth's primitive technology. Well, after his processing in the hotel room, Bowman will be master of new sciences. The narration *tells* us this—the visual effects prepare us emotionally—but we will only *believe* it when the room vanishes, and he is alone on the skyrock with the ship of the super-race—and his own Model T space pod. The ship is man's new tool—the equivalent of Moon-watcher's weapons. It symbolizes all the new wisdom of the stars. Bowman—with one backward glance at the pod—walks toward it. As if in greeting, it rises a few inches as he approaches. Close up, the hull texture must be beautiful (soft? warm?). He stands thoughtfully beside it, looking up at the sky and the road back to Earth. As he does so, he strokes it absentmindedly, almost voluptuously. (Appeal of fine sports car, camera.) "Now he was master of the world, etc . . ." The End

It would pack "a hell of a punch," he concluded, "and I'm sure it solves all our problems."

We don't know what Kubrick may have thought of Clarke's soft, warm Freudian spacecraft that rises "a few inches" in expectation of being stroked, but the director's terse response only extended an emerging split between their two approaches: "I prefer present nonspecific result for film. Maybe this can work in book but it won't on film."

He didn't do risible.*

. . .

Sets, meanwhile, were being assembled at a furious clip. Masters and Archer coordinated with art director John Hoesli and one of the best construction managers in England, Dick Frift, to build spacecraft interiors of unprecedented complexity, and these were starting to dominate Stages 2, 3, and 6. The floor of Stage 4, meanwhile, was being torn up and reinforced so as to bear the weight of the immense centrifuge set, then being prepared offsite by UK aviation firm Vickers-Armstrongs—makers of the famous RAF Spitfire, a representative of which was parked in the back lot, waiting for the Luftwaffe to return.

By now, the design team first assembled in New York was working together seamlessly and collaborating with Roger Caras, who'd remained in Manhattan for the time being, where he could interface more easily with major American corporations and seek their input into technological concepts and designs until the shooting started. After that, he would move to London. Masters's staff of forty architects, set decorators, model builders, and prop designers was spread across a network of workshops. Together they designed, built, furnished, and finished a unified vision of the future.

One key to the film's extraordinarily believable mise-en-scène—its overall "look"—was this coordination between technology-producing industry—which included leading industrial designers such as Eliot Noyes, the architect of IBM's integrated corporate visual identity and also its innovative Selectric typewriter—and the

---

* In a 1972 story titled "The Big Space Fuck," Kurt Vonnegut names his spaceship with "eight hundred pounds of freeze-dried jizzum in its nose" the *Arthur C. Clarke*, "in honor of a famous space pioneer." Its mission is to impregnate the Andromeda Galaxy (*Again, Dangerous Visions*, Harlan Ellison, ed., New York: Doubleday, 1972).

production's tripartite design leadership, with Ordway ensuring absolute technological plausibility based on corporate and governmental research and development, and Lange combining that with his extensive knowledge of space technologies to police the look of the sets and models and bring a sense of style to it all. Lange, said Masters, "put the authenticity into it."

To the extent that a division of authorship existed in such an intensely collaborative project, it was Masters who came up with kinetic concepts such as the rotating sets and designed the interiors and props, and Lange who did the vehicle and space station exteriors, as well as the film's extraordinary space suits. But in fact everyone worked on everything together, with lots of cross-pollination during meetings and discussions. Special effects pioneer Wally Veevers, who'd done the miniatures in *Dr. Strangelove*, would supervise the actual model production and filming, among other things. "The strange thing was that we all worked together for so long that we began to design in the same way," Masters recalled in 1977. "Just like Georgian or Victorian is a period, we designed *2001* as a period. We designed a way to live, right down to the last knife and fork. If we had to design a door, we would do it in *our* style."

On top of this, Masters and Kubrick weren't above using compelling work by outside designers, which simply had to conform to their vision of what the early twenty-first century would look like. These included dozens of futuristically curvaceous scarlet Djinn chairs, designed by Olivier Mourgue in 1963, which livened up the otherwise monochrome interior of Space Station 5, and several examples of Eero Saarinen's 1957 Tulip table. Geoffrey Harcourt's chrome-and-leather lounge chairs furnished the Clavius Moon Base conference room.

Complicating everything was the endlessly mutable story line. Despite the gargantuan nature of the undertaking—the giant, complex sets, the big budgets, the risk MGM was taking, the fact that a major studio complex with thousands of employees was almost entirely devoted to realizing his vision—Kubrick was winging it. The project was all in his head. In a normal head, of course, this would be a recipe for disaster. What began to emerge from *his* head, however, was actually a form of refinement. For all the seeming chaos, dross was methodically being stripped away and messages refined.

"We weren't working to any particular schedule or even any particular script," said Masters. "We had a basic idea—which was, of course, Arthur Clarke's story—but we never had a finished script. We worked through the movie, changing every day our ideas of what we were going to do tomorrow."

> We'd get together with Stanley in the evening and talk about what we were going to do the next day—and as a result, the whole thing would change. The production department was suicidal. Because, daily, absolutely *everything* that had been planned would be thrown straight out the window; we'd do something completely different. And that was how we worked from day to day. Having got through that particular problem, we'd then say, "Well, what are we going to do tomorrow?" but, in fact, when we came to it, he'd say, "Oh well, the hell with that, let's do something *really* interesting." And out of it—out of talk and talk and talk came something really much better than we thought we were going to do.

To add to the confusion, despite being a profoundly visual thinker, Kubrick was nearly incapable of actually picturing visual concepts when they were described verbally. He also didn't necessarily know what he liked until he saw it, and so he needed to be given several choices. These characteristics served to delay and sometimes confound many an idea conceived by the design department.

With most of the sets ostensibly depicting vehicles in a weightless environment, and with so much of the action centered on the turning, artificial-gravity-producing centrifuge at the center of *Discovery*'s spherical front end, several of the sets needed to rotate with a flawless smoothness. Masters conceived of a design permitting movement of the astronauts from the weightless part of the spacecraft to the centrifuge. In the final shot, they would proceed down a corridor, at the end of which was a rotating wall with a ladder. On reaching the end, they would seamlessly transition to the turning element—and, from an audience perspective, magically proceed to spin 360 degrees as they climbed out of the frame, thus "descending" into the centrifuge.

In order to accomplish this sleight of vision, however, both the entire corridor and the rotating wall and ladder assembly had to be mounted on external frames such that each could turn smoothly within armatures packed with ball bearings. "To capture that, we had a camera at the end of the corridor, screwed to the deck," Masters recalled. "The men walked away from us toward the revolving set, and as soon as they stepped onto the drum, we stopped it and began revolving the corridor."

> But because the camera moved with it, you couldn't see any change at all. On the screen, it looked like the end was revolving all the time and the men were going around with it . . . and they just went through a hole in the bottom of the thing. So that was two revolving wheels—one was revolving at the end, and then that one stopped and this one had to start revolving. And the mechanics of that, to get it to just the right moment, you had two levers, and as [they] stepped on, you had to go *arrrgh* with these great things—these were huge, great, massive sets revolving and rumbling around—it really worked well.

When all this was first explained to Kubrick, however, for all his legendary acuity, he just didn't get it, and Masters spent a marathon session one night in the fall of 1965 sketching away at a blackboard in the director's office, in an increasingly desperate attempt to convey the essence of the two elements. Finally, at about one in the morning, Kubrick said, "I think I see it! I think I know what you mean."

"Sweat was now running off, you know," the designer recalled. "Thank God for that—I won that one. But oh, was it hard going."

In the end, Kubrick loved the shot, which, despite the complexity behind it, unfolds with a seemingly effortless, antigravitational simplicity.

. . .

Harry Lange's transition from German to English life via a two-decade interlude with von Braun in the American Deep South didn't come without bumps and jolts. Nothing he'd seen in Alabama particularly disturbed certain attitudes that he'd absorbed in the fatherland, and though he was universally well liked by those who worked directly with him, that left most of MGM's substantial staff, who were not without memories of the Luftwaffe's blitz on major English cities in the early years of World War II. What they saw was a German man showing up to work in a quasi-militaristic "Janker" hunting jacket with stag-antler lapel details—the kind of "Tracht" traditional garments associated with Hitler, Austria, and Bavaria. As though willfully seeking to accentuate this effect, Lange—a man of some means following his marriage to a Huntsville heiress—had plunged into equestrian pursuits soon after his arrival. As a result, he also favored riding boots that, minus the Janker gear, would have

Harry Lange in his studio at Borehamwood.

looked a bit less like Prussian jackboots and more like standard British foxhunting attire. With them, however, they bore certain unfortunate connotations.

How conscious all of this was is open to debate. Christopher Frayling, author of an excellent monograph on Lange, believes the designer may have been "making a joke of it . . . perhaps subtly trying to 'lay the ghost' . . . by bringing the culture clash out into the open." If so, the designer forgot all subtlety when he brought a scale model of a V-2 rocket into his office one day and positioned it on his desk, where it existed in uneasy resonance with the Confederate flag he'd already pinned to the wall. Von Braun's "vengeance weapon" had killed more than two thousand people in London alone during the closing months of the war, and word of Lange's provocation spread rapidly, triggering a walkout of the British design staff. A swift intervention by Kubrick made the model and flag disappear, however, and gradually MGM's designers returned to their desks, still muttering and casting glances.

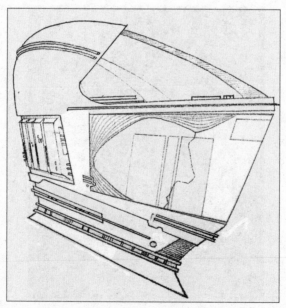

Lange's space helmet design.

Whatever his politics, one thing that you couldn't take away from Lange was the impressive new space suit designs he'd finalized that fall, just in time for the aptly named Manchester company, Frankenstein & Sons, to produce them. What Frankenstein delivered, however, still required accessorizing with all the details necessary to bring them to life, so to speak—for example, the addition of backpacks, front maneuvering controllers, button panels for the arms, and, of course, the helmets—which were produced either at Borehamwood or by AGM, the London company also busy manufacturing "Daleks": the cylindrical alien cyborgs seen in *Doctor Who*, the cult BBC-TV series.

Lange's suits came in two models. The silver, lunar versions were slightly different from *Discovery*'s multicolored deep-space EVA (extravehicular activity) suits. The latter had large silver air hoses looping, not unlike scuba regulators, from their backpacks to the base of their helmets—a design flaw thoughtfully added at the last minute as Kubrick and Clarke refined the nature of astronaut Frank Poole's upcoming accident.

Lange would spend the rest of his life working as a film designer in the United Kingdom, and it was *2001*'s extraordinary space helmets that perhaps best symbolize his transition from an Austro-Bavarian to a more acceptably British frame of reference. After multiple drawings of rounded designs—all of which ended up looking too similar to motorcycle crash helmets—his final design was based on the oval, forward-thrusting shape of a certain species of British hunting cap, the men's Ascot riding hat.

. . .

A good sense of Roger Caras's emerging role as Kubrick's most trusted advisor and confidant—a kind of New York counterpart to his LA fixer Louis Blau, only with professional PR smarts—can be found in a telling exchange in late July concerning IBM's advice on how to present *Discovery*'s talking computer, Athena. Kubrick had been expecting ideas concerning computer input and output devices and was unprepared for what he actually received from the company. In a cover letter to a document from Eliot Noyes's influential design

bureau think tank—one that included drawings of astronauts floating within a kind of "brain room"—Caras wrote, "As you can see,
they say that a computer of the complexity required by the *Discovery*
spacecraft would be a computer into which men went, rather than a
computer around which men walk. This is an interesting idea, and if
the plan for the *Discovery* will accommodate this, you may want to
consider it very carefully."

The communication caught Kubrick in an uncharacteristically de-
featist frame of mind. "[T]he IBM Athena drawings are useless and totally
irrelevant to our needs and what I must presume were Fred's discussions
with IBM," he responded, referring to Ordway. "I'm extremely bored
and depressed with all this." He proceeded to list what IBM should *really*
help with—including "detailed design concepts" (although the company
had provided just that). "There is absolutely no time to waste," he concluded. "Even having to write this letter adds chips to what seems to me
to be a completely lost hand. I know this is not your fault or Fred's, and
don't take this as criticism of yourselves. It is merely a total fuckup which
not only fails to do what was hoped for but costs time." He signed off in
caps: "Annoyed and depressed but lovingly, S."

In fact, IBM's recommendation couldn't have been more salient,
and would eventually result in the construction of *Discovery*'s Brain
Room—location of perhaps *2001*'s most extraordinary scene, when
astronaut Dave Bowman lobotomizes the ship's computer, by then
named HAL. But first Kubrick had to chill out, reconsider, and let the
dramatic possibilities form in his imagination.★

Other letters transmitting advice to Kubrick and his design trust
in London also tapped the thinking of some of the finest futurists
then working, and were equally critical to the film's final look. As a

---

★ Those tempted to observe that by the actual turn of the century computers
were already far smaller than what Eliot Noyes and *2001* envisioned would do
well simply to do an online search for pictures of contemporary supercomputers.
They will discover just how right Masters, Ordway, Lange, and company were,
based on IBM's idea.

by-product of Kubrick's and Clarke's questing, cerebral commitment to scientific and technological accuracy, a big-budget Hollywood production had been transformed into a giant research and development think tank. A sustained campaign by Ordway and Caras to interest leading American technology firms in participating—largely in exchange for product placement within the film's sets and mention during its promotion—was paying off.

A sampling of their letters written in July 1965 alone reveals they covered such diverse topics as mammalian hibernation; space suit designs; lunar charts; lunar photography from major observatories; interplanetary nuclear propulsion; information about Jupiter, Saturn, and their moons and rings; communications systems of many kinds; scientific equipment to be used on the Moon and at Jupiter; photos of the Earth from "balloons, missiles, and satellites"; and countless other themes. Apart from IBM, companies approached included Hilton Hotels, Parker Pens, Pan American World Airlines, Hewlett-Packard, Bell Labs, Armstrong Cork, Seabrook Farms, Bausch & Lomb, and Whirlpool. The number of firms consulted ultimately topped forty.

In the summer of 1965, Kubrick received two detailed Bell Labs reports written by A. Michael Noll, the digital arts and 3-D animations pioneer, and information theorist John R. Pierce—coiner of the term "transistor" and head of the team that built the first communications satellite. They recommended that *Discovery*'s communications systems feature multiple "fairly large, flat, and rectangular" screens, with "no indication of the massive depth of equipment behind them." Flat screens were, of course, unknown, at least outside of movie theaters, in 1965. Incorporating them within the production's set designs helped ensure *2001*'s futuristic sheen and gives it an eerily predictive, contemporary feeling even today. A single piece of advice from two of the best minds in the business helped bring about the film's prescient portrayal of our screen-based future.

As Kubrick's conception of Athena's role in the story evolved, his advisors' communications with IBM started hinting at the computer's transformation from a reliable but somewhat dense crew assistant to something rather more complex. Noyes and company's suggestion that crew members could physically move about *within* Athena had also suggested certain possibilities—at least, once Kubrick got over his initial aversion. Conflict being the essence of drama, as of August 20, the director was already contemplating that Athena cause the death of one of the hibernating crew members, V. F. Kaminsky.★

On August 24—the same day Clarke questioned the need to keep the waking crew ignorant of the mission's true purpose—Ordway wrote a somewhat guarded mail to IBM executive Eugene Riordan. In a tone indicative of the need to extract information without risking corporate alarm, he sought advice on "quasi-independent" actions that "conceivably could be taken" by the ship's computer. "Let us look down the road to the year 2001," Ordway wrote. "Do you think that by that time computers will be able to think, so to speak, somewhat to themselves and to initiate any actions that are not strictly according to the programming?"

Suppose in the interest of secrecy, that some important mission information was made known to the vehicle commander and not to the rest of the crew. Because the crew had access to Athena, this information could be stored in her. Yet she would be aware that certain flight procedures were being taken in a manner inconsistent with the information available to her. Take another possibility that occurred to us. Could Athena become slightly hypochondriacal, reporting rather more than necessary—not to be overdone—that this or that circuit or device should be checked for malfunction or impending malfunction? It would have to be very consciously presented at a threshold that would barely raise the suspicions of the mathematicians and computer

---

★ Given that he was to be murdered by an artificial intelligence, the name is an interestingly backhanded sort of tribute to one of Kubrick's top advisors, Massachusetts Institute of Technology cognitive scientist and artificial intelligence pioneer Marvin Minsky.

specialists aboard. Athena might also exhibit aggressions, of a mild type, that would somehow be manifested.

IBM's cooperation thus far had been predicated on an agreement that the company logo be visible on various of the film's technologies—and indeed it's seen throughout, including on *Discovery*'s iPad-like tablet computers. By October, however, those "slightly hypochondriacal" symptoms had clearly worsened to the point where another mail to Riordan was in order. This time Ordway informed the executive that *Discovery* had been "evolving" into "a considerably more experimental vehicle than originally visualized." Several "interesting plot points" had become feasible involving, in some cases, malfunctioning equipment. "Naturally, we do not want to present any IBM equipment in such light," Ordway wrote. Accordingly, they had decided *Discovery*'s computer would be labeled "as an experimental research and development type, recording only its number and the name of the sponsoring government agency."

The new name Kubrick and Clarke ultimately settled on compounded two terms, signifying a *H*euristically programmed *AL*gorithmic computer. The terms behind the acronym, and possibly the acronym itself, were originally suggested by Marvin Minsky, cofounder of MIT's Artificial Intelligence Laboratory. "Heuristics are, of course, rules of thumb; tricks or techniques that might work on a problem, or often work, but aren't guaranteed," Minsky commented in 1997. "Algorithmic implies inviolate rules, such as *If A then B, and A, therefore B*. HAL was supposed to have the best of both worlds." This duality between algorithms and heuristics—between dogmatic rules and interactive, trial-and-error paths to a solution—already hints at the core conflict that HAL would experience when asked to keep the mission's true purpose from the waking astronauts.★

★ Despite their efforts to distance *Discovery*'s "experimental" supercomputer from the Big Blue, when the film was released, someone noticed that if you advance the acronym HAL by a single letter, you arrive at IBM. Although the chance of this happening accidentally is vanishingly rare, both Kubrick and

On October 12, HAL was still Athena, but Clarke alluded to the emerging dramatic possibilities in a note to Kubrick: "If we wish, we can make the 'accident' an integral part of our theme, not just an episode inserted for excitement. After all, our story is a quest for truth. Athena's action shows what happens when the truth is concealed. We should delicately underline this at the appropriate point." The computer's internal conflict when asked to lie to the crew would be made more explicit in Clarke's novel than in the film, though certainly it's there in later script drafts and also in dialogue Kubrick filmed but ultimately didn't use.

An intriguing set of notes from late October survives in Kubrick's hand, a kind of Rosetta stone in which the director grapples with how to convey the computer's dilemma: that of being programmed to deceive the crew despite having been created expressly to provide them with flawless, objective data. "One evening (or whenever Poole's sleep cycle is), he brings up a rumor he heard prior to departure. He is semiserious," Kubrick wrote. "Rumor was that there was a secret aspect of the mission that only the sleepers knew, and that was why they were trained separately and brought aboard already asleep." The two crew members discuss this "fantastic possibility"—which the director compared with "rumors of high-level CIA men being involved in [the] Kennedy assassination"—and finally decide to ask the computer. "They do so as a joke (but obviously covering real interest) . . . The computer says, 'No.' There is a devilish, perverse humor about it . . . the imp of the perverse."

The imp of the perverse. That Kubrickian signature, surfacing as a note to self.

He also hit on the idea of chess—stylized combat, after all—as a way of conveying the computer's emerging deviancy. Both Bowman and Poole could play Athena, he suggested, and realize gradually that they

---

Clarke strenuously denied this was intentional, and there's no reason to doubt their sincerity. If not entirely coincidental, it may have been unconscious—much like their realizing that the name Bowman was connected to Odysseus months after deciding on it.

never win, "even when they play over their heads with chess books." Since she was programmed to lose half the time, Kubrick wrote, this "should immediately be recognized as a minor programming error; not serious but needs to be watched."

Aspects of these scenarios found their way into the final film, but one further intriguing element was discarded. Bowman was envisioned working with Athena when "suddenly computer asks about paradox all Cretans are liars. Or puts up an illusion and asks Bowman to define an illusion."★ And he wrote Bowman's response: "You seem very interested in illusion. You asked me this several times during the past week." Kubrick also envisioned that both crew members would gradually become aware they were being monitored at all times by the computer.

On the fourteenth page of his handwritten notes, under the heading "Killing the Computer," Kubrick arrived at a moment of sheer intuitive insight. It appeared as a sentence fragment so formally empathetic with its subject, it is as though he'd momentarily assumed the role he was conjuring: "Computer tries to talk Bowman out of erasing—incapacitating it—slowly become more + more"

There's no ellipsis, and the sentence never ends. He'd already moved the searchlight of his mind on to the next thing.

· · ·

Doug Trumbull had been suffering acute withdrawal pangs that Los Angeles summer. After painting the rotating galaxy for *To the Moon and Beyond*—the experimental Cinerama short film that had impressed Kubrick at the 1964–65 World's Fair—he'd worked for a few months on exactly the kind of thing that he loved: namely, drawings of moon

---

★ Kubrick is referencing Epimenides's paradox: "Epimenides the Cretan says that all the Cretans are liars, but Epimenides is himself a Cretan; therefore he is himself a liar. But if he be a liar, what he says is untrue, and consequently the Cretans are veracious; but Epimenides is a Cretan, and therefore what he says is true; saying the Cretans are liars, Epimenides is himself a liar, and what he says is untrue. Thus we may go on alternately proving that Epimenides and the Cretans are truthful and untruthful." (Thomas Fowler, 1869)

bases, spacecraft, and landing pads, and it had been like a steady dopamine drip to the brain. But Kubrick had moved farther east, cutting Graphic Films loose in the process, and with less work to do, the company had been forced to let him go. Trumbull had been into science fiction since he was a child, and although he'd started a little furniture company in Malibu to augment his income, it wasn't really where he saw himself.

He called his erstwhile boss, Con Pederson. It had been really exciting working on the Kubrick space project, he said. How could he contact the producer? Pederson explained that a confidentiality clause in his contract made it hard for him to say. There was an awkward pause. They'd worked well together on a couple projects, including *To the Moon and Beyond*. "Look," Pederson said finally. "If you went down to the office, you just might, maybe, find Mr. Kubrick's phone number penciled into the corner of the bulletin board."

Trumbull had gone in the back door of Graphic, bypassing the receptionist, so many times it was practically a reflex. Now he wasted no time repeating the move, found the number, took it home, calculated what time it was in London, and dialed. With no secretarial intervention, the director picked up immediately, and Trumbull introduced himself. "I'm one of the illustrators that's been working on the drawings you've been getting, and I want to go to work on your picture," he declared. Apart from his work on *To the Moon and Beyond*, he detailed his other qualifications as well. While that didn't take particularly long, his previous projects had all been seemingly perfectly attuned to *2001*'s subject matter and requirements.

"Absolutely great," Kubrick replied. "You're on, you've got a job—come on over . . . I'll pay you four hundred dollars a week." Hawk Films would arrange plane tickets for him and his wife, and find them accommodations as well. Welcome aboard. Was there anything else?

On arrival in mid-August, Trumbull, a fresh-faced California kid in a cowboy hat on his first trip out of the country, discovered so much activity going on at Borehamwood that he was afraid he'd gotten there too late. "I was twenty-three at the time and didn't know anything, really—just a little bit about animation and background painting. I didn't even understand photography very well. Before I left for England, I bought a Pentax camera, and when I got there, I started fooling around with a little black-and-white darkroom at home, just learning the fundamentals," he recalled.

He needn't have worried. There was plenty left to do—live-action production hadn't even started yet—and in the next two and a half years, Trumbull would rise from grunt animator to become one of *2001*'s four leading effects supervisors, leaving a distinctively innovative visual stamp on the film in the process.

■ ■ ■

2001: A Space Odyssey was an analog-to-digital undertaking from the beginning. The projected ubiquity of computer motion graphics decades in the future had been internalized by Kubrick and his designers, but because the kind of processing power needed to drive the incoming information age wasn't yet available, it would all have to be done by hand.

Trumbull had been a painstaking painter of cells for animations at Graphic Films, and Kubrick and Wally Gentleman gave him a day to shake off the jetlag before they put him to work. He was charged with making the computer readouts that would wink, jostle, and whizz purposefully about on all the flat screens being built into the sets, each of which had room behind it to hide a "massive depth of equipment," as Bell Labs had put it, in the form of a Bell & Howell 16-millimeter film projector. These included the ostensibly handheld tablet computers, which would actually be permanently affixed to the tabletop they appeared casually placed on top of, with projectors concealed behind them.

The screens themselves were opaque ground glass, perfect for rear projection. The Honeywell corporation had advised that what was to unfold on these ubiquitous displays should include headings such as Computer, Life, Radiation, and Propulsion—a cycling of systems critical to *Discovery*'s functioning. Otherwise Trumbull was given wide latitude to determine their contents; it simply needed to look authoritative and credible. Some animations would be more central to the shots,

Doug Trumbull at Borehamwood.

however, and would have more precisely programmed content, such as a navigational screen in the *Orion* space-plane cockpit, which would depict a rotating three-dimensional graphic display of Space Station 5's central rectangular docking port.

Friction had been building inexorably between Gentleman and Kubrick ever since they arrived in England. The former was alarmed by what he saw as a needless waste of resources, money, and time, and critical of what he saw as the director's pigheaded insistence on testing techniques Gentleman already knew wouldn't work. The latter was increasingly irritated at his collaborator's resistance to his ideas and what he regarded as a know-it-all manner. "Wally was a very scientific, linear, very elegant, very gentlemanly, very erudite, very well-trained, and experienced guy," Trumbull recalled. "And Kubrick was very free-wheeling: 'Whatever everybody's doing, I'm doing just the opposite' kind of thing. And I think it just kind of rubbed Wally the wrong way."

For his part, Trumbull was honored to be working with one of the visual innovators behind *Universe*, which he'd seen at Graphic Films, and which had been screened for him again immediately on his arrival at Borehamwood almost like a form of freshman orientation. Gentleman noted with approval that Trumbull was absolutely unintimidated by the director. "Doug would sort of wander onto the set and, with all the ingenuity and ingeniousness of a young man, would state exactly what he was thinking at the time," he said. "And this kind of nettled Stanley to begin with, but he gradually got used to it. But Doug was really working from a strongpoint of authority because his work was so very good. And I think he made the greatest contribution, finally, to the film of anybody."

Their early conception of a visual effects production chain had been to farm out the animation and model making to external companies, so Trumbull set about using traditional numbered animation cue sheets and creating artwork for the displays. It took only a few days of this for him to understand that conventional techniques were simply too labor intensive, and that their subcontractor, an animation service near the studio, was simply too slow. "Wally said, 'We can figure out a better way to do this,'" Trumbull recalled. "Because he was completely fearless. And he said, 'Let's build our own animation stand. Screw these guys. We're going to do it our way.'"

They cannibalized an old Mitchell single-frame movie camera from an optical printer, built a stand from a scaffolding made of pipes and clamps—the ubiquitous Borehamwood rigging material also used for sets and scenography, lumber being expensive in England—and started animating at a rapid clip. Trumbull recruited an energized teenager, Bruce Logan, directly from the animation service they'd just dropped and put him to work. Their telemetry displays were taken from "a million random sources of neat-looking graphs": *Scientific American* magazine, NASA manuals, and the like. These were xeroxed and then photographed as high-contrast stats, or negative transparencies.

Trumbull also devised an ostensibly high-tech language of letters and numbers to convey *Discovery*'s supercomputer data-stream, laced with "strange little acronyms and funny little three-letter fake words and things," typed it up on an IBM Selectric, and made more high-contrast transparencies. All this was funneled onto the glass plate in front of the Mitchell as he and Logan blasted rock 'n' roll at all hours of the day and night. "We'd sit there on the animation stand cranking out enormous volumes of footage, one frame at a time," he remembered in 1976. "We'd shove this stuff under the stand and line it up by hand, put a red gel over it, shoot a few frames, then move it around and burn in something else or change some numbers. It was all just tape and paper and colored gels."

Hitting their stride, they managed to produce ten minutes of read-outs a day—a pace practically unheard of in animation—all of it intended to provide the impression that the computer was processing an "autonomous, twenty-four-hour-a-day deluge of data." It was a triumph of youthful enthusiasm over the old, time-consuming way of doing things, and they brought it off without visible compromise. But "In reality, it was a kind of high-tech window dressing," Trumbull said.

Maybe so, but their evocation of endlessly hosing data—a nonstop stream of seemingly authoritative info-graphics, equations, acronyms, and letters—gave pulsing, almost capillary life to the computer running *Discovery* and the scrupulously engineered world in which *2001*'s story unfolded. No less critically than the design team's creation of a futuristic period equivalent to Georgian or Victorian, it conjured an entire upcoming chapter of civilization.

. . .

The studios at Borehamwood were stratified into what amounted to a rigid, union-enforced caste system, in which designers weren't supposed to pick up a screwdriver in case they threatened machine shop jobs, and set builders weren't to be seen applying a pencil to a blueprint for the same reason—at least in theory. As a faux cowboy from California with a genial manner and disarming grin, Trumbull was conspicuously sui generis, however, and he soon discovered that his outsider status came with huge advantages. American film workers were able to work at MGM in the first place because of a loophole in a British film subsidy program called the Eady Levy—essentially a tax on box office intended to keep the local industry alive by luring US studios with substantial tax breaks. As a result, UK production was cheaper than in the United States—one reason, of course, why Kubrick was there and why he'd shot his previous two films in England as well.

The Eady cap on non–UK citizens or legal residents limited the number of Americans who could work on *2001* to 20 percent. But because Trumbull didn't need to qualify for union membership under Eady, he discovered that he could go directly to department heads without bureaucratic obstacles—for example, to the guy running the machine shop, to request that parts be made for his new animation stand. He would use this freedom to fullest advantage over the next two years.

Kubrick, of course, had his own bigger-picture workaround. Simply by virtue of geographic distance, he'd seen to it that MGM had very limited abilities to intervene in his production. "There was no studio management interference at all," Trumbull recalled. "We were doing new things all the time. It was a free-wheeling situation where there was no schedule, no budget, no delivery date, and we were just going to solve all these problems, one after the other, as they came up. You know, the movie's trundling ahead. They're building sets. They're building all their props. They're building their miniatures. And every time something would come up, Stanley would say, 'Well, Doug, what can you do to help me fix this thing?'"

One thing he helped fix concerned miniature work. Soon after Trumbull and Logan got their animation stand rigged, a London company called Mastermodels delivered a three-foot-long Moon Bus—the first of several spacecraft they'd been contracted to make. At a glance, Trumbull realized it simply wouldn't do. The thing looked like a fiberglass travel agency display, not a credible piece of engineering. Having been essentially an illustrator up until that point, he was skilled with an airbrush, so he deployed a multimedia technique on the model that would subsequently be applied to all of *2001*'s spacecraft.

Using frisket masks—small pieces of plastic sheeting with adhesive on one side to protect areas from unwanted paint—he set about airbrushing sections of the fuselage, creating a nuanced, seemingly factory-assembled new exterior evidently composed of multiple joined metal panels with slightly different textures and shades. He also went to a hobby store in Borehamwood and started gluing small parts from plastic model kits in various places, augmenting the Moon Bus to create the effect of a meticulously engineered piece of technology. He scratched in fine lines, made new small panels, and constructed antennas. By the time he was done, *2001*'s lunar transportation workhorse had become palpably flight ready.

．　．　．

Part of Gentleman's mounting exasperation with Kubrick had to do with all the hairpin turns he kept making as he refined the film's concept. Up until the end of August, *Discovery*'s destination was Jupiter. But by early September, the director had grown increasingly intrigued by Saturn—certainly the solar system's most spectacular planet, but exceedingly difficult to represent convincingly on-screen. He was particularly provoked by the giant world's symmetrical rings, which are made of innumerable fragments of ice.

One day Kubrick took a knife to Czech illustrator Luděk Pešek's large-format illustrated book *The Moon and Planets*, cutting out a four-page gatefold depicting a stream of ring fragments suspended in front of Saturn and pinning it to a bulletin board in his office. Kubrick imagined a giant monolith floating just beyond the outer rings, and that the rings themselves might have been made four million years ago, when the titanic forces that his extraterrestrials had deployed to produce the Star Gate had also shattered one of Saturn's moons.

Most of those privy to the director's new obsession hoped it would go away. They could foresee all kinds of difficulties in attempting to depict this complex planetary system. But on September 5 Kubrick was having dinner with a group that included Ordway, and asked his scientific advisor how he'd react if they substituted Saturn for Jupiter. Ordway replied that it might be a bit late to make such a change, but the director persisted, his advisor recalled, "pointing out the beauty of the Saturnian ring system and spectacular visual effect of *Discovery*'s traveling near or even through it." Kubrick asked Ordway to write a document covering the visual aspects of Saturn—something duly produced after considerable research.

Despite initial objections from his effects staff, and in particular his small team of illustrators, Kubrick's new twist served as the basis for many months of intensive work focusing on how to depict the Saturn system. It may have been the last straw for Gentleman, who on November 29 complained to a friend, "We are supposed to wind up this production by next September, but to arrange a reliable schedule in the face of continuing radical changes is going to be exceedingly difficult." Years later, he expanded on his issues with the director. "My entire career, I had spent doing everything meticulously and precisely," he said. "But when I ran into Kubrick, I found a man who had an absolute *obsession* with the finicky details of *everything*, and I wasn't comfortable with that excess."

> I enjoyed Kubrick the man very much; the experience with him was vital and interesting. But I learned that one doesn't work *with* Kubrick—one only works for him—and I found that rather difficult. As a filmmaker, he was a little paranoid and certainly obsessive. He surrounded himself with really good people and then proceeded to dissipate their talents. Eventually I got fed up with the sort of autocratic methods that were being applied, often seemingly arbitrarily and at variance to the useful application of the associated talent.

Visual effects assistant Brian Johnson provided his own insight into their rift. "If Stanley thought someone was slightly insecure, he could be quite a bully, and he bullied Wally Gentleman quite a lot; gave him a hard time," he said. Gentleman, who was only thirty-nine at the time, had also been experiencing health complications, and by the end of the year had returned to Canada for treatment. (He lived to see the year 2001.) His loss was offset partially by the arrival of Con Pederson, who'd extricated himself, not without difficulty, from Graphic Films.

Discovery's new destination pleased Clarke, however. While researching planetary geometries, he'd discovered that Jupiter and Saturn were destined to line up like billiard balls in the year 2001. Accordingly, he built a Jupiter flyby into his book draft, and Saturn became *Discovery*'s goal in the published novel.★

But after months of attempts to produce a realistic set of rings, Pederson and the film's visual effects staff finally rebelled at the challenge of reproducing Saturn's complex environment. They were only halfway through the live-action shooting of *2001: A Space Odyssey*, and they simply faced too many other challenges. As Ordway remembered it, Kubrick reluctantly acceded to Jupiter's return as *Discovery*'s destination following an "altercation" with them in mid-March 1966.

It was a rare example of the director backing down.

■   ■   ■

Kubrick had watched innumerable actors' show reels throughout the summer and fall of 1965, and word had gotten around that *Dr. Strangelove*'s inscrutable mastermind was doing something Hollywood could understand: casting his next film. Rumor had it that Warren Beatty's agent had been lobbying on his behalf for the lead role. If so, Kubrick had other ideas. By late September, the film was definitively cast. It didn't include any "name" stars.

With Keir Dullea already slated to play Bowman, the director had met with Gary Lockwood weeks before sailing for London. A former college football star at UCLA who'd been expelled for brawling, Lockwood would play Dullea's second-in-command, Frank Poole. In addition to carousing and getting into fights, the actor was a hard-core gambler, but he'd had the discipline to score various second-string roles on Broadway and in Hollywood films, and had also played the lead in a popular TV series, *The Lieutenant*. An impressive physical presence with a quick wit and an unlikely ability as a painter, he'd found employment as a film stuntman immediately after his ejection from college, even working briefly on *Spartacus*, though he'd never spoken to the director before. At their meeting, Kubrick told Lockwood he

---

★ This alignment would, in fact, be used that year by NASA's unmanned Saturn-bound *Cassini* spacecraft, which flew past Jupiter on January 1, 2001, on its way to the ringed planet.

loved football, and asked why he thought the game was so popular. "I think it's a combination of chess and violence," was the response. The comment produced a great surprised belly laugh from the director.

Although surrounded in Elstree by some of Britain's top actors, Kubrick wanted to stick with North Americans for his principals and had selected two for relatively significant roles. Robert Beatty, a Canadian who'd handled numerous parts on the London stage, would play lunar base commander Ralph Halvorsen. For Dr. Heywood Floyd, presidential science advisor and chairman of the National Council of Astronautics—a key role carrying the film's middle section—Kubrick had chosen expat Californian William Sylvester, a regular presence on the London stage since shortly after World War II. Sylvester had played one of two leads in the BBC radio drama *Shadow on the Sun*, the production that had caught the director's attention while he was shooting *Lolita* in 1961.

Other players of note included English actor Leonard Rossiter in the bit part of Russian scientist Dr. Andrei Smyslov—a role he'd play with such impressively oily unctuousness that Kubrick returned to him a decade later and cast him as Captain John Quin in *Barry Lyndon*. For Elena, another Russian scientist—and the only adult female in the film with more than a few lines—Kubrick selected accomplished stage actress Margaret Tyzack, whose prior London theater roles had included Lady Macbeth.

Although Kubrick's UK company, Hawk Films, which he'd formed in 1963 for *Dr. Strangelove*, would announce most of these decisions in early January 1966, the director got steamed when MGM presumed to issue a press release in September announcing Dullea's lead role. This prompted a heated cable to Caras on the twenty-fifth, in which the director demanded that his VP find out why the announcement hadn't been left to Hawk. By now, Caras was in effect Kubrick's ambassador to MGM, the core leadership of which was largely based not in Los Angeles but in New York, and specifically to its publicity department, whose US chief was Dan Terrell.

Another cable followed the next day after Kubrick had stewed some more. "It is most inappropriate for Bob O'Brien to make casting announcements, as it makes me seem like a contract director," he complained. "Diplomatically tell this to Dan Terrell and ask for a policy agreement on this point. I, in turn, won't make Otto Premingerish announcements about distribution." (Preminger, a leading Hollywood director during the forties and fifties, was notorious among studio bosses for throwing his weight around.) The messages crystallized something of the ongoing transition from the old-school Hollywood way of doing things, in which hired-gun directors simply followed orders, and the emerging new paradigm of freelance directors and production companies making distribution deals with studios that granted them unprecedented levels of control.

Other personnel were being signed on as well, and on September 30 Victor Lyndon—a man well known in British film circles for his snazzy attire—informed Caras that Savile Row haute couture designer Hardy Amies was going to create the film's costumes. Famous as the dressmaker to the Queen, Amies was something of a queen himself but, as acting head of the Special Operations Branch for Belgium during much of World War II, had acquired a reputation as a cold-blooded plotter of assassinations, responsible for numerous successful schemes to kill Nazis and their sympathizers.

Although not as celebrated as some other alliances between noted directors and clothiers—Yves Saint Laurent's outfitting of Catherine Deneuve in Luis Buñuel's *Belle de Jour* comes to mind—Amies and his design director, Ken Fleetwood, would play an important part in establishing *2001*'s look. His simple, somewhat understated men's costumes, which included slim-fitting suits with idiosyncratic details, were designed to look futuristic without inappropriately calling attention to themselves. His female attire—which included bubblegum-pink space station receptionists' uniforms and cocoonish stewardess costumes with beehive shaped, shock-absorbing headgear—was a different matter.

"We were dealing with a period thirty-three years ahead in time," Amies recalled in 1984. "To try and get a perspective on this, I looked back over the previous thirty-three years to see what had happened in the world of fashion. I realized to my surprise that there had been less change than one had imagined. I didn't foresee, therefore, that clothing in the year 2001 would be dramatically futuristic. Mr. Kubrick accepted this."

Set decorator Bob Cartwright, who'd joined Masters and Kubrick in New York and was now working at Borehamwood, attended a production meeting with the director in which future trends in clothing and other potential social transformations were discussed. "I was skeptical about the major changes they were implying would have happened to ordinary people and ordinary life," he remembered. "I said that my daughter was very keen on animals. I was sure in 2001 when she's forty-six that she would have a daughter that also would have pets, and those things won't change." Cartwright believed the scene in which Heywood Floyd calls his young daughter from Earth orbit and discovers that she wants a pet for her birthday probably originated from his observation.

．　．　．

On September 10 Clarke visited Borehamwood and "emerged stunned," he wrote Wilson, then in final preproduction for his James Bond parody in Colombo. Stage 3, the biggest at the studio, was now dominated by a smoothly curving, 150-foot-long, 30-foot-wide structure floating under rigging suspended on chains from ceiling beams five stories above. Containing catwalks and innumerable rows of 1000-watt Fresnel lights, all pointing downward at diffusion gels set in the space station's ceiling, this nest-like pipework service structure matched the smooth curvature of the set below.

Clarke drew a diagram of it and compared it to a ski jump: "Actors at the far end will have to learn to walk tilted." It wasn't even a vital scene, he observed: "This is no longer a $5,000,000 movie. I think the accounts department has thrown up its hands; the guess I heard was $10,000,000."

Seen with all lights blazing, the visual impact of the set's exterior was indeed powerful. Visitors were permitted entry from the side and would then find themselves within a brilliantly lit rim section of an ostensibly wheel-shaped space station 750 feet in diameter. A strange

Exterior of Space Station 5 set.

phenomenon began to manifest itself as they approached either end. Although the furniture and walls indicated that gravity should pull *thisaway*—namely, in the direction that all visible cues indicated it should—in practice, it pulled *thataway*: toward the invisible studio floor. While this optical-versus-gravitational illusionism was somewhat tolerable when walking at an increasing tilt toward either end of the curve, visitors inevitably experienced an unsettling vertigo upon turning around. The perspective switch was too much, and a spinning sensation kicked in. With the trickster floor now sloped nauseatingly away, many had to sit down rapidly to avoid falling.

"In about six weeks, things should be very exciting," Clarke continued. "Shooting now set to December 1. Release slipped to Mar '67, tho' Stan still talks '66 . . . Unfortunately he has no time to finish the script (!!!) so can't finalize the novel. So I can't sell it, while Scott chews his fingernails. Never mind."

Shooting would slip further than that, and Scott Meredith's nails would suffer for much longer than either of them suspected.

. . .

In key ways, Clarke's futurism had been formed by that comment made by Russian spaceflight visionary Konstantin Tsiolkovsky—the one he'd quoted in his short story "Out of the Cradle, Endlessly Orbiting": "Earth is the cradle of the mind, but humanity can't remain in its cradle forever." The thought was contained in an essay the rocket scientist had published in 1912, five years before the Russian Revolution and less than a decade after the Wright brothers lifted their plane above the Kitty Hawk sand. In a sentence, Tsiolkovsky had laid out the ideological foundations for spaceflight. Even more than the work of British science fiction writer Olaf Stapledon, the pronouncement had influenced Clarke's worldview, in which Earth was but a stepping-stone to the cosmos.

His most influential novel, *Childhood's End*, was named in oblique reference to Tsiolkovsky's concept. In the book, a final recognizably human generation loses touch with its children, who begin to display signs of group clairvoyance and telekinetic powers. Under the supervision of a powerful alien master race, the Overlords, Earth's children merge into a single group consciousness comprising hundreds of millions of individual minds.* "To all outward appearances, she was still a baby," Clarke wrote of one of the children, "but round her now was a sense of latent power so terrifying that Jean could no longer bear to enter the nursery."

With production looming, pressure was building to finalize important aspects of the story, including, of course, the ending. Ideas continued emerging in a steady stream from both writer and director. "Stanley phoned with *another* ending," Clarke noted on October 1. "I find I left *his* treatment at his house last night—unconscious rejection?"

---

* For an evocative rendition of Clarke's telekinetic wild children, see Aubrey Powell's cover for Led Zeppelin's 1973 album *Houses of the Holy*, inspired directly by the book's ending. A dozen naked children climb the square boulders of Giant's Causeway, Northern Ireland, under an eerily orange sky.

He retreated to his apartment on the top floor of his brother's house, not far from Borehamwood, and began drafting alternate conclusions to *2001: A Space Odyssey*.

Their current final sequence relied heavily on voice-over. According to an omniscient narrator, the extraterrestrials had undergone an evolutionary transition of their own since decisively influencing prehistoric humanity four million years before—transitioning through an "age of . . . Machine Entities" and learning how to "preserve their thoughts for eternity in frozen lattices of light." Freed from "the tyranny of matter," they were now "Lords of the galaxy . . . But despite their godlike powers, they still watched over the experiment their ancestors had started many generations ago." On the final page, Bowman was supposed to be seen flying his pod toward a "great machine" in orbit of Saturn (or presumably Jupiter, when they reverted to that destination). What happened next was frustratingly opaque, however, and the narrator's final words didn't help much: "In a moment of time, too short to be measured, space turned and twisted upon itself."

Within a few days, the author had produced a selection of new endings, and on October 3 Kubrick called to fret about the existing one. It wasn't clear what it was supposed to *mean*. It was simply too inconclusive. Clarke went through the alternatives he'd been working on, and one "suddenly clicked: Bowman will regress to infancy, and we'll see him at the end as a baby in orbit. Stanley called again later, still very enthusiastic. Hope this isn't false optimism: I feel cautiously encouraged myself."

Kubrick's approval was likely rooted in his own appreciation of Tsiolkovsky's formulation,★ which he cited only a few weeks later in a draft response to a question from the *New York Herald Tribune*. Clarke's new ending may have been prompted by two other factors as well. One was an illustration reproduced in Robert Ardrey's *African Genesis*, the book already so influential to the film's prelude. A small pen-and-ink

---

★ He likely heard the quote from the author directly, but as already mentioned, it had been cited in one of the Clarke stories he'd optioned, "Out of the Cradle, Endlessly Orbiting."

rendering, it depicted a fetus floating in a bubble-like amniotic sac, surrounded by other featureless bubbles in a kind of pointillist void composed of black dots. It looked exactly like an unborn baby in space.

A second influence—certainly regarding Kubrick's receptivity to the idea—was Swedish photographer Lennart Nilsson's stunning color photographs of human embryos, which had appeared just a few months before in *Life*. As with the illustration, the cover of the magazine's April 30, 1965, issue depicted a fetus seemingly floating in cosmic darkness—ostensibly the inner space of the womb, though, in fact, all but one of Nilsson's pictures were of embryos that "had been surgically removed, for a variety of medical reasons," as the text explained. Nilsson's work caused a worldwide sensation and definitely caught the director's attention.

For a while, *2001*'s Star Child was supposed to detonate orbiting nuclear weapons in Earth orbit—a scene that remained in the novel but not the film, where it was judged too similar to *Dr. Strangelove*'s ending. In any case, after so many failed attempts, Clarke himself initially appeared less than certain of his new idea. But a few days later, he wrote, "Back to brood over the novel. Suddenly (I think) found a logical reason why Bowman should appear at the end as a baby. It's his image of himself at this stage of his development. And perhaps the Cosmic Consciousness has a sense of humor. Phoned these ideas to Stan, who wasn't too impressed, but I'm happy now."

Whatever Kubrick may have thought of Clarke's *post festum* rationale, his initial enthusiasm held. They had their ending.

◦ ◦ ◦

Kubrick's management methodology was simultaneously egalitarian and hierarchical. He was the boss, of course, but his door was always open to his closer collaborators regardless of their ostensible rank. Particularly when he felt comfortable with someone, he used that person as a sounding board at all hours of the day or night, seeking his opinion (and it was almost always a he) on matters well outside his official role. He was a good-humored presence, by and large, and in particular liked to banter with his younger assistants about their extracurricular activities, sometimes with prurient curiosity as the sexual revolution unfolded in Swinging London, directly down the new M1 Motorway from Borehamwood.

Still, he wasn't always happy when one of his subordinates piped up with unsolicited opinions or when he perceived that his more junior employees were freelancing in areas outside of their assigned domain— particularly if this hadn't been taken under his own initiative.

For his part, Doug Trumbull was feeling increasingly confident. Lately, he'd been experimenting with creating large slices of lunar topography and had devised an effective way of hosing down his large Moon surface model with water, climbing to a high catwalk, and dropping various-sized stones on it, thus producing convincing craters in the wet clay. He'd received unstinting support in all this and had already been invited to the director's home for dinner. (Although Kubrick would ultimately elect to use a more jagged landscape, Trumbull's softly rolling terrain proved closer to what Apollo astronauts actually saw when they started flying to the Moon in 1968.)

In short, he was feeling his oats. Another thing he'd been tasked with was producing animated content for the medical status screens mounted above each of the sleeping astronauts: six stacked horizontal waveforms indicating cardiovascular and pulmonary function. Like others within the production, he'd been receiving regular revisions to the film's ever-evolving script, each bearing a new date and printed on a different color of paper. Trumbull had read these with keen attention, and one thing he'd noted was that the story had evolved from Poole and Bowman going into the Star Gate, with the rest of the revived crew manning *Discovery*, to Poole getting killed and only Bowman making the film's final journey. In the latest draft, in fact, the rest of the crew had seemingly been left entombed in their hibernaculums, alive but still in suspended animation, thereby playing no role in the story whatsoever. It occurred to him that they were effectively loose ends.

Having arrived at certain conclusions, early in the week of October 11, Trumbull chose his moment, walked down the hall from his office on the east end of Building 53, and appeared in Kubrick's inner sanctum. Finding the director no more distracted than usual, after a perfunctory update on his animation efforts, Trumbull got to the point. "Stanley, we've got these changing circumstances where it was supposed to be Poole and Bowman who go into the Star Gate," he observed. "Now it's just Bowman, and he's alone, and this whole trip thing is going to happen. Isn't it kind of messy to leave all those guys back on the ship? Isn't there some way to kind of get rid of them? I don't think this story's going to work if you leave them behind, unresolved."

As he spoke, the magnitude of Trumbull's transgression mounted on Kubrick. Finally, he rose from behind his desk and pointed at the door. "Trumbull, get the fuck out of my office and pay attention to your own goddamn business," he said in the icy tone he sometimes used when irritated. "*I'm* the director of this movie."

"Okay, leaving now," said Trumbull, beating a hasty retreat.

Although it was the closest they'd ever come to a confrontation, Trumbull didn't sweat it afterward. So what if he'd been a bit presumptuous? It was well meant, and at the service of the project. And indeed, on October 15 Tony Frewin distributed a new round of script revisions to the production's inner circle. As Clarke noted in his journal, "Stan has decided to kill off *all* the crew of *Discovery* and leave Bowman only. Drastic, but it seems right. After all, Odysseus was the sole survivor."

．　■　．

With the space suits ready to wear by mid-November 1966 and preparations under way to begin filming at Shepperton's vast Tycho Magnetic Anomaly 1 set—located inconveniently far away on the other side of London from Borehamwood—Kubrick kept delaying *2001*'s start date. By now, set dresser Bob Cartwright, who'd been deeply involved throughout the preproduction period, had become something like the director's personal assistant on design matters. In practice, this meant he was endlessly running messages and schematics back and forth between Building 53 and Tony Masters and Harry Lange, whose studios were farther back in the sprawling Borehamwood complex. This was not by choice, and Cartwright felt that his actual job—that of arranging and decorating the sets—was suffering as a result. He was also growing increasingly exasperated by Kubrick's delaying tactics.

"My upbringing in films was if a date was set for a shoot on Friday the fifteenth, then that's the day you had to be ready," Cartwright recalled. "Not so with Stanley, 'cause Stanley was in such control he'd say 'Don't worry,' and could push it to three weeks later. That wore on me." Like Gentleman, who'd felt that his views concerning the right way to do things were being ignored, Cartwright began to stew. He'd also grown tired of being woken up in the early hours of the morning by phone calls from Kubrick—another point of friction with Gentleman—and finally, one day in early November, he'd had enough.

"Stanley, you're not paying attention to me," he said. "How do I get your attention?" To Kubrick's amazement, he climbed directly onto the director's desk and proceeded to stand on his head. "Pay attention, Stanley," said Cartwright, now upside down and facing him from only a few feet away. "If you don't make a decision today, they're going to have to cancel the start of shooting."

Kubrick soon conveyed the essence of this unusual occurrence to Masters, who felt he had to intervene. "Look, are you upset?" the designer asked his colleague. Cartwright explained that he hadn't been able to get Kubrick to focus, which meant they couldn't manufacture key set components—already difficult enough, given that everything then had to be broken down and transported to Shepperton—which, in turn, meant they wouldn't be able to start filming on schedule. Masters's mediation failed to ameliorate the situation, however, and although Cartwright remained in an upright position during his next meeting with Kubrick, he was no less exasperated. "I find that I'm failing," he observed. "And further, I'm going to have to leave and not continue."

"You can't do that," the director protested.

"Well, I can give you five or six weeks' notice," said Cartwright.

"No, that won't work," Kubrick replied. "We can work it out."

Cartwright, who'd truly had enough, refused. "After that, he was very cross," he remembered. "He said I'll never work on films again.

I can understand his frustration. It was unfair of me to give in; I really should have stuck it out. I could feel myself getting more and more wound up. You know, you get to a state where you can't finish anything, can't do anything."

Despite his threat, a decade later Kubrick offered Cartwright the position of set decorator on *Barry Lyndon*—a significant credit. "I had to say no," Cartwright recalled. "He said, 'Well, what are you doing?' I said, 'I'm not doing a film, I'm doing a reconstruction of a large house.' If I hadn't been occupied, I might have gone to do it. Ken Adam worked on it and actually had a nervous breakdown. It's not . . ." he hesitated.

"He's such a nice guy, Stanley is. It's not that he's nasty. I was used to working on deadlines . . . Stanley didn't even bother about deadlines."

. . .

In late November Kubrick delayed the start of production yet again, this time from early to late December—true brinksmanship, as Stage H would need to be broken down and handed over to another production within a week. With typical exactitude, he'd ordered that multiple grades of sand be brought to Borehamwood from the British Isles' various coasts, but none was deemed the right color, so 110 cubic yards of a particularly fine sand had been dyed a dark ash grey—90 tons of it, in fact, all subsequently trucked to Shepperton, dumped on Stage H, and carefully raked around the 120-foot-long, 60-foot-wide lunar excavation site. Its base was the stage floor, with the Moon's surface represented by a rectangular platform constructed thirty feet above, and accessible by two broad steel-mesh ramps leading up its reinforced walls.

Following Masters's suggestion, an eleven-foot-high black hardwood monolith had been prepared. After much discussion, the ratio of its dimensions had been settled as the square of the first three integers, or 1:4:9—a determinedly logical formula worked out shortly after Masters's visit to the Perspex fair. Subsequently, a kind of monolith assembly line had been set up in the carpenter's mill directly behind MGM's power station. Various kinds of wood had been tried. Graphite was mixed with matte black paint and sprayed on, which, after many coats, produced a darkly metallic surface luster. In the end, about fourteen had been made, with Masters forced to throw them away almost as fast as they were produced due to small flaws that robbed them of their desired aura of powerfully inscrutable perfection.

Even the one that ultimately passed muster—*the* monolith—wasn't immune to directorial criticism. Although its form and finish were impeccable, dusty fingerprints could sometimes be seen on its surface with extraordinary clarity. "He'd stand like this," Masters said, rising to demonstrate. "And then he'd say, 'I can see finger marks on it.' 'I know you can, Stanley, I don't know what we can do about that.' And he said, 'Well, there can't be any finger marks on a thing from outer space. Everybody who handles it will have to wear gloves, and nobody's going to come anywhere near it—it's not going to be marked.'"

To further complicate things, while in transit to Shepperton, the monolith acquired a static charge, and when exposed to the set's powdery lunar regolith, "*Foomp*, it'd be covered with dust," Masters recalled. "We had to use air blasts to keep the dust off the damn thing. Then we'd put the lights on it, and after three or four hours, I'd be biting my nails, because as it got hotter and hotter, you could see slight bumps starting to form along the surface. 'Oh Christ! I wonder if Stanley will see that.' And if an electrician happened to move alongside and put his hand on it—'Stop the shooting! Back to the paint shop for a respray!' It was unbelievable what went on to protect that thing."

With the monolith in place at Shepperton, the set prepped and ready, Lange's suits delivered and accessorized, all the actors' signatures dry on their contracts, and another shoot nipping at his heels, Kubrick nevertheless waited until the last possible minute. Then he announced, "We've got to do it."

Production was on.

CHAPTER SIX

PRODUCTION

Never let your ego get in the way of a good idea.

—STANLEY KUBRICK

Production of *2001: A Space Odyssey* began on December 30, 1965, at a thirty-foot-deep excavation in the Moon's surface within the prominent southern crater Tycho, a location where Earth was at a permanently low inclination above the lunar horizon, one advantageous for cinematography. Actually a 120-by-60-foot quadrangle of interlocked, fretted metal plates rising from the concrete floor of Stage H at Shepperton, the Tycho Magnetic Anomaly set (TMA-1, to Kubrick, Clarke, and company), location of the first alien artifact ever discovered by humankind, was as sizable an affair as could be housed under a studio roof within the borders of the United Kingdom. Twenty-one months had passed since Kubrick wrote Clarke proposing that they make "the proverbial 'really good' science fiction movie."*

It's difficult to exaggerate the magnitude of the transition between preproduction and production. Preproduction is aspirational and notional. It's where filmmakers project an idealized vision of what they want to make and then endeavor to set up the best possible conditions to achieve it, be they financial, logistical, or conceptual. The right collaborators, the right equipment, the right schedule. It's not unlike planning a battle.

Production is where one discovers what percentage of those aspi-

---

* The start of production has otherwise been widely cited as December 29, but that day was spent in camera rehearsals and lighting, with no film shot.

rations will actually be achieved. It's where the rubber meets reality's road. Any military historian will tell you that even the most meticulously organized offensive deviates from its objectives the moment it's launched. To quote boxer Mike Tyson, everybody has a plan until they get punched in the mouth. Still, achieving even a reasonable percentage of those original aspirations can frequently yield something a good deal better than okay.

Kubrick's ambitions were much greater than that, though he might not have admitted as much. Despite the seeming chaos—the whiplash switches of plot and concept, the last-minute design changes, the ongoing winnowing of dramaturgical chaff—his preproduction process had delivered an entire major studio complex's wealth of resources to his fingertips. Behind it stood a significant budget, with the possibility of going back for more, should he go over—something the accounting department was already warning he would.

His recruitment had been impeccable. Kubrick had arrayed around him a team of exceptional talent and ability. He could count on the unstinting support of studio boss Robert O'Brien. His cinematographer, Geoffrey Unsworth, was among the best in the business. Much the same could be said for the rest of his collaborators. And he had a world-class intellectual interlocutor in Arthur C. Clarke, who'd accepted the reality of ever-lengthening story development—something by no means over yet, despite their best efforts, despite the completed sets, despite the wide-gauge 65-millimeter film stock even now being threaded into those fat Panaflex camera magazines.

Even if his tendency was always to aim higher than yesterday's iteration of his vision—a process producing a constant upward ratcheting of effort and aspiration, sometimes at the expense of his collaborators' sanity—Kubrick had achieved conditions more than adequate to vault beyond his original goal. What he was after now was a film spanning "nothing less than the origins and destiny of Man"—something he hadn't allowed Clarke to state so explicitly in his draft article for *Life*, the one they hadn't published.

Everyone present for *2001*'s first production day cites one sight above all that transmitted the thrill of the event: Kubrick hoisting the twenty-

Kubrick with a heavy Panaflex camera on his shoulder at Shepperton.

four pound bulk of the Panaflex on his own shoulder and filming his space-suited Moonwalkers from behind as they descended the ramp toward the monolith below. Assistant director Derek Cracknell flanked him on one side, lugging the battery case. Camera operator Kelvin Pike brought up the other, monitoring focus and aperture. It was simultaneously the most vivid demonstration of Kubrick's hands-on approach possible and something not unlike the opening move in a chess game.

A young assistant to Roger Caras, Ivor Powell, remembered the moment vividly. "This was Panavision, these were big buggers, and to see him just handle this camera and *do* it, it was just incredibly exciting," he said. He was particularly struck by Harry Lange's space helmets. "Jesus Christ, they were good," he said. "Shivers up my spine. They were so beautiful, unbeaten until today."

Actually, Kubrick's handheld shot was made on Day Two: Friday the thirty-first, New Year's Eve. It made it into the final film, providing a subjective first-person sense of being among the astronauts as they descended toward the mysterious rectangular object. December 30 had been spent getting a number of static high-angle wide views, including from a tall rostrum platform, with plenty of empty black space left around the well-lit Tycho Magnetic Anomaly excavation at the center of the frame.

The first main unit film exposed for *2001: A Space Odyssey* was thus devoted to simple, powerfully evocative, stage-setting master shots. In them, six astronauts in silver space suits walk to the edge of the excavation and pause to look at the monolith at its center. On this opening production day, they were filmed such that the pause permitted a short exchange between base commander Halvorsen and space agency official Floyd:

"Well, there it is."

"Can we go down closer?"

"Certainly!"

Perhaps recognizing the dialogue's banality, Kubrick would redo the establishing shot on January 2 with no words spoken. In those takes, one of which is seen in the final film, the astronauts simply pause to observe the monolith from above in silence. Rather than being rushed to the lab for development, this footage was couriered to cold storage at MGM, where it became "held takes." The visual effects crew would eventually fill in the black areas around the brightly lit central part of the set with rugged lunar topography, complete with a blue-white Earth hanging in the starry sky.

It's a measure of Kubrick's confidence in the process he'd worked out with his two visual effects Wallys—Wally Gentleman and Wally Veevers—that for each shot requiring that a lunar landscape be added later, only two takes were held for that purpose. This left little latitude for error. They certainly couldn't be redone later in the event of a mistake: the set would have long since been struck. And because they'd been filmed so early in the production, those Shepperton shots requiring additional effects work would languish undeveloped in cold storage for almost two years before finally being run through an animation-stand camera to add the surrounding lunar terrain. It was a risky procedure, and far from the usual practice with precious first-generation camera negative stock, which was customarily developed immediately. As an opening example of the complexity of the game that had just commenced, it required both nerve and luck.

Prior to the Tycho Magnetic Anomaly rehearsals on the twenty-ninth, a number of second-unit sequences had been filmed as well. (In film production, a second unit is typically tasked with capturing shots or sequences that don't require a director's physical presence, while the main or first unit works under his immediate supervision.) Most of *2001: A Space Odyssey* was shot on 65-millimeter film, but all of this material, shot just before Christmas, was filmed on standard 35-millimeter stock and would later be seen on various screens embedded in the sets. It was intended to simulate TV programs or other video content. Scenes produced before Shepperton included a passport girl, a judo match, and a young couple chatting in a futuristic, javelin-shaped General Motors concept car. All three are visible in the final film, with the latter glimpsed briefly on the flat screen mounted to the seat in front of a sleeping Heywood Floyd as he is vaulted toward the space station in his Pan Am shuttle.

So to be truly accurate, production on *2001: A Space Odyssey* actually began on December 17, 1965, with second-unit sequences shot on 35-millimeter. That day already featured lines written by Kubrick that subtly differentiated his stance from Clarke's more optimistic take on the space age. A young woman in a brick-red uniform welcomes a space station visitor to "voice-print identification." Having instructed him on procedures necessary to pass through her automated portal, she continues:

Despite an excellent and continually improving safety record, there are certain risks inherent in space travel and an extremely high cost of payload. Because of this, it is necessary for the Space Carrier to advise you that it cannot be responsible for the return of your body to Earth should you become deceased on the Moon or en route to the Moon. However, it wishes to advise you that insurance covering this contingency is available in the main lounge. Thank you. You are cleared through voice-print identification.

Like so much of the film's scripted dialogue, it wouldn't make it to the final cut.

. . .

Not technically a soundstage because of its lack of soundproofing, the Shepperton venue was big enough to contain its own ecosystem. Flies buzzed through the floodlit camera rehearsals. On January 1 Kubrick noticed that one had landed on a helmet during a take, alerted the script "girl" (actually an adult woman, Pamela Carlton), and the event was duly noted in the *Daily Continuity Report*—a document intended to help ensure that a film's scenes flow seamlessly. Another day, a bat kept on flitting down among the astronauts and past the monolith. No doubt seeking lunar flies, it was hunted in turn by a propman with a net. Finally captured after bringing production to a halt all morning, it bit its tormentor on the hand, drawing blood before being hauled outside and released into the frigid air. A much-touted "magnetic induction" communications system, installed in the venue at great expense, was supposed to allow two-way communication between Kubrick and his space-suited actors. Worthy of its own press release, it failed to work as advertised.

Despite these problems, shooting concluded ahead of schedule on January 2. In part, this was because so much dialogue had been cut just prior to showtime and transferred to another scene: the Moon Bus ride between the lunar base and the Tycho excavation site. Script drafts from late November had featured a conversation between the astronauts in the presence of the monolith, during which they debated its purpose and also made clear that it would soon be exposed to sunlight for the first time in four million years. Asked about the object's color, one was to have replied, "At first glance, black would suggest something sun powered. But then why would anyone deliberately bury a sun-powered device?" This, in turn, would have clarified why the monolith emitted "a piercingly powerful series of five electronic shrieks" at the end of the scene—sending its visitors into paroxysms as they tried to protect ears unreachable under their helmets. But Kubrick, evidently already intent on inserting calculated measures of ambiguity into his film, elected to convey this part of the story purely through image and sound cues.

Dramatic films are usually shot out of sequence, and a series of scenes depicting Floyd's journey to the Moon followed over the next few days. With Shepperton's flies and bats behind them, the troupe of actors, technicians, makeup people, wardrobe assistants, grips, electricians, set dressers, visual effects people, and note takers made their way back to the more sterile environs of Borehamwood's Stage 2, where a set representing the circular, elaborately padded passenger area of the spherical *Aries* lunar transit vehicle had been constructed. There a stewardess in one of Hardy Amies's crisp white Pan Am uniforms attempted to bring food to a sleeping Heywood Floyd, and in a series of takes on January 4, the *Aries*'s captain, played by Ed Bishop, tried to squeeze information out of his VIP passenger. "Rumors about some kind of trouble up at Clavius" were circulating, he said, referring to their Moon base destination. Meanwhile, Floyd's rectangular tray of slurpable liquid food floated up from his lap in the zero gravity—an effect achieved with a simple but effective fishing line rig devised by Wally Veevers.

He was politely rebuffed by Sylvester, who feigned ignorance. If it had made it into the film, their dialogue would have furthered a story element made explicit in a space station scene to be shot later that month but meant for earlier in *2001*'s narrative arc. The next day, January 5, they shot one of the film's few overtly humorous moments: a deadly serious Heywood Floyd, intently reading the zero gravity toilet instructions ("Passengers are advised to read before use"), all ten points of which were helpfully located *outside* the toilet.

∎ ∎ ∎

It was the morning of January 6, and Vivian fidgeted and squirmed. She wanted to do what Daddy asked her to, and she was wearing her new red blouse with the frilly sleeves, but the lights were very bright in her eyes—why did they have to be *so-ooo* bright?—and he was over there, beside the camera, not here, closer to her. And that man was holding a big thing over her head—he'd said it was to record her words, the ones she'd been trying to remember—but why was it called a "boom" if it was supposed to listen? It was all very confusing. Her sister had warned her it would be strange. Anya had already gone through this yesterday,

and even though her mother was hovering in the background, occasionally offering words of encouragement, and even though everyone had been very nice and friendly, there were all these strange men
around, doing strange things with strange pieces of equipment.

They'd done three "takes" already, but nobody had taken anything
away, as far as she could see, and they said this was the fourth. Daddy
had said, very quietly, "*Cut,*" in the middle of the first, but nobody cut
anything that she could see, and then he asked her not to look off to
the side but instead to look right at him—there, where he was sitting,
right beside the camera, with the lights in her eyes all around, so she
couldn't really see him. Now he asked if she was ready, and she said
yes, and they were starting again. First she heard somebody say, "Turn
over," and then somebody else said, "Turning." Then the first person
said, "Quiet, everyone!" and then "Roll sound," and somebody *else*
said, "Speed." And then a fourth person came up to her with a small
rectangular blackboard with a kind of stick on top that you could raise
and lower, and raised it and lowered it—it made a small sound, like a
clack—and then he went away very quickly, and Daddy said, "Hello,
sweetheart, how are you?"

And she'd waited for a bit, trying to remember the answer. Finally,
she said, "All right"—but he'd already started to say something on top
of *that.* So he said, "I'm sorry, let's start again," and they did, this time
without all the cutting and taking. "*Ding-aling-aling,*" Daddy said, like
a telephone. "Yes," she said. "Hello," he said. And they went from
there.

She remembered everything, all the words, until they got to Rachel
going to the bathroom—that was the name of her *real* housekeeper:
Rachel—but then she'd forgotten what to say after that. So Daddy said,
"No, sweetheart, you ask me something, if I'm coming to your party,"
and she'd said it, but very quietly, and he said, "Do it again. Remember, the party is tomorrow, and say it right to me, real loud." So she'd
asked him, Was he coming? And he'd said he was sorry, he couldn't
come, he was away, and he wouldn't be coming home for about a
year—a year!—but he'd get her a present anyway. And he'd asked if
there was anything she'd like, and she'd thought about it but couldn't

remember, so she said, "A telephone." And daddy had said, very patiently, "You say 'A bush baby,'" and so she'd said *that* immediately, in case she forgot it again—"A bush baby," very fast—and he'd said, "Wait till I ask you the question again." She waited, and then said it, but then he'd said, "Say it again, but wait till I've finished speaking now." And she had, and it had been okay.

And finally after she stood and said "Bye-bye," and after Daddy had said "Bye-bye," and after he'd said "Cut" again—but this time it was "Cut, print," not just "Cut"—after all that, everybody clapped, and Daddy came to give her a hug, and told her it was very good, and he was proud of her, and Mommy hugged her, too, and took her home.

. . .

On Monday, January 10, Ivor Powell was delegated to go pick up Keir Dullea at Southampton, where his ocean liner, the SS *United States*, had docked that midafternoon. Like Kubrick and Dullea's costar, Gary Lockwood, the actor had a fear of flying. The most convincing film about space exploration ever made would be captained and crewed by groundlings.

Powell's mother had died on Saturday, so he was in a shaken state but had elected to fulfill his duties anyway. After greeting Dullea—a square-jawed, appropriately handsome type with strikingly blue eyes—they watched together as his Mercedes 250 SL was hoisted from the giant ship's hold and deposited on the pier. Dullea had been in England less than a year before, working on Otto Preminger's psychological thriller *Bunny Lake Is Missing*—a nightmare experience under a shouting Teutonic despot. The actor had high hopes that this would be a different experience. At age twenty-nine, he was still up-and-coming, but to Powell, seven years younger, he was glamour personified—a bona fide movie star—and his arrival with a two-seater sports car gave a kind of James Dean frisson to the moment. Powell soon discovered that Dullea was the opposite of snooty—on the contrary, he was personable and open, very sympathetic when he heard about his escort's loss, and as they drove into town in the Mercedes, they hit it off.

Lockwood had arrived in London a few weeks earlier, and then gone on a jaunt to Rome and Paris, where he'd spent time with Jane Fonda—a friend ever since he'd played a supporting role alongside her in *There Was a Little Girl* on Broadway in 1960. He'd returned to London on the sixth, subletting a Bayswater apartment from Sean Connery's wife, actress Diane Cilento. Starting in the second week of January, both actors were regularly present at Borehamwood, being fitted with costumes and undergoing makeup tests designed to make them look like they were in their midthirties. Their first shooting day was scheduled for the thirty-first.

As an assistant to the film's chief publicist, Powell spent a lot of time with Dullea and Lockwood. They were good unpretentious company, always ready for a joke and a party—something that would soon be helped along by ample quantities of unstructured time between camera setups, which extended for many hours and even days as Kubrick and Unsworth lit the challenging sets. Although both were married—Keir to actress Margot Bennett, and Gary to Stefanie Powers—Swinging London was peaking, and the city had shed its bleak postwar torpor and emerged as a colorful center of style, sexual liberation, and rock 'n' roll. As good-looking actors working on a big-budget film with an internationally acclaimed director, they had major sex appeal, they knew it, and they were not averse to taking advantage of it.

Both were professionals, however, and knew how to learn their lines—even if they'd just been revised for the fifth time that morning. On Stage 2, a short walk from Hawk Films HQ behind MGM's central administrative building, Kubrick discovered that the same wasn't necessarily true of his middle-aged North American expats.

From the outside, the Moon Bus set looked less like the miniature of a lunar transportation workhorse that Doug Trumbull had added detail to, and more like a weirdly shaped amusement park hotdog stand: an oblong wooden box with multiple windows and an elongated snout, surrounded by film lights. With its padded walls, airplane seats, stowed equipment, and recessed ceiling lights, the interior was thoroughly credible, however, and associate producer Victor Lyndon had scheduled just two days of production there starting January 12 for what was expected to be an easily shot sequence. Their dialogue had been filled out with lines taken from the Tycho Magnetic Anomaly scene, and was intended to bridge Heywood Floyd's Clavius Base conference room talk and the arrival of the astronauts at the monolith. The bulk of the scene was actually shot on the thirteenth and fourteenth, and featured three principal actors: William Sylvester as Floyd, Robert Beatty as Halvorsen, and Sean Sullivan as Dr. Bill Michaels.

From the very beginning, there were dialogue problems. The first two takes—more than five minutes of film—were unusable. Soft-spoken and solicitous as usual, Kubrick had little to say beyond "Let's try it again," with assistant director Derek Cracknell conveying camera and sound cues in his distinctive East End twang. Typical of Kubrick, he'd set the frame, tweaked the lights, and only then allowed operator Kelvin Pike behind the camera. Unsworth stood just outside the cramped set, keeping a keen eye on developments.

After the standard UK film production litany—"Turn over," "Turning," "Roll sound," "Speed," "Slate it," "Action"—they tried again with a third take, out of sequence in the scene's middle section; a wide shot with Sylvester and Beatty seated on the frame's left and right sides, respectively. Sullivan approached at frame center, offering sandwiches to his colleagues. "Looks pretty good," Sylvester offered. "Well, they're getting better all the time. Whoops," said Sullivan, dropping his sandwich on the floor. When he didn't hear the expected "Cut" from Kubrick, he squatted and scooped it up.

"You know, that was a . . . an excellent speech you gave us, Heywood," said Beatty, flubbing the line. Again Kubrick refrained from cutting, and they rolled onward. Sylvester complimented his colleagues on how they'd handled "this thing"—the monolith's discovery and the subsequent cover story concerning an epidemic at Clavius Base—whereupon Sullivan put down the pesky sandwich, retreated to the bus's front end, and returned with photographs of the excavation site.

After some more discussion, it was back to Beatty: "When we first found it, we thought it might be an outcrop of magnetic rock, but all the geological evidence was against it. And not even . . . um . . . big nickel iron meteorite could produce a field as intense as that. So, we decided to have a look."

Once again, Kubrick desisted from cutting.

Beatty: "It seems to have been deliberately buried."

Sylvester, mildly incredulous: "Deliberately buried? And your tests don't show anything?"

Beatty: "We've only been able to make some preliminary checks. We're still waiting for security to get in a . . . or, to clear in a special evaluation team to go over everything. What we have found is that the surface is completely sterile, it's completely inert, and we haven't detected any vibration, radioactivity, or any other energy source apart from some magnetism."

Despite the new flub—making four in all, including the sandwich and "Whoops"—Kubrick went to the planned end of the shot. At 480 feet of film, or about four minutes in total, it was a mistake a minute. He spoke reassuringly to Beatty—an experienced character actor now in his late fifties who'd once been a B movie heartthrob—and they tried again. Once more he flubbed several lines, in Takes 4, 5, and 6. So they broke for lunch and reconvened that afternoon to try again.

After lunch, Beatty's shipwreck really started. Kubrick's new strategy was to chip away at the problematic parts with shorter takes. Notes taken on set tell the tale. Slate 99, Take 1, almost three minutes long: no good, dialogue. Take 2: cut short at less than a minute due to Beatty's flub. Takes 3 and 4, just over a minute of film: no good, dialogue again. Take 5: to Beatty's relief, Kubrick said, "Print that." But the reprieve was short; the next two takes were no good, dialogue again. Then Beatty began crashing repeatedly into the dialogical rocks. Fourteen takes followed, making a total of twenty-seven that day, with sixteen no good, dialogue—or twenty minutes of tossed negative stock. Those few the director deemed worthy of printing

William Sylvester, Sean Sullivan, and Robert Beatty in the Moon Bus.

were filled with flubbed iterations of his "preliminary checks" speech. In all, Beatty attempted the three lines about thirty times; fifteen in the eleven takes that were printed, and an unknown number in the sixteen that weren't.

Throughout, Kubrick had been as cool and collected as the mid-January drizzle that had been pelting the studio roof all day. He didn't like confrontations—they were both counterproductive and antithetical to his way of working—but he'd been laboring intensely without break for many months trying to master this exceedingly complex production, and he expected others to hold down their parts. He occasionally said of acting, "Real is good, interesting is better." This was clearly neither, and having shot forty-five minutes of film with almost half of it going straight into the trash, he was quietly fuming.

He hid it well, though, and calmly told Cracknell to wrap. They would try again tomorrow.

. . .

The Clavius Moon Base conference room scene of *2001: A Space Odyssey* isn't remembered as particularly remarkable, given the more fancy and complex spacecraft sets. At first glance, it is nothing but a rectangular room largely indistinguishable from corporate meeting areas today. But although perfectly bland and anonymous, as such spaces usually are, it represents a fascinating example of both Kubrick's perfectionism and his sophisticated understanding of photographic technique.

The director had insisted not only that the three large, unadorned walls visible to the viewer be the only light sources for the room but also that this trio of oblong rectangles have absolutely no brightness variations whatsoever. They needed to be completely uniform: uninterrupted sheets of pure white light. This was not nearly as easy to achieve as it might sound. They couldn't simply use a brute-force approach and flood the walls with so much light that they were overexposed. That would only blow out the scene and make it unfilmable. And as soon as the walls were brought down to reasonable light levels, it was extremely hard to ensure that hot spots or other subtle tonal variations wouldn't show up when they were captured on film.

After discussing the problem with Kubrick, Unsworth worked with Masters to create a simple but highly effective light-diffusing structure. The conference room itself was essentially just a ceiling, built on the bottom of a wooden frame suspended by steel cables from Studio 5's roof beams, and a carpeted floor made of dense blue wall-to-wall nylon fiber. As for the walls, despite their apparent solidity—in the film, they look like featureless blocks of glowing marble—they were simply large sheets of translucent polyester light-diffusing gel, stretched tight over the gap between ceiling and floor. The room's corners were mostly hidden by drapes.

These gels alone wouldn't have ensured uniform illumination, however, and Masters had constructed another external structure, surrounding the entire conference room, intended solely to diffuse light. Giant floods, each powered by a single industrial-strength tungsten bulb the size of a cantaloupe, were mounted on top of the conference room's external ceiling frame—a rectangular phalanx of light artillery pointing counterintuitively *away* from the set below and at the surrounding structure. Made of reflective bounce boards, it served as a kind of diffusion moat. By the time all that firepower had ricocheted around and reached the gels, it was so dispersed that they finished the job, and the light was rendered completely flat inside. A small miracle of photographic problem solving, it was a remarkably efficient way to achieve Kubrick's goal of absolutely invariable illumination.

So far, William Sylvester had been entirely prepared and unruffled. His Moon Bus line readings had been relaxed and word perfect. But Beatty's travails four days previously must have spooked him. Now the three actors from that scene, plus nine extras playing Clavius Base officials, had moved into the flawlessly opaque wash of light bathing Masters's conference room, where Sylvester was to deliver the longest single monologue in the film. To make things harder, Kubrick was so pleased by the neatly geometrical precision of his master shot—which he'd framed, in what would become a directorial signature, as a symmetrical composition, with the speaker's podium at the center and the rest of the room in mirrored equalization—that he wanted to cover the whole thing in a single take.

Sylvester's monologue was about 280 words, broken into nineteen uninterrupted sentences, after which he was asked a question by Sean Sullivan, and then had another 100 or so words to say, totaling about 380. In principle, that many lines within a single take of four and a half minutes shouldn't be a big deal for an experienced actor. After all, in Stratford, Ontario, in 1959, Sylvester had played Orlando in *As You Like It*, a character with almost that much to say the moment the curtain went up—and that was in Shakespearean English. But he didn't have other characters to play off of now. It wasn't a dialogue, it was a speech. He was on his own.

Sylvester's first problematic sentence was, "It should not be difficult for any of you to realize the extremely grave potential for cultural shock and social disorientation contained in the present situation, if the facts were prematurely and suddenly made public without adequate preparation and conditioning." A bit of a mouthful, to be sure, but, in

Sylvester in the conference room.

any case, he found himself unable to say it. He blew the phrase over and over, with the number of takes growing to a total of twenty-one, only five of which were deemed good enough to print. The last nine constituted an uninterrupted, building humiliation of unprinted flubs and "Let's try it agains" and "Turn overs" until Sylvester, who'd started visibly trembling and was bathed in flop sweat, said finally, "I can't do it anymore. I've had it"—and was helped from the podium by one of the nurses always on call and taken away.

To make matters more humiliating, a far larger crowd was watching than had been present during the cramped Moon Bus scene: camera assistants, grips, sound crew, and costume and makeup people. Visual effects assistant Brian Johnson recalls witnessing the unfolding scene. Sylvester, he said, "never got it; he tried all day. He just lost his nerve. He had, virtually, a nervous breakdown. He shook. He shook, and he had to be taken off the set because he was shaking." Throughout this ordeal, Kubrick had been seated in his chair directly beside the giant

Panavision camera just outside the open fourth wall of the conference room. He'd been quietly furious after Beatty's failings a few days before, and now he was utterly uncompromising. He wanted his master shot, and he aimed to have it.

Although Johnson professes great respect for Kubrick, whom he remembers with some warmth, he characterized the incident as an example "of how cruel Stanley could actually be with his actors." Asked about the director's demeanor, he responded, "He wasn't nasty. But he just wouldn't let go."

■   ■   ■

Roger Caras had played multiple roles in the film industry prior to being a PR man and then VP of Kubrick's production companies. He'd been assistant director of story and talent for one studio, and casting director for another. He'd been Joan Crawford's press secretary, no less, and he'd dealt with many a problematic actor in his time—something Kubrick knew very well. One day in mid-January, the director summoned him after a long, aggravating shoot. It's unclear if he was upset following Beatty's failings in the Moon Bus scene, or Sylvester's in the Clavius Base conference room.

In any case, Kubrick was "beside himself," remembered Caras. He vented about how many takes he'd just blown and said he never did get the scene, which was easy enough to do.

"Is he on drugs?" he demanded about the actor in question.

"Yes," Caras said.

"Are you certain?" asked Kubrick.

"Yes, he is."

"Well, do what you have to do. I don't want to ever have a day like that again. Don't tell me what you have to do."

Unsure exactly how he should handle the situation, but in no doubt that the problem was now his, Caras went off to find the actor, whom he located in his dressing room. After brief opening pleasantries, he announced the reason for his visit. "We're going to a press conference tomorrow at ten o'clock," he said cheerily. "Stanley won't be shooting. Please come to a press conference."

The actor looked surprised. "What's it all about?" he asked. "It's unusual to have a press conference at this stage of production."

"Well, yeah," Caras conceded, suddenly serious. "We're gonna announce that we're gonna be replacing you, because you're a junkie, and you can't learn your lines anymore, and you're useless as an actor."

At this, the actor in question stared disbelievingly for a moment, and then burst into tears. After a wordless pause punctuated by sobbing, Caras extracted a promise concerning the rest of the shoot. "Which was kept," he remembered. "I made no such statement. There was no press conference. But I scared him to death." From then on, the man knew his lines.

A few days later, Kubrick took Caras aside. How had this miraculous transformation been brought about? "I told Stanley exactly what I'd done," Caras recalled. "And Stanley just kept saying, 'Jesus, oh Jesus,' and he was writhing that this was said. I said, 'Stanley, you said, "Do whatever you have to do." Are you having trouble getting your film in the camera now?' It was that kind of thing. Whatever Stanley needed."

. . .

Throughout December and January, a strangely inwardly focused predecessor to the London Eye Ferris wheel was rising and being accessorized in Stage 4. With an interior designed by Masters and Lange in New York and constructed in sections at MGM, and a hulking exterior frame engineered and built by Vickers-Armstrongs, *2001*'s centrifuge set represented a mechanical way of producing artificial gravity for long-duration space missions. At thirty-eight feet in diameter, ten feet wide, and thirty tons, it was among the largest kinetic film sets ever built. It would doubtless have been bigger still, but its frame took it to the maximum height the tallest soundstage at Borehamwood allowed.

Constructed at a cost of over $750,000, the centrifuge was supposed to be the Jupiter-bound astronauts' main living quarters, and would gobble up about one-eighth of the film's original budget, which MGM had adjusted upward to $6.5 million—though after all the dust settled and the accountants had a chance to do their work, it eventually

The centrifuge set dominated Stage 4.

worked out closer to one fourteenth of the film's final price tag of between $10.5 and $12 million. The central part of the studio floor had to be jackhammered open so that a new concrete base could be sunk into the ground to accommodate the wheel, with dual pyramidal I-beam support structures bolted directly into it. The set's two sides were perfectly matched drums, their external parts made of mirrored steel-fretwork circular walkways with curving wooden floors and a ring of film lights, all pointing outward from the hub, or "down" toward the outermost rim—meaning in the direction of the set's hamster-wheel floor, invisible from outside. A disc-shaped chain of diffusion gels ensured evenly distributed light within the cloistered interior.

Arranged at various points along the external wheel on both sides were batteries of Bell & Howell 16-millimeter projectors, each pointing at one of the centrifuge's flat-screen displays. Twelve were mounted on a rectangular frame backing HAL's console alone. With the additional medical displays for the three hibernating astronauts, a total of fifteen projectors ran simultaneously. Getting all of them to run together and in synchrony was a nontrivial accomplishment, and it provided an example of a management technique Kubrick sometimes used: playing his talented staff off against one another. Rehearsals had been scheduled to commence on January 31, but by midmonth, a satisfactory way of mounting the projectors within the rotating structure still hadn't been devised. Kubrick's lead on this, as on all mechanical effects, was Wally Veevers.

Veevers was a fascinating character. He'd handled the filming of *Dr. Strangelove*'s B-52 bomber, among other things, and had visual effects credits extending as far back as the seminal 1936 British science fiction movie *Things to Come*—the film that had caused Kubrick to exclaim "I'll never see another movie you recommend!" to Clarke in New York. Bald and wide, with a physiognomy similar to Alfred Hitchcock's, only with a pug nose, he was well loved by his coworkers, extremely capable, and notorious for his short fuse, which periodically caused him to turn bright red. Veevers had constructed an unwieldy rig of steel joists for the centrifuge projector arrays, such that the projectors themselves were arranged at different distances from HAL's tightly clustered face-panel screens—a design that, in turn, necessitated lenses with different throws to compensate for the distance discrepancy. Unfortunately, this had introduced a new problem: the screens supplied by the more distant projectors were dimmer than the others.

Kubrick was aware of all this, and, unbeknownst to Veevers, had summoned Brian Johnson and told him they were running out of time. "I want you to redesign it, 'cause it's not going to work the way Wally is designing it," he said. Johnson had been hired at the age of twenty-six straight from producer Gerry Anderson's successful *Thunderbirds* TV series, where he'd been responsible for constructing and filming the program's rocket planes and spacecraft. He'd been working on *2001*'s miniatures, which by now were being made entirely at the studio. After his discussion with Kubrick, he hiked over to Stage 4 to study the situation.

Johnson quickly realized that the projectors could be arranged differently. Each upright one could be matched with one bolted below it. Although nominally upside down, the lower projectors immediately took care of the main problem introduced by Veevers's design, namely some screens being dimmer than others because of varying projector distances, and since the whole set would be turning, all the projectors would be going through full 360-degree rotations: there was no up or down here. Of course, in Johnson's design some of the projectors *were* upside down in relation to the set inside. But in those cases it was easy simply to run the film upside down and in reverse—something that immediately oriented their footage "up" vis-a-vis HAL's "face" inside the wheel. It was an elegant solution.

He rigged four projectors in this way just a few feet from Veevers's more complicated array, alerted Kubrick, and a side-by-side test was arranged. The massive wheel had been turned so that HAL's multi-screen console, which centered on the computer's cyclopean red eye, was at the bottom, easily accessible through one of the rectangular floor hatches.* Arriving with Veevers, Kubrick said, "Right, show me what you've got." Trumbull and Logan's animated films depicting HAL's data stream had been printed on eight reels of 16-millimeter film and threaded by assistant film editor David de Wilde through the various projectors. Veevers, Johnson, and Kubrick climbed inside the set through a trap door in the floor. Veevers's array was behind the four screens on one side of HAL's eye, and Johnson's behind a similar set of screens on the other.

When de Wilde rolled the films, it was immediately clear which system was better: Johnson's. They climbed silently out of the centrifuge. "Wally, file your stuff. We're using Brian's," Kubrick said, departing. A furious Veevers picked up a light stand with both hands, smashed it onto the floor, and stalked away, his head a navigational beacon fading in the gloom.

He didn't talk to Johnson again for weeks.

---

* HAL's eye wasn't a prop but was actually a Nikon Nikkor 8-millimeter wide-angle lens lit from behind. It was reportedly requisitioned occasionally by Kubrick, affixed to a camera, and used for filming—though not for HAL's point-of-view shots, interestingly enough, which were made with another lens, a Fairchild-Curtis 160-degree ultrawide angle.

When they did their first test run with a full slate of fifteen projectors and the centrifuge spinning simultaneously—thirty film reels in total, all rotating within the gargantuan set's mega-reel—they'd somehow forgotten to take into account the gravitational forces at play, and the reels all fell off, clattering two by two onto the circular wooden flooring of the centrifuge's external framework.

·   ·   ·

If a map were to have been made of Kubrick's on-set peregrinations during the making of *2001: A Space Odyssey*, the orbits inscribed therein would always have centered on the camera. He might go on long cometary trajectories to outer regions, the better to consult with Geoffrey Unsworth on lighting issues, or Tony Masters on the sets. But he always returned reflexively to the mammoth Panavision, which was almost invariably outfitted with Zeiss lenses, contrary to Panavision's contractual understanding. He usually had one or more still cameras around his neck as well.

If the shot was even slightly complex, the director usually operated it himself—no reflection on the abilities of operator Kelvin Pike but simply Kubrick's way. This was one of the many reasons why the director couldn't and didn't last in Los Angeles in the late 1950s. Union rules prohibited anybody but the cinematographer and his assistants from operating the camera in Hollywood, and a Kubrick unable to handle his own camera was a Kubrick artificially hobbled. Even when he didn't operate it, he always set the compositions, which were invariably things of beauty.

Alongside Kubrick was Unsworth, the lighting cameraman, aka the director of photography. As with Pike, in practice it was a role he shared with Kubrick—something this avuncular, balding, soft-spoken professional had adapted to with the egoless assurance of a master who firmly believed that cinematography should always serve directorial intent. A versatile, intuitive craftsman, Unsworth had shot thirty-five features before *2001*. Veterans of the production remember him and Kubrick in ongoing quiet dialogue throughout the long production weeks, sometimes flanked by Pike and camera assistant John Alcott,

Geoffrey Unsworth.

sometimes off by themselves as they hashed out photographic approaches to some of the most complex sets and setups in filmmaking history.

If Kubrick wasn't on set throughout the lighting process, he invariably tweaked the results on his return. As with the Clavius conference room, most of *2001*'s scenes were characterized by a chilly, almost preternaturally uniform illumination. This wasn't easy to achieve, and he and Alcott had established an effective way to determine if their lighting ratios worked. Their solution wasn't based on light meters but by taking innumerable black-and-white Polaroid stills.

Late in the preproduction period, Alcott had conducted many tests to calibrate the relationship between the Polaroid's aperture settings and those of their lenses. After this had been definitively established, those small rectangular instant photographs—which at the time required peeling away from their negative and coating by hand with a wet fixative—were the most important on-set guide to how their filmed shots would actually appear. This master-slave relationship between a mass-production camera and their expensive 65-millimeter Panavision

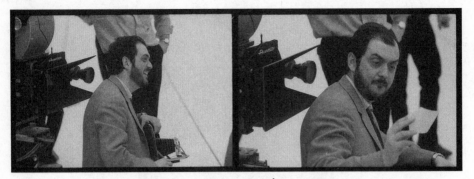

Kubrick with Polaroid camera and print in Space Station 5 set.

was remarkably effective, and somewhat comparable to using a cheap car radio speaker to test a mix in a recording studio.

Kubrick and Alcott's innovation was part of why the director frequently had a boxy Polaroid Land Camera around his neck while on set, its black bellows extended. But their replacing of traditional light meters with Polaroids was also an outgrowth of Kubrick's prior practice of checking compositions with stills. "I think he saw things differently that way than he did looking through the camera," Alcott commented.

> When you're looking through a movie camera, you're seeing a three-dimensional image, so you're getting a feeling of depth. But when Kubrick looked at this Polaroid still, he would see an entirely different, two-dimensional picture—it was all one surface and closer to what he was going to see on the screen. And many times, just before we were going to shoot, he would change the setup because he didn't like what he'd seen in the Polaroid.

As a result, throughout the production, the director shot an estimated ten thousand Polaroids as the lighting was adjusted, then readjusted, and the camera positions were modified. They ended up strewn everywhere, a kind of instant-image confetti piling up in the corners and being swept away by janitors every evening. Only those stills judged of direct utility for continuity or composition, or in determining key

Polaroid lighting test shot of HAL's "eye,"
most likely taken by Kubrick himself.

exposure settings, were retained by Alcott, who kept them neatly in an
album near the camera.

Kubrick's attention to cinematographic questions was the legacy of
a lifelong engagement with photography predating even his five years
as a photojournalist for *Look* magazine in the late 1940s. His understanding of aperture, focal length, depth of field, film stock, frame
rates, and exposure times was unparalleled. Asked about this in 1980,
Alcott—who would shoot *2001*'s Dawn of Man sequence, and later
serve as director of photography on Kubrick's *A Clockwork Orange*
(1971), *Barry Lyndon* (1976), and *The Shining* (1980)—said, "If he were
not a director, he would probably be the greatest lighting cameraman
in the world." Kubrick's mastery in this regard sometimes resulted in an
on-set dynamic where, when finally ready to roll after hours of lighting, Unsworth would call out to Alcott, "Make that a five-six"—an
aperture of 5.6—and Kubrick would countermand, "No, a six-three,"

whereupon Unsworth would give Alcott a discreetly good-humored wink, and they would split the difference.

Or not. At the time they worked together, Unsworth was fifty-two and Kubrick thirty-eight. One day late in the production, Caras invited Unsworth to lunch in the MGM restaurant. They were making their way past the studio's slab-sided soundstages to the commissary building when he noticed the cinematographer squinting meditatively down at the pavement as he walked, his brow furrowed, clearly lost in thought. "Geoffrey, why so deeply contemplative?" asked Caras.

Unsworth considered the question for a minute. "You know, Roger," he replied, "if anyone had told me six months ago that I had anything of any substance to learn about my profession at this stage of the game, I would have told them they were mad. I have been a top British cinematographer, a top man, for twenty-five years. In fact, though, I have learned more about my profession from that boy in there in the last six months than I have in the previous twenty-five years. He is an absolute genius. He knows more about the mechanics of optics and the chemistry of photography than anyone who's ever lived."

He stopped and turned to look Caras in the eye. "Are you aware of this?"

. . .

With the giant wheel almost finished, MIT artificial intelligence pioneer Marvin Minsky paid a visit to Borehamwood, and Kubrick proudly took him to demonstrate its capabilities. One issue the Hawk Films contingent had been facing throughout its construction was that the MGM workmen installing lights, panels, and bulkheads would simply forget the nature of the object they were working on, drop their tools, and go to lunch. With Minsky standing directly at its base, the better to appreciate its magnificent kineticism, Kubrick ordered that the centrifuge be taken for a spin. With a low moan of power and a portentous whirr, thirty tons of steel heaved into motion. At the half-turn mark, a heavy pipe wrench shifted with an ominous rasping sound at the twelve o'clock position and plummeted forty feet—slamming into the ground with a sickening *blam* directly at the scientist's feet.

"I could have been *killed!*" Minsky remembered. "Kubrick was livid and shaken and fired a stagehand on the spot." Once the dust had settled, the scientist's near-death experience created an interestingly paradoxical yin and yang of fact and fiction—wheels within wheels again. It was Minsky who'd recommended the terms behind HAL's acronym, and he'd also been the one who confirmed to Kubrick that computers thirty-five years in the future might be advanced enough to suffer breakdowns when faced with apparently irresolvable conflicts. And Kubrick had named one of *2001*'s hibernating astronauts Kaminsky in tribute to the creator of the first self-learning neural network, SNARC.★ An astronaut in turn destined to be bumped off, during the course of its mission, by a murderous supercomputer.

.　.　.

Following the Clavius conference room scene—which Kubrick was finally forced to break into shorter takes in response to Sylvester's problem—all the Space Station 5 material was shot relatively uneventfully in late January. On the nineteenth, the first actual words spoken in *2001*'s running time, twenty-five minutes after the film's opening title—"Here you are, sir, main level, please"—were enunciated by the elevator operator (played by model and actress Maggie London) as the station's circular lift rotated into view. On the twenty-fifth, Christiane brought little Vivian back into the studio, this time to watch her father film Heywood Floyd calling her on the station's videophone.

By the end of January 1966, all attention was focused on the giant wheel. The first two weeks of February were spent on camera, camera mount, video assist, and lighting tests, as well as actor rehearsals. It was the most complex shooting environment anybody on the highly experienced crew had ever encountered.

In practice, two categories of shots were planned: those with the set turning, and those with it stationary. The centrifuge could rotate

---

★ For stochastic neural analog reinforcement calculator. But you already knew that.

Stanley and Vivian Kubrick watch
William Sylvester at the videophone.

in either direction at speeds sufficient for an actor to stroll along at a
normal pace at the bottom. When turning at its maximum speed, three
miles per hour, he could break into a jog. Shots where the wheel wasn't
in motion were relatively straightforward and were handled in a regular way, with auxiliary film lights sometimes brought in to augment
the diffused light cast from the wraparound ceiling.

Filming while the set was turning was another matter. The kinetic
shots were also divided into a further two categories: those in which
the camera remained near the bottom of the wheel, and those where
the camera (and its two-person crew) rotated all the way around as the
centrifuge turned. In both cases, the actors usually remained at the
bottom—though in one spectacular shot, Lockwood was strapped to
his seat at the dining table and then rotated to the top, where he was
seen upside down eating lunch when Dullea entered and climbed down
a ladder from the center—thirty-eight feet below and 180 degrees away
from Lockwood; then he walked around to join him, with the turning
wheel effectively bringing him around to his deputy. Both actors ended

Centrifuge conference with, from left to right:
Stanley Kubrick, John Alcott, Kelvin Pike, Keir
Dullea, Gary Lockwood, and Geoffrey Unsworth.

up in a heads-up position by the end of the shot. It was a bravura piece
of work.

These two species of camera position—if one can use the term for
a set in motion—had necessitated the construction of a unique pair
of shooting platforms. For shots in which the camera remained near
the bottom, it was mounted on a dolly that rolled along on four large
rubber wheels—fixed in relation to the studio outside, but in motion
vis-a-vis the rotating set. On the other side of the centrifuge floor,
the dolly was mounted to a platform external to the set by a thin steel
blade, which permitted the camera to be hoisted up to 40 degrees from
the very lowest extension of the wheel—or about twenty feet up the
sloping floor from the actor. This was necessary because when walking
or running, Dullea and Lockwood always stayed at the bottom of their
high-tech hamster wheel. Once a given angle was chosen and the dolly
fixed, the giant set turned around it, and the blade knifed between
rubber matting on the floor, which fell back into place after it passed.

The second species of centrifuge camera position required an even more ingenious camera mount: one in which *it* was now fixed in re-lation to the centrifuge, but in motion vis-a-vis the studio outside. According to a production memo that January, a rig permitting the Panavision "to be rotated like a clock, as well as being turned over and over in the other plane of movement" was required. MGM engineer George Merritt soon devised an ingenious gyroscopic framework per-mitting the camera—as well as camera operator Kelvin Pike and cam-era assistant John Alcott—to ride around and around as the set turned, all while bolted securely to the wall. A nested pair of stainless steel hoops, the spinnable mount allowed its passengers full freedom to pan and yaw the camera no matter what their orientation. In 1984 Alcott remembered the experience of riding Merritt's rig.

> We would go up *with* the wheel itself. It was a little hairy being up there thirty-five feet. It was like going to a fairground. There was a massive gimbal mechanism to hold the camera. It was a 65-millimeter camera—a sound camera, too, so it had a big blimp on it. The camera was attached to the base, and we could sit on it; and as we went up, the movement would be counter-acted by the movement of the wheel. It *was* a bit hairy. We used to have lifelines up there, you know, which we could jump on anytime we needed. Gyro lines—you'd just have to step onto them, and you'd come down.

Since the lens was always pointed at the actor down at the bottom even as the camera crew rotated all the way around, Pike and Alcott had to perfect the art of keeping their feet out of the frame. Once that was accomplished, however, the handmade technologies devised to shoot in their topsy-turvy environment would enable them to produce some of the most blatantly original footage in filmmaking history.

. . .

With all thirty projector reels now affixed antigravitationally to their stubs by retaining clips, and each projector housed in thick-walled sound-insulating blimps, the first actual takes in the big wheel were made on February 16—the so-called roadwork sequence in which Gary Lockwood jogs along, shadowboxing as the wheel turns around him. In the final film, those four shots of astronaut Frank Poole working out constitute an extraordinarily innovative display of filmmaking. The first extended take alone served to catapult the audience into a wondrously strange realm in which up and down no longer had their accustomed meanings.

In this first interior view of the giant Jupiter-bound spacecraft *Discovery*, the elongated rectangular aspect ratio of the 65-millimeter film frame is used to full advantage, and we see Poole coming toward the camera from a horizontal position, face left and feet right. He's running along on what transpires to be a kind of endlessly circular floor, and as he jogs by, his horizontal position changes rapidly to a vertical one, as seen from directly below—but this, too, transforms quickly, becoming a kind of flip-turn from his first position: now Poole's running *away*, sneakers to the left, head to the right. Meanwhile, without a cut or change of camera position, our point of view continues to morph seamlessly from what may initially have seemed a shot from midway up a wall, to a view clearly looking *up* from the floor, and finally a vertiginous look straight *down*, from the ceiling—unless *he's* running across the ceiling? It all has a Möbius-strip, M. C. Escher, WTF quality, a disorienting tour de force in which Lockwood runs circles around the audience and prior cinematic verities. Nothing like it had ever been seen before. All that problem-solving had paid off. Kubrick had made it new.

Because only Pike, Alcott, and Lockwood could be inside the set while shooting—and sometimes, when the camera was on its dolly, Lockwood was alone—a small video camera was mounted directly beside the Panavision's lens. This allowed Kubrick to monitor events from a closed-circuit TV monitor in front of his command chair outside—among the first times such a video assist had been used in film production. Seated at the centrifuge base, the director communicated with Pike and Lockwood through two microphones, one of which hooked up to the wheel's internal PA system and the other to their earpieces.

Kubrick's bulbous TV monitor had been masked with tape to define the borders of the rectangular film frame. It was a far cry from the ersatz high-tech flat screens inside the wheel, but it worked. Apart from the microphones and monitor, he had a turntable on the desk in front of him, and for Poole's workout session, he'd cued up an LP of Chopin waltzes. It was the first time those multiple rotations—spinning vinyl evoking dancing couples turning in three-quarters time; and gleaming white machinery rotating smoothly in space—had been combined within the wheeling, weightless, rhythmic, ever-kinetic context of *2001: A Space Odyssey*.

On their first day of shooting, they managed seven takes, exposing more than twenty minutes of film as they turned the wheel twenty-two times. Impressive for a first try, particularly given that they managed both kinds of shot: those with Alcott and Pike riding around the wheel in their gyroscopic camera rig, and those with Lockwood alone inside the centrifuge with a "fixed" camera riding a dolly as the set turned around it. In the latter scenario, after a flurry of barked commands from assistant director Cracknell, Kubrick cued Lockwood, who reached forward and started the camera himself before everything else began turning. In effect, the familiar filmmaking litany had changed from "Roll camera, roll sound," to "Roll camera, roll *set*"—with "Spin Chopin" somewhere in between. A nicely multivalent set of rotational energies.

Not everything went flawlessly. On February 21 they were filming with an ultrawide lens when the Panavision detached from the dolly and smashed to the floor, breaking the video system in the process and delaying the shoot by several hours. The Grundig video camera was primitive, and being clamped to the Panavision with an adjustable bracket, was easily nudged away from its correct orientation. The communications system between deep space and ground control wasn't wireless but required cables that fed through slip rings and gradually wound onto the centrifuge drum, requiring that the whole thing be reversed and unwound at the end of long takes. The second the finicky video camera went out of alignment—a regular occurrence—Kubrick, desperate to know what was happening inside, fired rapid questions at Pike: "Kelvin! What are you doing? What are you doing! *It's getting away from you!*" Pike's voice would echo back, as distant as an astronaut from the Moon: "It's fine, Stanley, it's okay, it's okay!" Houston, no problem. Apart from everything else, a Nikon F single lens reflex, one of Kubrick's favorite still cameras, had been mounted on the Panaflex, and Pike periodically squeezed off shots.

After a few days of jogging, Lockwood had run so many miles in new sneakers that he emerged from the centrifuge limping from blisters. Returning downtown, he started to notice that no matter which direction he turned, London's streets seemed to curve upward in front of him, causing him to lean slightly forward in compensation. The set's circular logic had warped his visual perception. Heat from all the lights and a buildup of exhaled carbon dioxide made it hard going inside. Outside, the wheel turned with a groan of applied power accompanied by the constant ominous tinkle of fallen nails and shattered tungsten bulbs. "It was a scary place to work," Trumbull recalled. "Film lights don't like to go upside down when they're hot. There was a constant barrage of exploding glass. Every time the centrifuge went around, there was this *Pow!* as lights blew up."

In fact, Kubrick's Mission Control area, jammed with speakers and wires, had a large, rectangular chicken-wire frame around it to shield the crew from a perennial rain of broken glass and other debris. Anyone leaving the protected zone was supposed to put on a hard hat.

With the actors—and frequently the camera crew—sealed inside, the wheel was also a significant fire hazard. Exit through the floor hatches was neither fast nor easy. A fire department team was always on alert at the peripheries of Stage 4 during shooting. Given the amount of electricity coursing through fat cables; the popping bulbs and falling objects; the snaking, overlapping air-conditioning conduits and projector wires; and the flammable materials comprising many of the interior set elements, the risk of a sudden flash fire was real.

But they were lucky.

■　■　■

Throughout the shooting of *2001*, Gary Lockwood provided a textbook case of a well-documented filmmaking phenomenon in which the goings-on behind the camera are sometimes more interesting than those in front. Despite his deltoid quarterback's body, maintained by working out several times a day and by his rigorously self-enforced diet of veggies and fish, Lockwood's presence within the film was curiously anodyne. This was intentional. Kubrick and Clarke had conceived of the spacemen in charge of *Discovery* as well educated and acutely self-controlled, a pair of dual PhDs so unflappable as to be almost android. Like Charlie Chaplin's bolt-tightening factory laborer in *Modern Times* (1936), spinning along with the very cogwheel he's working on—only with less humor and more pathos—they were ghosts in the machinery, component parts of their chilly, refrigerator-white mother ship.

In real life, however, Lockwood was a charismatic trifecta of charmer, hustler, and rogue, a gambler, pool shark, and ladies' man; not the brightest bulb on the set but by no means unintelligent; and possessing a highly attuned antenna for imprecision and bullshit almost comparable to Kubrick's own. Although he'd been a rabble-rouser in college, and although some on the set remember him as never seeing a recreational drug he didn't like, the actor kept his head together and was a disciplined presence throughout.

To his own surprise, Kubrick found himself intrigued. The director had more than a streak of hustler himself, albeit in a more cerebral way, and, like Lockwood, had on several occasions supported himself with gambling of one kind or another. While famous for having earned just enough money to survive by playing chess for quarters in Lower Manhattan's Washington Square Park in the late 1940s, it's less well known that a decade later, Kubrick kept food on the table by playing a weekly high-stakes poker game with Hollywood fat cats. Having completed *Paths of Glory* in 1957, he returned to Los Angeles from Germany accompanied by his glamorous new wife, Christiane, her daughter, Katharina, and a large debt. The latter was the result of loans made over the previous several years by supporters, including his partner Jim Harris, who'd produced three of his early films—money that Kubrick had used to complete his second and third features, *Killer's Kiss* (1955) and *The Killing* (1956).

As a result—and despite the success of *Paths of Glory*—the couple hit the ground broke. Christiane has vivid memories of discomfort with her new husband's gambling. "I felt this was so not-solid, so Wild West.

I just felt, 'This is going to end badly,'" she recalled. "He sits there in a circle of people, there are these big, fat cars parked outside our house, and it made me so nervous." She soon realized, however, that his approach to gambling mirrored his on-set discipline, producing exactly the results he'd predicted. "He played, and he never won big, and he never lost, he made very small irregular [bets]," she said. He told her, "I don't want them to think I have a pattern or certain technique or I play in a certain way . . . so it has to be irregular. I'm only trying to get two hundred bucks for the supermarket next week. No more."

"And he did it," said Christiane. "We survived on it."

Throughout his time as *Discovery* subcommander Frank Poole, Lockwood engaged successfully in various forms of gambling as well, both in the London casinos that had been legalized in 1960—providing an excellent opportunity for organized crime figures such as the Kray brothers—and in relatively small-stakes games with the Borehamwood staff. At one point, he was owed so much by his lighting stand-in from gin rummy winnings that the man failed to show up for work the next day. Evidently he didn't want to pay up, and thereafter simply went AWOL. Hearing of this, Kubrick felt he had to intervene. Summoning Lockwood to his office, he asked him to quit gambling with his film crew. "It was very simple," said Lockwood. "It wasn't like I needed it. And I just said, 'Sure.' We didn't make a big thing out of it."

"Everybody was so in awe of Kubrick, they just constantly were like sycophants," Lockwood remembers. "And that's not my nature. I'm a quarterback. I'm an alpha male, aggressive. I'm a bar fighter. I was smart in school. I'm a piece of shit, you know." The actor's success with women fascinated Kubrick, as did his knowledge of sports, particularly football. Kubrick had been flying in films of NFL games from New York since his arrival, and he and Lockwood began watching them on Friday nights at the Kubrick residence at Abbots Mead, the early-nineteenth-century house just south of the studio that he bought in 1965, periodically commanding projectionist Eddie Frewin—Tony's dad—to stop the film so they could discuss plays.

When they tired of this, they played a lot of snooker. "He had the most beautiful snooker table I ever saw," Lockwood recalled. In an

Gary Lockwood and Stanley Kubrick take a break in the pod bay.

inverse of the chess-dominance tactic Kubrick had engaged in with unwary actors on set—most famously George C. Scott, whom he'd repeatedly checkmated while shooting *Dr. Strangelove*—Lockwood never failed to thrash him at the game. When Kubrick noticed that the actor was fully ambidextrous, using his left and right hands with equal skill and shooting without a rake, it was the last straw. "How did you get the way you are?" he demanded. Lockwood shifted his attention from table to director. "Life," he replied. "How'd you get the way *you* are?"

Asked years later if Kubrick ever showed signs of irritation at being unable to beat him on his own table, the actor denied it. On the contrary, he said, Kubrick always seemed to enjoy their games. "He was the epitome of cool," Lockwood observed appreciatively. "I mean, Steve McQueen was called 'Mr. Cool.' But it was really Mr. Kubrick."

. . .

The question of how to transmit the essence of HAL's internal dilemma to the audience had preoccupied Kubrick and Clarke for many months, and it hadn't been resolved even as the film went into production. In late January Kubrick wasted almost a week filming space agency ground controllers, including actor Neil McCallum, looking straight into the camera and giving lengthy, verbose explanations about the type of malfunction HAL "may be guilty of." McCallum advised Bowman and Poole that they were conducting a three-day "feasibility study" to look into the possible transfer of spacecraft control from HAL to an Earth-based computer. The material was intended to play as a video uplink received by HAL and played for the astronauts in the centrifuge, presumably upping the computer's paranoia in the process.

The problem with this approach was that the astronauts were essentially passive receivers of someone else's decision making. Plus, it was a telling, not a showing—an elementary dramaturgical error. None of this sat particularly well with Lockwood, who instinctively sensed something was wrong, and by mid-February, when the two jump suited actors were preparing to shoot the sequence at HAL's centrifuge console, he was convinced that a mistake was being made. The scene was "a little bit redundant and, candidly, just a little bit on the corny side," he remembered. "I really didn't like it, and I just felt it wasn't up to where we had been. And two guys as brilliant as Kubrick and Clarke, they were trying to solve a problem verbally. And we were making a very nonverbal type film."

For his part, Kubrick sensed that Lockwood's heart wasn't in it. "You're not showing your normal enthusiasm today," he observed. His cover blown, the actor launched spontaneously into a weirdly off-color story. "You know, Stanley, I grew up in a redneck town with a lot of rough people and rednecks and stuff like that, and I was always disruptive in class," he said. "And I sat in the back of the room, and this one girl would come down the aisle that had the greatest ass on her— callipygian, if you will—and she would hand out test papers.* And every time she got to me, I would look up and say, 'Hi, Peggy, how about a piece?' And then I took my right elbow and I went"—he jabbed his elbow at Dullea, who was watching the whole performance with

---

* Callipygian (kal-*uh*-pij-ee-*uh* n) adjective 1. having well-shaped buttocks. Origin: 1640–50; Greek kallipyg(os): with beautiful buttocks, referring to a statue of Aphrodite.

opaque amusement—" 'Yuck yuck yuck.' Well, that's what I think of this fucking scene."

Kubrick quietly absorbed the information. He wasn't used to being spoken to in this way. "Derek?" he said. "Yes, Guv?" Cracknell replied. "It's a wrap." It was only eleven in the morning.

Lockwood withdrew to his dressing room, unsure if he'd blown it. While he felt bad that he'd sassed Kubrick, his conviction that the scene wasn't right remained unchanged. "I guess there was a side of me that wanted to protect this great movie," he said later. Although worried about the potential consequences, he'd started his daily exercise routine when a knock sounded on his door, and Cracknell stuck his head inside. "The Guv wants to see you," he said.

"Right," said the actor. "Hey, Derek, before you go, am I fired?"

"Don't know, mate."

Lockwood showered rapidly, toweled himself off, threw on some clothes, and went to Kubrick's private lair, a second-floor dressing room off Stages 6 and 7 that he used for rewrites and solitary brainstorming. "You're Polish, aren't you?" the director asked, inviting him in.

"I'm Polish and German," Lockwood responded.

"You want schnapps or vodka?" Kubrick asked.

"Candidly, I like tequila," said Lockwood.

"How do you take it?"

"In a snifter, like brandy."

Kubrick poured him a shot of "not a great tequila but a so-so tequila," as Lockwood remembered it, and walked over to "a wall of records; an absolute wall. 'What do you want, Chopin?' 'Yeah, Chopin's fine. Stanley, before we get into what happened, am I fired?' "

No, Kubrick replied, Lockwood had been very much a team player, and he'd learned to pay attention when somebody who'd been working conscientiously had a problem. He referred to an incident in which John F. Kennedy, visiting NASA headquarters for the first time, saw a janitor at work behind the assembled brass. Circumventing the suits and walking over, he introduced himself and asked what the man was doing. "Well, Mr. President, I'm helping put a man on the Moon," was the response. You never know who's going to have a good answer,

Kubrick observed. He asked the actor to explain what the problem was. Lockwood said that the series of little scenes that Kubrick and Clarke had written to "tighten the screws on the computer" didn't "cut to the marrow." In his opinion, they were too diffuse. He said that while he thought Kubrick was the best director in the world, there had to be a better way to trigger HAL's paranoia.

Kubrick listened carefully, and when they finished their drinks, he said he knew Lockwood liked deli food. He would give Eddie Frewin some money, and Eddie would drive him to the best deli in Golders Green—London's small kosher nexus—for lox, whitefish, and bagels. "Then I want you to go home, and the next time I hear from you, I want to know how you would solve the problem."

Back at home, Lockwood finished his bagel and got out a spiral notebook. He listed everything he liked and disliked about what they'd been filming, writing down what he thought Clarke and Kubrick were trying to accomplish, how they were trying to accomplish it, and where it had gone awry. At some point late that afternoon, he remembered that just the week before, he'd gone to visit a new set being constructed on Stage 1 called the pod bay. With typical attention to detail, Kubrick had asked him to go there, don a space suit, and see if it was awkward to climb in and out of one of the space pods that aviation firm Hawker Siddeley would soon stuff with buttons, screens, and controls.

In doing so, Lockwood had, in fact, noticed a small problem and had asked that a handhold be installed directly inside the pod door. If Kubrick hadn't asked him to reconnoiter the situation in the first place, the actor wouldn't have known about the space pods, let alone that they could hold two people. The pod bay scenes were months away.

By nine o'clock, Lockwood had developed an idea. He called Kubrick and said he thought the astronauts should find a pretext to climb into one of the pods, thereby cutting themselves off from HAL entirely so that they could talk in private. Their discussion could then include disconnecting the computer—and HAL could stay a step ahead of them by figuring out how to overhear them. That way, he said, the audience would discover everything they needed to know, and HAL could be made to experience a very human persecution complex.

Hearing this, Kubrick grew excited. He sent a car to pick up Lockwood. It was a subzero February night, but by the time the actor got to Abbots Mead, a fire was roaring in the fireplace. They sat in front of it until the early hours of the morning, drinking, discussing, and, finally, improvising the scene together.

▪ ▪ ▪

Kubrick rarely raised his voice on set. That was Cracknell's job. Sometimes he didn't even interact with his actors at all, particularly if they were young, attractive women—such as *2001*'s stewardesses, several of whom were well-known London models.

Heather Downham, who played the Pan Am stewardess that picks a floating pen out of thin air and then gently inserts it back into a sleeping Heywood Floyd's pocket, described how she would show up on set in the morning and relay any questions concerning her role to Cracknell, who would ask Kubrick for his thoughts, which would then be transmitted back to her via Cracknell again—even though they were all standing in the same tightly constricted space. While this may sound bizarre—and it says something about Kubrick's shyness when confronted by female beauty—it's actually not that unusual. A film director is constantly barraged from all sides by questions and demands. If he's not to deplete his energy reserves, he needs to rely on assistants to serve as filters. Making a film, the director sometimes observed, is "like trying to write *War and Peace* in a bumper car in an amusement park."

Keir Dullea's relationship with Kubrick was somewhat different from Lockwood's, and ultimately more distant. The actor arrived still shell-shocked from Preminger's tyrannical behavior on *Bunny Lake Is Missing*—a sadistic daily shout-fest—and so Kubrick's tranquil ways were a great relief. On the other hand, Dullea was very much in awe of Kubrick, whose work he'd admired for years. He regarded the director as a genius, and so was initially a bit starstruck. It hobbled his first week of work.

Despite Sylvester's experience of Kubrick's implacable will during the Clavius conference room scene, most who worked closely with Kubrick

Keir Dullea as David Bowman, testing
HAL's erroneous fault prediction.

describe him as a caring and empathetic presence, soft-spoken but with a wicked sense of humor. Dullea was no exception, but if anything, this made the situation worse: these very qualities made the director practically godlike, particularly in contrast with Preminger. "I was projecting my awe, and being in awe is not freeing," Dullea reflected.

Kubrick intuited what was going on, and after several days of awkwardly tentative takes, he dismissed the crew and took aside the actor. For several hours, they calmly discussed the problem. Kubrick told Dullea he thought he was one of the finest actors working. He hadn't been chosen by accident, he said, but because he was perfect for the role. He felt badly because it was his own fault—*he*, Kubrick, was doing something wrong, and was personally responsible for the actor not feeling more relaxed. Somewhat dismayed at this, but relieved to be discussing the matter so frankly, Dullea assured him it was entirely his own problem; in fact, Kubrick had been great. He felt better, however, and soon transitioned from a state of trepidation to one of trust. He would soon need it.

Of the two actors, it was Lockwood who knew from the beginning that they were involved in "the greatest film ever made." The sets were so brilliantly lit that the actors wore sunglasses between takes. Occasionally Lockwood would survey their environment, turn to Dullea, lower his shades, and say, "Can you *believe* this?" He was just as impressed by Kubrick's ability to improvise within all this grandly conceived futurism.

"Stanley was simply much more intelligent than other directors, and in a nonlinear way," the actor observed. "You found that anything could happen at almost any time. Despite all the careful preparations, he designed *2001* with an air of flexibility, and that's what made the picture brilliant. Directing is all management of human relations, logistics, details, and so on. On top of all that, Stanley still found room for a kind of danger—even a kind of bravery or recklessness. He was always trying things where there was a high risk of failure. He wasn't the obstinate, solitary genius of popular imagination. He needed people to bounce off, and he would often turn around and ask if he was doing the right thing."

As with other actors who'd worked with him, however, Dullea and Lockwood sometimes experienced Kubrick's inability to articulate what he wanted. At the end of takes, they would frequently hear a terse "Okay, let's do another." When they asked what he wanted done differently, he would respond with "Let's just see what happens."

As Tony Masters had discovered, he needed to be presented with options, and only then could he decide.

. . .

Clarke had returned to Ceylon in early February 1966, where he immediately found himself in a ludicrous position. Despite having put more than five thousand miles between himself and the daily struggle to shoot *2001*, he discovered an all-new, high-stakes set of film production issues confronting him. He'd poured a significant amount of money into his partner Mike Wilson's Sinhalese James Bond parody,

Jamis Bandu. But Wilson, who'd been behaving increasingly erratically, had devoted only fitful attention to the project. After shooting just over half the film, he'd simply dropped everything and left the country, bound for England.

Forced to protect his investment—which would soon rival the total sum he'd made from his involvement in *2001* thus far—Clarke had no recourse but to assume the role of film producer. If he could get the remaining scenes shot, with luck he might recoup at least some of his outlay from ticket sales.

Apart from his drug use, Wilson's distraction had undoubtedly been triggered by a competitive response to Clarke's involvement in big-time filmmaking, with its promise of the imminent arrival of the world's first blockbuster science fiction film. Arriving in London, he quickly used Clarke's entrée to insinuate himself into the Hawk Films group at Borehamwood. Meanwhile, his partner was pouring his last savings into the film he'd abandoned—and simultaneously fielding a steady stream of urgent cables from Kubrick. In the midst of trying to persuade the Ceylonese government to lend him a detachment of army troops for *Jamis Bandu*'s grand finale—something it flatly refused due to a recent military coup attempt—Clarke tried to help Kubrick find a plausible way for HAL to listen in on Bowman and Poole as they discussed the computer's fate within a hermetically sealed space pod.

He had one foot planted in an escapist, derivative Third World popcorn feature and the other in the most sophisticated evocation of human origins and destiny Hollywood had ever attempted, with the latter funding the former.

■ ■ ■

Centrifuge sequences dominated February and March. With Stage 4 being a cavernous, noisy, chilly place, Kubrick had a large trailer brought directly onto the premises. Following Lockwood's space pod breakthrough, he invited his actors into this heated, sound-insulated bubble between setups, and they improvised what they might say to each other, ostensibly out of HAL's hearing. After taping their exchanges, Kubrick had them transcribed. Then he edited the transcripts every morning, gradually whittling their dialogue down to the essentials.

But he hadn't yet figured out a plausible way for HAL to overhear them. They still had six weeks before moving to the pod bay set, but the director was unwilling to defer a solution. Clarke's proposal—that HAL access "geophones in the soft lander probes"—risked incomprehensibility. How to explain what a "geophone" is, let alone a "soft lander probe"? The film wasn't for geeks alone.

One afternoon, Victor Lyndon arrived in the trailer with some documents. Kubrick's associate producer was looking increasingly pale and tense these days. A large part of his work had devolved into the thankless task of filing a seemingly endless succession of insurance claims, with the quality of Kodak's film stock and that of the Technicolor lab's development regularly exchanging leading positions of dishonor in the director's eyes. He was relieved to take a short break and watch the actors run through their lines. On hearing that they still hadn't figured out how HAL could overhear them, Lyndon looked at Dullea and Lockwood as though it was the most obvious thing in the world. "He could just read your lips," he said. There was a moment of thunderstruck silence. "God, that's a great idea!" Kubrick exclaimed. They had their answer.

In late April they finally moved to the pod bay set for a couple weeks of filming. Possibly the purest expression of *2001*'s collective design genius, *Discovery*'s parking garage with its three space pods poised in front of circular air lock doors was among the most finely realized of Masters's collaborations with Lange and Ordway. Everything in the set was purely utilitarian—a Bauhaus marriage of form and function. The total effect was as pure as physics, with even the vivid colors of its racked space suits justifiable as the fastest way for spacewalking astronauts to identify each other from a distance.

On May 6 Kubrick shot Bowman's and Poole's conspiratorial space pod discussion in thirty-five takes. At almost three minutes, it would ultimately constitute the longest single dialogue in *2001: A Space Odyssey*. Lockwood still considers it his biggest single contribution to the film.

Dullea and Lockwood
rehearse their discussion.

▪ ▪ ▪

Another aspect of the story that hadn't been adequately solved when shooting started concerned how to get Bowman back into *Discovery* after his commandeering of a space pod to recover Poole's lifeless body. Following his request to "Open the pod bay doors, HAL," and the computer's famous response—"I'm sorry, Dave, I'm afraid I can't do that"—*Discovery*'s sole surviving human crew member had to find another way back inside. Complicating matters further, he'd left his space helmet behind in a distraught rush to recover Poole.

In another example of their quest to ensure maximal technical accuracy, Clarke and Ordway had both reached out to various air force researchers, seeking confirmation that a human being could, in fact, withstand unshielded exposure to the hard vacuum of space for the short period necessary to move between the pod and ship. While they'd verified it was indeed possible, Bowman's method of ingress hadn't been worked out.

Harry Lange had designed 2001's spherical, anthropomorphic pods, but their ingeniously contrived servo arms had been taken directly from Marvin Minsky's published work—another instance of the scientist's influence on the film. But even if the same arms that enabled HAL to kill Poole via remote control could also be used to open Discovery's emergency air lock when under Bowman's command, how would he vacate his pod and get inside the ship fast enough to survive? On February 10 Kubrick wrote Clarke that he had a "completely believable" solution.

> After the pod arms open the air lock door . . . Bowman backs the pod up, so that the rear pod door, which is smaller than the air lock door, is perfectly aligned and almost touching the open airlock door. This rear pod door will have explosive bolts which Bowman will explode without bleeding the pod air. The result of this should be that if Bowman is in a curled-up position and pointing toward the door, he should be shot like a cannonball directly into the open air lock. If he then manages to bang the control button, he would then, within a matter of seconds, have made the exit from the pod into the air lock and have the air lock door closed. I think this should cover even the worst skeptics. What do you think?

Clarke responded that he'd already thought of that gambit and "it's quite okay. Also crawling with Freudian symbols, as you are doubtless aware." Bowman's explosive reentry into Discovery became one of 2001's most memorably kinetic moments. It interrupts the film's stately rhythm with an eruption of smoke out of which Dullea vaults like a cartoon figure in a Roy Lichtenstein "Pow!" painting. There was no question of using a stuntman. The actor needed to be recognizable, since he'd left the ship without his helmet—a mistake legitimating the scene in the first place.

The emergency air lock set was a padded, vaguely vaginal tube flooded in pink light. Constructed not far from the centrifuge on Stage 4 and oriented end-up, it stood twenty-four feet high—almost three stories—directly above which a space pod had been suspended by chains from the ceiling. The pod was stabilized and made accessible by a raised platform, its back to the set. The sequence required some old-school theatrical effects. To his trepidation, Dullea needed to climb to the top of the platform, then be tethered to a thin wire attached to the small of his back by a leather harness fastened around his waist and hidden under his space suit. After about twenty feet, the wire was attached to a thick rope, which had a knot tied in a position just prior to Dullea's maximum extension—meaning, as close as they wanted him to get to an encounter with the camera at the bottom, before he was made to ricochet back to the air lock's entrance.

Hidden from the camera by Dullea's body, the wire and rope were attached to another person: an expert in theatrical flying from Eugene's Flying Ballet, a company famous for the technique. The company roustabout was poised on the platform outside, ready to jump as soon as the knot hit his gloved hand. He would provide the counterweight, smoothly reversing Dullea's fall and returning him back to the entrance, where, if everything worked as planned, he would float convincingly toward the "emergency hatch close" lever—the mechanism of his salvation.

When they shot the scene on June 16, Unsworth had camera operator Kelvin Pike roll thirty seconds of static shot with the pod doors closed—material for Kubrick to edit seamlessly with what was to follow—and then they cut and opened the pod door. His heart pounding, Dullea was lowered into place and braced himself with both hands at the open door. At Kubrick's command, he released his grip, extended his hands forward, held his breath, and Pike rolled the camera again. At "Action," a team on the platform blasted a thick dollop of smoke through the air lock entrance, and the actor was released, his wire rapidly unspooling in a headlong plunge toward the camera. A split second later, the roustabout saved Dullea from being impaled, with all movements rendered jolt-free by a system of geared aluminum drums and "ball-races"—

Dullea, dangling from a wire, closes the emergency
air lock door. Polaroid most likely by Kubrick.

ring-shaped components stuffed with ball bearings. Rebounding to the
air lock top, Dullea swiftly reached around to the lever, closing the door
and smiling in relief as oxygen ostensibly flooded the chamber.

On the twentieth, a Monday, they continued covering the scene.
Now Unsworth had Pike undercrank the camera—setting its frame
rate at eighteen and not twenty-four frames per second, thus speeding
up the action—and shot it again. Dullea, an actor so nervous of heights
that he refused to fly, willingly repeated his emergency air lock swan
dive five times during the two shooting days.

Reflecting on the experience decades later, he said, "People ask,
'Why were you willing to do it?' And my answer is, I totally trusted
Stanley."

. . .

In March Kubrick shot two quickly executed scenes with Lockwood in
the centrifuge that on first viewing may have seemed unimportant—as

close as the director ever got to filler—but have since become practically legendary, at least for some. One was Poole's birthday greeting from his parents. The other was the chess game.

The question of who was to play HAL remained unresolved throughout the actual live-action shooting. At first, Kubrick brought in UK actor Nigel Davenport, who for the first week was present to read the computer's lines. His delivery was soon judged too British, however. Kubrick's next plan was to record American actor Martin Balsam, but that would happen only in postproduction anyway. Meanwhile, HAL was played either by Derek Cracknell—whose cockney accent, Dullea observed, made the computer sound a bit like Michael Caine—and sometimes even by Kubrick himself.

In practice, Kubrick usually played HAL when working with Lockwood. On March 7, they filmed Poole's centrifuge tanning bed scene with a small crew. A birthday greeting from the astronaut's parents had been filmed weeks before and was cued up on a projector behind the solarium screen. Kubrick usually didn't provide his actors with much direction, relying on them to feel their way into a scene and shooting many takes. This time the utterly dispassionate way Lockwood played it resonated with him immediately, however, and they nailed the sequence in just eleven minutes of film. The director introduced an element of improvisation, throwing in an unscripted curveball to see how Lockwood would react.

In the final take, Kubrick suddenly said, "Happy birthday, Frank," after Poole's parents sang to him, and following his father's concluding line, "See you next Wednesday."* Lockwood responded with an impassive, "Thank you, HAL. A bit flatter, please"—commanding the computer to lower his automated bed. The actor's opaque reactions to both his parents and HAL were indistinguishable: a kind of

---

* The seemingly throwaway line concerning Wednesday was used by film director John Landis as the title of his first, unproduced script, and subsequently deployed as a recurring gag in most of his films, including *The Blues Brothers* and *Twilight Zone: The Movie*. It then memed its way into innumerable other films, TV shows, and video games.

numb flatness. More than any other scene in the film, Poole's birthday transmitted a quietly excoriating message about the desensitizing effects of technologically mediated communications. Those who criticized the film for its emotionless performances missed Kubrick's point entirely—in this case, as channeled through an underrated actor who knew exactly what he was doing.

The chess scene, shot later in March, immediately follows Poole's birthday in the film's running time as well. This one wasn't even scripted and appears to have been a pure improvisation—though its dialogue is based on a game Kubrick had chosen in advance and asked Trumbull to animate. While the sequence's significance is so understated as to be practically invisible to anyone but hard-core chess players, it's nevertheless consequential, because it presents the first hint that something's not quite right with HAL.

Kubrick based the game on an actual documented match played in Hamburg in 1913. In the part we see, Poole decides to resign when HAL—again voiced during filming by Kubrick—says, "I'm sorry, Frank, I think you missed it. Queen to bishop three. Bishop takes queen. Knight takes bishop. Mate." This early in *Discovery*'s trajectory toward Jupiter, it didn't occur to Poole that the computer might be cheating or making an error. But that's how the director played it, because it should have been "queen to bishop six"—something the astronaut failed to notice, despite its being on the screen in front of him. Kubrick was subtly transmitting that HAL's first victim had his guard down.

There's another subtlety to their game as well. Of filmmakers then active, Kubrick valued Ingmar Bergman above all—so much so that he wrote the Swedish director a fan letter in 1960, praising his "unearthly and brilliant contribution," and stating, "Your vision of life has moved me deeply, more deeply than I have ever been moved by any films." In view of Poole's eventual fate at HAL's remote-controlled hands—so remote as to convert the pod that killed him into one pawn taking another—there's a parallel between the brief, seemingly inconsequential scene where Poole effectively plays chess with Death, and Max von Sydow doing the same in his incarnation as a medieval knight playing the same grim adversary in Bergman's *The Seventh Seal*.

. . .

Kubrick's mastery of the goings-on at Borehamwood and his almost preternaturally keen attention to detail were legendary, as were certain personal peccadilloes that his young assistants sometimes satirized, always behind his back.

His ability to see through obfuscation was usually immediate and unsparing. Kept waiting one afternoon due to an alleged problem with a piece of recording equipment, Kubrick took in the mumbo-jumbo technical explanation offered by Derek Cracknell, digested it, waited a few more minutes, and then said, "The next time the sound mixer's late back from lunch, just tell me."

Regularly stopping assistants such as Ivor Powell in the hallway, the director would pull out a set of index cards and ask for status updates as he shuffled rapidly through them, plowing through innumerable bullet points covering various aspects of the production. Powell quickly learned that if he didn't know something, he should say so rather than risk bluffing.

When Kubrick asked for a technical explanation, he homed in mercilessly on fuzzy thinking or partial accounting by people who should know better. "If you can't explain something to me, you don't understand it," he said to special effects man Bryan Loftus, who had to admit he was right. What he wanted was the truth, observed Loftus, and if somebody didn't know something, that was acceptable to him, because it, too, was a truth, one that could then be tackled.

Quickly spotting technical anomalies, Kubrick attacked them with intuitive intelligence. After days of subpar images from an optical printer that manifested as misaligned color layers, he grew impatient of failed attempts to fix the problem and came to inspect the machine himself. Evaluating the device with its keeper, Loftus, he asked if the printer was, in fact, rock steady. "As far as I know," Loftus replied, pointing out that it was bolted securely to its concrete base. At this Kubrick, who happened to be holding a plastic cup of water, placed it on the machine. The two of them leaned forward to observe as tiny wavelets started coursing across its surface. "Look at that, Bryan," the

Few portraits of Kubrick have captured
his alert, contemplative essence as
well as this one by Dmitri Kasterine.

director said. "It's rippling. It must be moving." Having established the
source, they could now address the problem. There were many such
stories.

As to his peccadilloes, those close to him might wave them away
as overblown or even manufactured, but they existed. Fastidious about
personal hygiene, he couldn't stand to be around anybody who was
ill. If this was unavoidable, he insisted that they—and everybody else
on the set or in the office—wear surgical masks to lessen the danger
of germs spreading. The sight of blood sent him speed walking in the
opposite direction. Flies and other insects were anathema, and despite
his crush of responsibilities, he found time to pay close attention to
the studio's window screens so as to keep bugs out of his working
environment. Like General Jack D. Ripper in *Dr. Strangelove*, he insisted on drinking only bottled water—something one assistant earned
quiet backstage laughs circumventing by amassing empties of Malvern

Water, surreptitiously filling them at the tap, and providing the director with "new" bottles on demand. "Stanley drank tap water for two and a half years," said Brian Johnson.

A heavy smoker since the age of twelve, under pressure from Christiane he'd convinced himself that he was quitting, and so never bought cigarettes. As a result, he was the most reliably shameless bummer of smokes in the British film industry, to the point where poorly paid crew members took to keeping empty packs in their pockets so they could demonstrate they were out. Discovering what was going on, an embarrassed Christiane brought a carload of cartons to the studio one day and saw to it they were distributed, in partial compensation.

Although stories that he only drove at slow speeds have been dismissed by some, credible accounts have it that during the production of *2001*, Kubrick operated under a self-imposed limit of twenty-nine miles per hour, having deduced that this was the top survivable speed should an accident occur in his Mercedes or Rolls. Attempting to disguise what he knew must sound strange, he explained to cabbies and car service drivers that he was recovering from back surgery, and they'd simply have to keep their speed down, ostensibly for medical reasons.

Concern for his personal safety also lay at the root of Kubrick's fear of flying, which originated in a traumatic near-death experience while piloting a plane in the late 1950s. One day he observed to Roger Caras that being brave was stupid; why would you risk the only chance you have at life, simply to prove something? Commenting on Gary Lockwood's physique while playing touch football in the Kubricks' backyard one day, Jeremy Bernstein told the director that the two of them together might manage to bring him down. "Never get into a fair fight," was the response.

.   .   .

Three months into the production of *2001*, financial and logistical pressure had been building so severely on Clarke that he finally erupted at Wilson, then still in London. "I am now fighting to finish the movie," he wrote from Colombo on March 12—referring not to *2001* but to Wilson's Bond parody. "I am absolutely sick and disgusted with the whole situation, now I know what has happened. I don't trust myself to write about it."

> Last week I borrowed another 10,000—making 35,000—so that we could continue. I now hear that the Rocket Publishing Co. [Clarke's UK company] has spent almost £3,000—half of its income—on the movie—and now has nothing in the bank. I cannot even help my own family, and there is nothing to cover the next tax demand. Though I was anxious not to do it, I have now had to start borrowing from Stanley. Because of all these worries, I have not been able to sleep or to do any work on the novel. Whether I am ever associated with you again in *any* project depends entirely on satisfactory answers to my earlier letters and the arrangement you are making to settle your liabilities.

Wilson's profligacy had caused Clarke to cross a line he'd sworn he wouldn't. By borrowing from Kubrick, he was now literally in his debt—something that, in turn, reduced his leverage when it came to publishing the novel. Despite this, wrote Clarke, "I am still anxious to help you, even though I have practically ruined myself to do so." With striking frankness, he continued: "Everything I have earned in the last 10 years has been poured into your projects." But like a long-suffering parent, he concluded, "Meanwhile so that you can carry on until I get back to London on the 5th . . . I have asked Stanley to advance you a couple of hundred if you need it." Kubrick had become a banker so that Clarke could continue funding his partner of seventeen years.

The financial pressures on Clarke made it that much more imperative that he sell the book. As of March, he and Meredith were still operating under the assumption that Kubrick would shortly allow them to sell hardcover rights, thereby giving them sufficient lead time to monetize a paperback edition as well, with the whole program ostensibly timed to the film's release. A communications nexus had developed in which Clarke bitched to Kubrick about the novel and Meredith pressured Kubrick's lawyer Louis Blau regarding the same, with plenty of pushback from both and cross-communication among all parties.

Clarke and Mike Wilson in the *Aries* lunar vehicle set, below
a sign reading "Caution: Weightless Condition."

Although he masked it with various excuses, the director simply wasn't
giving permission. His loans to Clarke took off some of the pressure.
Those close to Kubrick understood that he simply didn't want the book
coming out before the film was released. To be straight about that
would be to break his word, however, so he said he needed time to
suggest revisions—time he didn't have at the moment.

As Clarke's intimate and confidant, Caras was aware of the financial
strains on him and the resulting pressure to get his novel into print after
two full years of work. Uncharacteristically for the prolific author,
apart from a smattering of magazine articles, he hadn't managed to do
any significant writing outside of *2001*. Caras had few illusions about
the source of his friend's pecuniary problems. He was carrying Wilson's water. "His relationships, one after the other, the most important
relationships of his life, have been founded strictly on his homosexuality," he told Clarke's biographer Neil McAleer in 1989. "It has cost him
millions of dollars . . . He has spent millions of dollars on these rela-

tionships." Clarke, said Caras, "was victimized by Mike Wilson. Severely." Adding to the burden, the author wasn't just supporting Wilson, who was evidently bisexual, but also the man's family: his wife, Liz, their kids, and even Wilson's mother in England.

Apart from that, his collaboration with Kubrick had long since moved on from its happy perambulatory brainstorming phase. Caras was well positioned to evaluate Clarke's state of mind throughout the production, and although the author tried to mask it, what he saw was not good. "I think he found it all very frustrating because he had no control. Stanley had no time for him," Caras said. "He was never *not* in control of his own project before." Asked to elaborate, he continued, "Arthur was not happy . . . I have every reason to believe . . . knowing Arthur's look on his face, things he said, that Stanley was being unreasonable. Just expressions he dropped, and knowing that he rarely voices a negative response . . . Arthur has a positive opinion about anything—he's a very positive fellow. I have reason to believe he was desperately unhappy about the whole thing." Concerning Clarke's struggle to complete and sell the novel, Caras observed:

> He couldn't really finish the book . . . until the picture was made because he never knew what Kubrick was going to do next on the set. At any moment, Stanley could have gone off on some tangent, and the book would be nonsensical as a representative of the film. He was held up there. And he was very perturbed about Stanley holding up the pub date of the book. That bothered him. He talked about it quite a bit—more than any other single negative element in all the years that I've known him.

Despite this, Clarke showed immense loyalty to Kubrick. In response to repeated cables and letters from the director, who was handling the colossal strains of the ongoing day-by-day production schedule and simultaneously the stresses of attempting what amounted to a coherent jazz improvisation with critical plot points, Clarke kept a nervous breakdown at bay—something he admitted to Meredith—and proceeded to greatly tighten and improve the film's plot.

As Lockwood had intuited when he spontaneously told his off-color story to Kubrick the previous month, Bowman's and Poole's reactions as they become aware of HAL's incipient disorder had retained, as Clarke put it in a letter from Colombo on March 11, "a lot of unnecessary dead wood and fossil material from previous versions. (Sorry about mixed metaphor.)" His detailed communication to the director that day constituted a decisive intervention, further reducing and compressing the story following Lockwood's space-pod conspiracy idea by listing nine actions he proposed Kubrick either film or use existing material to illustrate. At this key moment halfway through the shoot, Clarke's nine-point plan became a blueprint the director largely followed, though he cut them down a bit further, in part by incorporating some into the astronauts' pod dialogue, which hadn't been shot yet.

In much the same way that Lockwood's and Dullea's dialogue improvisations had methodically been reduced to their essence, Clarke's suggestions stripped down and clarified the story. HAL predicts the failure of *Discovery*'s antenna's guidance unit, necessitating Bowman's spacewalk to go retrieve it. Tested in the pod bay, the unit checks out okay, however, and Mission Control "hints that the fault may be in HAL." Bowman and Poole discuss this with HAL, who "sticks by his diagnosis and claims he is right. (Your pod-bay eavesdropping here?)" Kubrick indeed put it there. Poole then goes on his fatal space walk. "Unless I am suffering from chess blindness and can't see an obvious move, this seems a vast improvement," Clarke wrote. "It is also more logical."

Clarke's contribution from Colombo in March 1966—a time of great personal and professional turmoil—redefined *2001*'s middle passages, producing essentially the film we see today. It serves as a clear testament to the pivotal role he played even during production.

• • •

Although as cool as they come while on set—cooler than Steve McQueen—Kubrick frequently showed the result of the enormous pressures on him after returning home at the end of long, supercharged production days. His constant modifications of story and concept were taking their toll. All eyes were on him at the studio, and he couldn't afford the slightest sign of weakness or indecision. Safely back at Abbots Mead, however, sometimes he melted down. "I don't know what I'm doing, I have no idea!" he'd exclaim, his face pale from the strain. At times he struggled "with this feeling of being the dumbest animal walking the earth," Christiane recalled. Wrestling with how to approach a scene, he'd ask, "Does that sound right? No. It sounds really pompous! Stupid!"

Discussing the project with her, he observed that he was working on "one of those films where you think any minute, 'I've got the answer,' when, in fact, I know I don't." He'd list people he'd previously found ignorant or worthy of scorn, and say, "I'm just like that. Why am I critical? Because I think I'm better! That's really pathetic." Weary and seemingly at the end of his rope, he'd obsess over an actor, a bit of dialogue, a production mishap, or a technical problem. "This isn't doable," he'd vent. "Why'd I think I could do that? I have no idea! I don't even know how . . . I'm going to cut the whole scene!" To this, Christiane—an accomplished actress who'd read his script changes, was tracking developments closely with astute intelligence, and understood that he was exhausted and that this was part of his process—would sometimes push back, replying, "No you're not! It's really good." Then he'd either reconsider, or stick to the decision, or table the thought—sometimes concluding the whole thing with "I'm gonna have a hamburger."

"And then, five o'clock in the morning, he would suddenly say, 'Well, maybe if I let *him* come in first and say *that*, then maybe it would work,'" Christiane recalled. "He'd find an intellectual game worth playing. I think he often got himself out of it like that."

Some days, after returning home thinking the whole thing didn't make any sense, and rewriting until sunrise, and then returning to the studio the next day to announce the changes, Kubrick struggled with a profound sense of shame that he'd gotten it so wrong to begin with and had deployed dozens of people and much money to realize a set or a scenario that went one way, when, in fact, he now realized, it absolutely had to go in another direction entirely.

Asked incredulously if this was really so—if the great Stanley

Kubrick actually felt shame over his many revisions and modifications—
Christiane affirmed it.

"He would feel ashamed?"

"Very."

"But he hid it very well in the studio."

"He tried."

"He succeeded. I've never heard the slightest—"

"He would hide it. He looked like he didn't doubt himself, yes. But he *so* did. He had these 'I'm just an asshole' moments all the time."

.   .   .

From the outside, the HAL Brain Room set on Stage 6 resembled a strange art installation: a four-story-high rectangular shape so encased in a boxy nestwork of bracing pipes and joints that its true form was hard to discern. The gridded exterior supported and was surrounded by some twenty evenly spaced 10-kilowatt film lights, producing a 200,000-watt glare so brilliant that the towering structure was almost impossible to look at from the outside without sunglasses.

Temperatures in the interior spiked well above 90 degrees. It was constructed of dark-grey sheet metal perforated by hundreds of evenly spaced rectangular slots backed by red-orange gels. With the lights blazing away outside, HAL's brain glowed from all sides—looking, camera operator Kelvin Pike observed, like the inside of a toaster. Dullea was to kill HAL by floating in through a hatch and methodically disconnecting the computer's higher-logic functions as it begged for its life. As with the emergency air lock scene, he would need to hang by a single strand of cable for many of the shots, but in this case nothing particularly acrobatic was called for.

"I want this to be a murder," Stanley had said to Christiane, who clearly remembers the origins and etymology of the scene. "It was Arthur's idea," she said. "Stanley wrote it. But he, Arthur, planted the concept of an intelligence as something that's alive. An intelligence is life. If you hurt an intelligence, it can't bear it. It knows you're hurting it." Recalling her husband's approach, Christiane said, "It was very important to him that the computer suffers when he takes these bits out and removes bits of the brain. That's why he lit it red, so it looked fleshy."

As for HAL singing "Daisy Bell (Bicycle Built for Two)," this, too, was Clarke's contribution, including the song's gradual devolution to near incomprehensibility at the end. The idea originated in a visit he'd made in 1962 to Bell Laboratories, where he'd heard John Kelly's voice-synthesizer experiments with an IBM 7094 mainframe, which had coaxed the machine to sing Harry Dacre's 1892 marriage proposal—the first song ever sung by a computer. Even as he expired, HAL was referencing a significant moment in computing history.

As with many of 2001's sets, the Brain Room was a hazardous place. On the morning of June 15, a "sparks" (electrician) had climbed to the very top of the structure and was moving one of the massive 10Ks when he lost his footing, fell almost thirty-five feet to the studio floor, and broke his back. An ambulance rushed him to the nearby Barnet Hospital, where his life was saved. Because the crew had been working on a simple insert shot—of a TV screen blinking fitfully to life with Heywood Floyd's pretaped message concerning the actual purpose of the mission, supposedly triggered by HAL's final throes—Kubrick had framed the shot and left it to Pike and Cracknell to film, and was working in his office. A young assistant, Andrew Birkin, was there when word came in that somebody had been badly hurt. "Gee, that's terrible," said Kubrick, looking concerned. "Did it wreck the shot?"

Dullea knew that as the stress of losing his fellow astronauts and dueling with HAL took its toll, his character's seemingly unflappable quality would need to deteriorate, and he prepared for the Brain Room scene by replaying in memory Burgess Meredith's powerful performance as George in the original 1939 film version of John Steinbeck's Of Mice and Men. At the end of the film, George, realizing that his mentally disabled friend Lenny (Lon Chaney Jr.) is about to be murdered by a vengeful mob, decides to shoot him himself—a mercy killing. But first they go over a familiar story: their mutual dream of owning their own ranch, with cows, chickens, and rabbits. Then, with grim determination and evident sorrow, George shoots Lenny in the back.

If this kind of channeling may sound a bit excessive, given Dullea's performance, it's only because Kubrick reduced the scene significantly in the editing. In the film as released, the actor has only two short sentences to say while lobotomizing the only other sentient being for two hundred million miles: "Yes, I'd like to hear it, HAL. Sing it for me." In fact, the scene was shot with much more dialogue. While the computer still had by far the most lines—four hundred words before Dullea's first—the actor originally had eight subsequent lines totaling more than fifty words.

The Brain Room sequence was shot across five long days between May 31 and June 29. By the last day, the tricky wirework was done, and Kubrick framed a medium close-up, a side view of Dullea's face through his helmet. The crew had rigged a seat for the actor to sit on, allowing him to sway backward and forward slightly, simulating

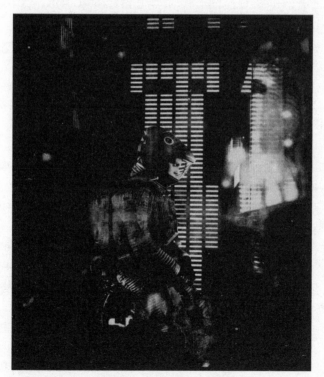

Side view of Dullea in the Brain Room. Partially
degraded Polaroid probably by Kubrick.

weightlessness as he used a small screwdriver to loosen the rectangular Plexiglas blocks representing HAL's logic and memory circuits.

Dullea's best performance came on the last take of the last day. "Dave, look. I'm really sorry about everything. I am genuinely sorry," said HAL, now channeled by Cracknell. Dullea continued working without comment—methodically, inexorably. "Dave, even a condemned criminal isn't treated like this," pleaded Cracknell. Swaying slightly under his flat-topped helmet, Dullea had a hooded, cobra-like quality to him—less an agitated Burgess Meredith than a determined assassin. "Please . . . Please, Dave, stop . . . Dave, you will destroy my mind, don't you understand . . . I will become nothing," said Cracknell.

"Shut up, HAL, you won't feel anything, just like when Poole and I go to sleep," said Dullea finally, his eyes panning over to return the computer's unblinking gaze.

"Well, I have never slept . . . I . . . don't know what it's like," said Cracknell.

"It's very nice, HAL. It's peaceful. It's very peaceful."

"What happens after that?" asked HAL plaintively.

"It will be all right," said Bowman, his attention flickering between his task and HAL's accusatory glass eye.

"Will I know anything? Will I be me?"

"It will be all right," Bowman repeated.

"Say, Dave, you know, I've just thought of something," said HAL. "The quick brown fox jumped over the lazy dog."

"Yes, that's right, HAL," said Bowman, relentlessly turning his screwdriver.

"I've thought of something else," said HAL. "The square root of pi is one decimal seven seven two four five three eight zero nine zero." (Had it been used, this would have provided evidence of the computer's accelerating decline—the number's incorrect.)

"Yes, that's right, HAL," repeated Bowman, working to finish the job.

In a shot Kubrick captured but didn't use, HAL's eye—previously alive with it usual glowing red cornea and yellow iris—had gone black and cold by the end of the scene.

Arthur C. Clarke and Stanley Kubrick in the *Aries* lunar lander set.

Pierre Boulat takes stills at dawn, assisted by Catherine Gire.

Scouting Dawn of Man locations in the Namib Desert.

Colin Cantwell "pivot point" test shot of monolith and African sky
before the sun and moon were added.

Dan Richter as Moonwatcher intuits a new use for a large thighbone.

Orion space plane matches speed with Space Station 5. This and most of the stills presented in this color insert section are from rare 65-millimeter footage not actually used in the film.

Right: Heywood Floyd's shuttle approaches the space station.

Below: William Sylvester as Floyd encounters an inquisitive group of Russian scientists.

Bottom: Aries spacecraft on its landing pad. The moon base windows haven't yet been filled with live-action footage.

Opposite middle: Accompanied by Clavius Base officials, Floyd surveys the lunar monolith.

Opposite bottom: Sylvester touches the monolith—an action Dan Richter will repeat a year later for the Dawn of Man sequence.

8

9

Kubrick works with his actors under the usual cloud of cigarette smoke.

Sunrise over the lunar monolith—Cantwell's second "pivot point" shot.

Right: The astronauts recoil from a powerful radio signal.

Below: The Jupiter-bound *Discovery* spacecraft.

Bottom: HAL's red eye was actually a Nikon 8-millimeter wide-angle lens lit from behind.

14

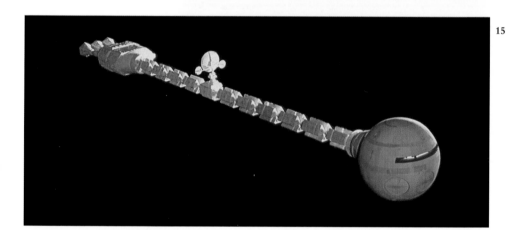

15

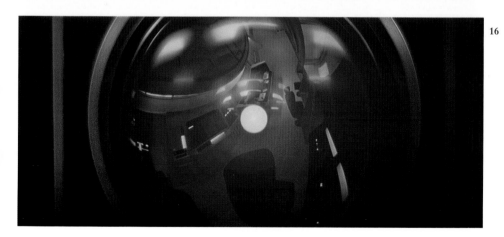

16

17

18

19

Left: Gary Lockwood as astronaut Frank Poole plays chess with HAL.

Left middle and bottom: Keir Dullea as Bowman being extruded from *Discovery* in his space pod.

Right: Keir Dullea and Gary Lockwood as astronauts Dave Bowman and Frank Poole, seated at HAL's computer console in *Discovery*'s astonishing centrifuge set.

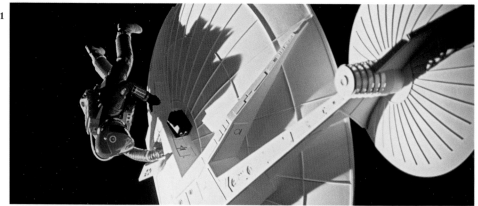

Top: Dangling from wires, stuntman Bill Weston plays Dave Bowman retrieving the antenna guidance unit. *Middle:* Mission controller Frank Miller delivers news that HAL is making mistakes. *Bottom:* Poole and Bowman discuss HAL, visible through the pod window, ostensibly in private—though he's reading their lips.

At thirty-eight feet in diameter and thirty tons, *2001*'s centrifuge, seen here from outside, was one of the largest and most expensive kinetic sets ever built.

Top: Seen here in a lighter moment, Kubrick frequently wielded
one or more cameras during production. In this shot he has two.
Above: Watching video feed from inside the centrifuge during filming.
Cinematographer Geoffrey Unsworth is to Kubrick's left and camera
assistant John Alcott is behind.

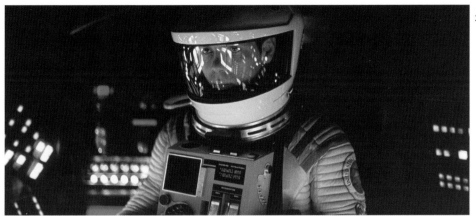

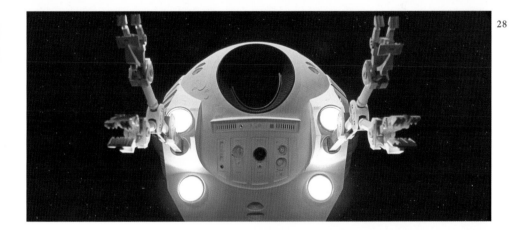

Top: Poole goes to retrieve a second allegedly faulty antenna guidance unit.

Above: Mimicking a pose from the 1931 film *Frankenstein*, Poole's pod attacks the spacewalking astronaut while under HAL's control.

Right: Bill Weston as Poole tumbles through space, struggling to reconnect his severed oxygen line. Given Weston's working conditions, the feeling was not unfamiliar to him.

Top: Bowman retrieves Poole's spinning body—one of Weston's most difficult stunts.

Left: Bowman marches grimly toward HAL's Brain Room.

Lower left: Following his disconnection, HAL's eye goes dead.

Top right: An early effects shot showed the waiting monolith silhouetted against Doug Trumbull's rendition of planet Jupiter.

Middle right: Unused Star Gate material in which the hurtling space pod (the white orb near center) was seen from outside.

Lower right: Kubrick frames a shot in the Hotel Room.

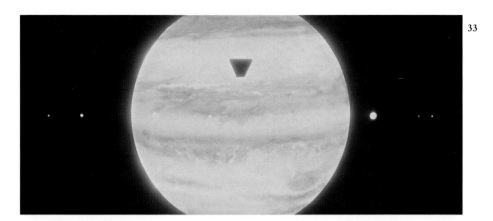

33

34

35

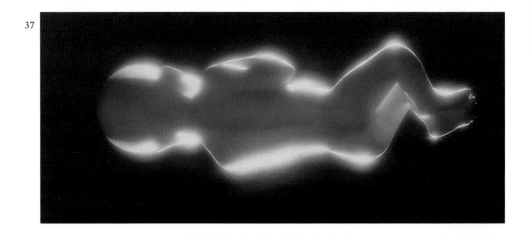

37

Top and above: Keir Dullea as Bowman dies and is reborn as a Star Child.

Right: Crop of an establishing shot in which the Star Child approaches planet Earth.

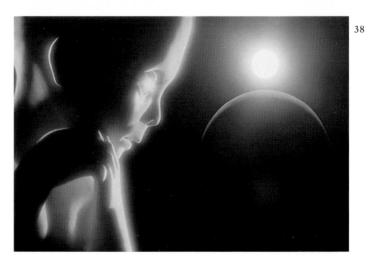

. . .

Early in preproduction, Kubrick had conceived of a documentary prologue, designed to protect *2001* from the kind of misunderstandings engendered by the then prevalent Buck Rogers and Little Green Men school of science fiction. Featuring eminent scientists such as Freeman Dyson, Margaret Mead, B. F. Skinner, and eighteen other leading researchers discussing such subjects as extraterrestrial life, space travel, and communication between alien civilizations, the prologue was entrusted to Roger Caras and was actually shot, on black-and-white 35-millimeter, throughout 1965. Caras even traveled to Moscow to film the eminent Russian biochemist Alexander Oparin, author of the influential "primordial soup" theory of life's evolution.

Kubrick had good reason to believe he was up against a perception that such subjects weren't to be taken seriously by serious people, and therefore "could hardly be the basis for what is commonly known to the film industry's barkers and shills as a *major motion picture*," as Tony Frewin put it in his book-length compilation of these interviews, published in 2005. (Transcripts are all that's readily available, though the actual footage may still exist somewhere in a Warner Bros. vault.) The documentary segment, Frewin pointed out, would have served much the same purpose as the pages of quotations on whales and whaling that Melville inserted at the start of *Moby-Dick*, prior to the immortal line "Call me Ishmael" that everyone remembers as the novel's opening. They recall it that way, of course, because the quotes weren't necessary to the story—something Kubrick ultimately realized about his documentary material. It was "one of Stanley Kubrick's few really bad ideas," observed one interview subject, Jeremy Bernstein.

Despite the director's distaste for Carl Sagan's patronizing manner at their New York meeting the year before, he'd readily agreed to his inclusion in the prologue. Nothing about Sagan's reaction to Caras's letter of invitation, however, was calculated to mollify Kubrick's aversion to the young astronomer. In his first response in February, Sagan had asked what he would be paid—the only subject to do so. Caras replied that they hadn't intended on paying a fee. In his second, on

March 10, Sagan asked for editorial control due to the "high frequency of quotations out of context and misrepresentations through
cuts and juxtapositions" in his "prior experience with the news
media"—a statement that couldn't have been better tailored to irritate
the director.

Observing, accurately enough, that the film was not a documentary
"but a commercial enterprise of some magnitude," Sagan wrote that
"the scientific introduction is clearly designed, among other things, to
gain respectability for the film; this is obviously a negotiable item." He
asked for 0.002 percent of the gross receipts of the film for every minute of his appearance in it.

Prodded vigorously by Kubrick, Caras responded that editorial
control would not be possible, and neither would a fee. Sagan was out.

. . .

2001's penultimate scene, set in a hyperreal hotel room following
Dave Bowman's epic journey through the Star Gate, had its origins in
the perception by Kubrick and Clarke that their surviving Odysseus-
astronaut would need a place to recover from his exposure to sights
the likes of which no human being had been exposed to before. Their
room was also supposed to have overtones of a holding tank or zoo
cage. "If you were having an extraterrestrial in your bottle in your
chemistry lab, and you wanted to make it comfortable so you can observe it," said Stanley to Christiane one day, "you would try to learn
what it likes. And if you milk the brain of a human being, it might like
a fancy art-book hotel room."

The concept originally stemmed from Kubrick's "robots who create
a Victorian environment to put our heroes at their ease"—as Clarke
had written in October 1964—and also from the author's own early
draft text in which he'd imagined Bowman seeking shelter in a building that could provide "mental security, for it would shut out the view
of that impossible sky." The Victorian element had evidently shifted
by July 1965, when Victor Lyndon specified to Caras that the room's
furnishings should be from the 2001 period. Only a week later, however, Tony Masters—who as production designer had better access to

Kubrick's thinking—wrote Caras that they really didn't know yet what the room would consist of.

Taking this rather inexact guidance, Kubrick's vice president proceeded to contact the Armstrong Cork Company, a leading American manufacturer of floorings and ceiling materials, proposing that they take the lead in designing the space. "The situation is roughly this," Caras wrote on July 29. "[W]hen the lead astronaut arrives . . . he is ushered into what is, in effect, a suite of rooms that is a direct duplicate of a suite . . . he sees on a television program coming from Earth. What has happened is, the superintelligence . . . has recreated this hospitality suite just as they saw it on television, and as they assumed everything on Earth would be."

Armstrong Cork's response was a model of industrial futurology. The company's designers proposed a set of rooms containing a type of inflatable high-tech furniture capable of disappearing into the floor when not needed. The ceilings, they wrote, could be made of "a series of lumps" that "move back and forth . . . giving a very restful, undulating type motion." If stairs were required, they should be invisible until stepped on, and then vanish again after use. "They propose that the floor will be very soft, and, in effect, padded," summarized Caras in a letter to Masters. "They say that this floor would glow, would give off light, give a very warm, indirect lighting effect."

By the time actual construction loomed, however, in the spring of 1966, it was clear that Armstrong's proposal, which amounted to a setting even more futuristic than the film's spacecraft interiors, might not have the desired effect. Such a set might, in fact, produce more of a sense that this environment "beyond the infinite" served as living quarters to the invisible aliens, rather than being created to put Bowman at ease. And the idea that the surviving astronaut flip through TV channels and discover a program featuring the very room in which he was sitting had long since been discarded.

By now, Kubrick's faith in Masters's judgment was about as strong as his "trust but verify" persona permitted. The designer had saved the day following the Perspex monolith fiasco, rapidly conceiving of what would soon be regarded as perhaps the most powerfully opaque object in film history. All his interiors had been exquisite as well. With the discussion now centering on that black monolith's last setting—and Bowman's transformation into a Star Child—Masters proposed another elegantly simple solution. "Why not have a French bedroom?" he said. "I mean, if you're going to have a bedroom, it could be anything. But one thing we can do quite well is a French bedroom. We'll do it all in nice soft grey-greens."

Kubrick evaluated the idea. "Okay," he said, nodding. "Let's do a French bedroom."

"Just like that," Masters said of the conversation in 1977. The decision had taken seconds.

When he drafted his rococo Louis XIV room, however, the designer did retain an important element from Armstrong Cork: the glowing, indirect floor lighting. Masters's set was constructed on Stage 4 using that ubiquitous Borehamwood substrate, steel piping, and it floated about twelve feet above the studio floor, giving space for lights amid the pipework. The set's floor was made of another familiar material: Perspex, in this case, three-foot-square tiles. The lighting was provided by 370,000 watts of firepower from below, creating a kind of Sahara for the crew inside, one that MGM stagehands tried to ameliorate with a giant flexible air-conditioning pipe that was shoved in through the bathroom periodically but didn't ever seem to cool the set enough. The lights gradually warped the tiles, which required periodic replacement.

Although to a viewer the effect was almost subliminal, Masters's joined bedroom and bathroom were entirely without exterior doors or windows. While the wall panels could be removed for filming, from an audience point of view, there was no *outside* to Bowman's extraterrestrial zoo cage.

The scene was shot during the second half of June, and for the first time required major work by Stuart Freeborn, then already recognized as among the world's best makeup artists. Freeborn had taken a cast of Dullea's face in January and since then had created a number of form-fitting foam rubber pieces, which when affixed and augmented during ten-hour makeup sessions, ushered the actor through several

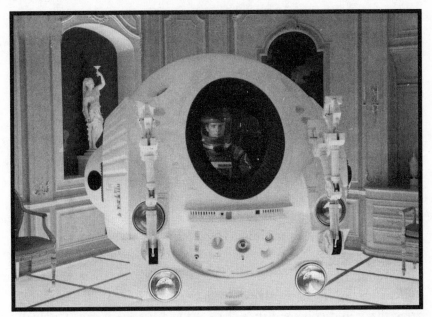

Dullea arrives at the hotel room in his space pod.
Polaroid lighting test probably by Kubrick.

stages between his original assumed midthirties vintage all the way to Bowman's deathbed. These included an octogenarian having supper, and a nonagenarian breathing his last.

A few weeks from the nominal end of main unit production for a film that by now had gone two months over its original shooting schedule, people were starting to drop away. Following the Brain Room scene, Cracknell left for another film and was replaced for several days by an inexperienced Ivor Powell, who'd transitioned in midproduction from serving as Caras's man in London to being one of Kubrick's assistants. While Geoffrey Unsworth would shoot the hotel room, he was wanted immediately thereafter on the set of a British musical and would not be available for the Dawn of Man, whenever it might be tackled.

The sequence depicted four transformations of *Discovery*'s sole survivor. Dullea first appears in his space pod seemingly undergoing a nervous breakdown after his passage through the Star Gate. He's visible again as seen through the pod window, standing wide-eyed in a

Ferrari-red space suit—evidently from the point of view of the shattered, quivering version of himself. Closer shots reveal this to be an older, wrinkled Bowman, perhaps in his midseventies, staring vacantly at the place where the pod no longer is. He walks into the bathroom, seemingly exploring his environment for the first time. Seeing himself in a large mirror, he numbly absorbs the sight of his own wrinkled face. Hearing the sound of cutlery, midseventies Bowman peers from the bathroom door to see octogenarian Bowman from behind, seated at a table beneath an oil painting recessed in the elaborate baroque walls—an arboreal scene. Finally, octogenarian Bowman sees the final version of himself on the bed.

The eight days spent filming the hotel room included a number of actions not used in the final cut. On June 23 midseventies Dullea was filmed walking to where his space pod had just been, then kneeling in disbelief to feel the floor where it had disappeared. Rising, he suddenly grew dizzy and collapsed in a chair, still in his space suit. Gradually he noticed that neatly folded clothing had appeared on the bed—an implicit invitation to shed the suit. Slowly rising, Dullea inspected the clothes. Then he walked into the bathroom—after which he sensed his older self eating at the dining table, as in the existing cut. Various props produced for the scene were also never used. These included a Potemkin phone book produced by Trumbull, which Bowman might have opened to discover nothing but blank pages inside.

The scene gave Dullea an opportunity to show off acting chops that hadn't really previously been called for. His quivering, catatonic state on arrival in the pod is utterly convincing. His octogenarian incarnation is a kind of understated symphony of stiffened, arthritic movements. In one particularly long take on Friday, June 24—the dolly shot in which midseventies Bowman sees his older self from the bathroom, with the latter seated and eating—octogenarian Bowman senses something, turns in his seat, rises, and approaches the camera to survey the now empty bathroom. Discovering nobody there, he returns creakily to his seat. Handwritten in blue ink, Kubrick's note in the *Daily Continuity Report* reads simply, "Very good acting!"

That day, Dullea had an inspired idea. The previous two transitions between his character's incarnations had been shot to permit straight cuts later in the editing, with each cut reflecting a jump from his younger to his older character's point of view—something Kubrick would subtly augment with the film's sound design. Seated now at the dining table with the camera on a high rostrum looking down at him from in front, Dullea surveyed the small tabletop in front of him, which included two cut-crystal goblets, and suddenly thought of a way to mark the final shift between his incarnations. "Stanley, do you mind if I knock this glass over?" he asked. "Let me find a different way of being in the moment of hearing something, or sensing something. Let me knock the glass over, and in the action of leaning over, let me be caught with that sense, right in midmotion, so that isn't a repeat of how I have done it up until now."

Intrigued, Kubrick contemplated the suggestion. "Okay, that's fine," he decided, dispatching the propman to fetch additional crystal. They proceeded to film Dullea's idea twice, in wide and medium shots from the rostrum. Then after multiple takes they lowered the Pana-vision to the floor and shot it a third time, now at table height from the side—the best angle to document the moment Dullea was looking for, as his attention shifted incrementally from the broken glass below him toward the bed, where his older self lay quietly breathing his—*their*—last, a sight greeted through narrowed eyes with scarcely repressed in-credulity. Once again, Kubrick registered approval in blue ink on the *Daily Continuity Report*: "Very Good."

While it may seem a less significant contribution than Lockwood's space pod breakthrough—without the latter's clarifying effect on the film's structure—Dullea's glass shatters during the most ideologically freighted final section of *2001*'s multipart structure. Every action in Masters's eerily echoing French period room resounds with potentially allegorical meanings. It's a percussive moment, that crystal smashing in the enclosed space, and it has provided grounds for plenty of inter-pretations in the decades since. In the end, it's probably as important as Lockwood's pod-conspirators scene. Resounding in a metaphysical en-vironment somewhere beyond the infinite, it's simultaneously a mani-festation of humanity's propensity for error, every glass ever broken in

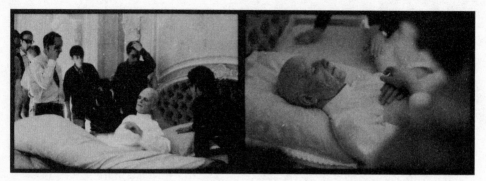

Keir Dullea as David Bowman on his deathbed.

all the Jewish weddings of history, a cymbal crash marking the death of God, a metaphorical token akin to the flash of insight produced by a Zen koan—and on and on.★

It's also, of course, another example of a onetime aspiring jazz drummer's innate ability to hear and adapt to the players in his group— a cinematic bandleader's real-time genius for listening, evaluating, and marshaling his players' talents to maximum effect.

On telling the story in 2014, Dullea referred to it as a "tiny contribution." It was more than that. Six months after his lead actor departed from Borehamwood, Kubrick sent a message to producer David Wolper. "Understand you are considering Keir for part in film," he wrote in the clipped diction of international cables. "He does unneurotic parts with same genius as neurotic ones. You cannot find a better actor or more cooperative or intelligent one. I think he is one of the best actors in the world."

He didn't get the role.

★ Consider, for example, this koan: Ikkyu, the Zen master, was very clever even as a boy. His teacher had a precious teacup; a rare antique. Ikkyu happened to break this cup and was greatly perplexed. Hearing the footsteps of his teacher, he held the pieces of the cup behind him. When the master appeared, Ikkyu asked, "Why do people have to die?" "This is natural," explained the older man. "Everything has to die and has just so long to live." Ikkyu, producing the shattered cup, said, "It was time for your cup to die." (From www.ashidakim .com.)

CHAPTER SEVEN

PURPLE HEARTS AND HIGH WIRES

SUMMER–WINTER 1966

> *Any sufficiently advanced technology is*
> *indistinguishable from magic.*
>
> —CLARKE'S THIRD LAW

With main unit production over at least for the moment, Kubrick was free to focus on two major areas of concern. One constituted the visual effects necessary to turn his film into a convincing simulacrum of our spacefaring future—exterior shots of the *Aries*, *Orion*, and *Discovery* spacecraft in flight, of the Moon Bus landing, and of his astronauts' free-floating space walks—as well as adding a more varied visual pizzazz to augment the neopsychedelic Star Gate material he'd shot already in that abandoned New York brassiere factory in early 1965.

The other was the Dawn of Man sequence. Originally scheduled for filming immediately after the hotel room scene, *2001*'s man-ape prelude had been giving him headaches for months, with shooting deferred repeatedly until a series of thorny problems was solved. These covered a spectrum of issues, from makeup (Stuart Freeborn's creation of a truly believable tribe of prehistoric savages that didn't look like swinging Londoners stuffed into ape suits), to location (shooting in South West Africa or a Spanish desert had been momentarily tabled by a director reluctant to travel, in favor of scouting locations in the damp northern climes of the British Isles), and finally, to dramatic construction: the specific actions our hairy ancestors were supposed to take as they fought for survival four million years ago, and in what order.

With production still under way in April, Kubrick had written Clarke—who, having finished shooting Wilson's film, was at the Chelsea Hotel during a US lecture tour—with suggested modifications to the Dawn of Man. He proposed that they abandon their prior idea that the monolith (still called "the Cube" in the letter) should be seen overtly transmitting audiovisual lessons to the man-apes. Rather, he wrote, the contents of the lesson should be unknown to their tribe-leader, Moonwatcher. "A secondary mystery is made of what the lesson is," he wrote. "Aside from possibly better narrative construction, it seems to me that not showing the visions in the Cube helps prevent a kind of silly simplicity of which I think we are presently in danger."

In any case, the director stressed, if they showed such lessons in the literal fashion they'd previously imagined, "we are at the risk of frustration at the end by *not* showing what the lesson is for Bowman. Whereas if we show the *result* of the lesson with Moonwatcher, then, when we show the *result* of the lesson for Bowman (the Star Child adaptation), we are still sticking to the rules that we have made." And he made another point: that the rival tribes they'd conceived for the scene—one led by Moonwatcher, the other by "One Ear"—should not be seen actually fighting until *after* the lesson has had its effect. Rather, "shrieks and howls" should be sufficient to indicate their territoriality. "I think it makes the introduction of weapons much more significant if you have a stalemate and a lack of fighting . . . prior to the introduction of weapons," wrote Kubrick. After that, "they not only make weapons, but they break the long-standing truce at the waterhole and invade the Others' territory." His changes "affect both the script and the novel," he wrote, adding that he hadn't yet managed to reread Clarke's novel draft.

This last point was not insignificant, since it amounted to a kind of continuing fig leaf intended to cover Kubrick's ongoing refusal to approve the novel's publication. He was, he insisted, simply too busy to give it the attention it deserved—and yet it couldn't be published without him making certain revisions. While no record exists of Clarke's response to Kubrick's suggestions, he evidently didn't agree sufficiently to change his text—the director's "silly simplicity" observation notwithstanding. Instead, he cabled that he was flying to London to make final arrangements for its publication.

Kubrick's response on April 19 was immediate, direct, and, in case that wasn't sufficient, delivered not once but twice: as a message transcribed by a Chelsea Hotel clerk and as a Western Union telegram. "Please do not come here for purposes of compelling me to abandon my position on the novel with which you are fully acquainted," he wrote. "No possible consideration can influence me. Sorry to be so intractable, but I feel I must do it this way." He concluded with, "Do not waste a trip here."

Clarke responded that he was coming anyway—and he did. A message from his agent underlined why he was increasingly alarmed. Apart from making some stark points about the cost of delaying any further, Meredith's letter contained estimates of the money they could make if a deal was signed by May 1—just a couple weeks away. Figures in Clarke's hand totaled his agent's calculations, which included magazine serializations and UK rights, arriving at a potential sum of $250,000. Such a deal would have solved the author's financial problems in a stroke, with plenty of cash to spare. It was worth a fight.

When Clarke arrived at the studio a couple days later—where he witnessed filming in the centrifuge ("A portentous spectacle, accompanied by terrifying noises and popping lamp bulbs")—Kubrick himself brought up the subject. Swearing that he didn't want to hold up the novel until the film's release, he pointed out that its *general* release wouldn't be until late 1967 or even 1968. (General release of *2001*, meaning in its mass-market, lower-ticket-price 35-millimeter incarnation, actually didn't happen until the fall of 1968.) Even if the film's first showing was in April 1967, said Kubrick, it would "be running only in a few Cinerama houses, which will give us some more breathing space." (In fact, the 70-millimeter premiere of the film took place a year after that, on April 2, 1968.)

Although the distinction may have been lost on Clarke, Kubrick was indicating that he felt the secondary, 35-millimeter release of the film, which would follow its so-called road-show 70-millimeter release by many months, was the appropriate time to publish the book. His seeking to minimize the impact of the Cinerama run, which took place in more than "a few" houses—in fact, the film would dominate some of the world's leading theaters in the spring and summer of 1968—was disingenuous. He was breaking a verbal agreement, and he was trying to put the best possible spin on it without backing down. He also required that Clarke and Meredith not show the manuscript to anybody until he signed off on it. Confronted once again by "the sheer North Face of Stanley's resolve"—in writer Michael Herr's phrase—Clarke had little choice but to accede. Meanwhile, however, he authorized Meredith to continue trying to make a deal, with the thought that an actual contract might change Kubrick's mind.

Always the good soldier, within a few days Clarke was off with Roger Caras and Mike Wilson to visit a private zoo in the town of Nuneaton, where Wilson filmed boisterous chimps and solemnly uncooperative gorillas as part of ongoing research for the Dawn of Man sequence.

．．．

Threading his way innocuously throughout all these activities, thus far an asterisk to a footnote in a film history afterword, was that ubiquitous daily provider of hot beverages and cold call sheets, the tea boy. Twenty-one years old, strikingly handsome—if not perhaps as stunning as his extraordinarily radiant sister Jane, then still an unknown—Andrew Birkin was somewhat analogous to the "young gentlemen" brought on board Royal Navy warships at the height of the empire, meaning that despite his lowly position, he was an officer in training. The product of an upper-class family, a descendant of "that dickhead George the Second"—as he put it, referring to the eighteenth-century monarch—he'd been a difficult student, and had suffered repeated canings and hazing before dropping out of his posh private school, Harrow.

As with most of the young assistants to be found running messages between departments at Borehamwood, Birkin's subentry-level position came via parental connections. Like his naval predecessors—

who were expected to leave home as teenagers, land on their feet, and immediately start issuing authoritative commands to grizzled seamen with decades of experience—Birkin simply had to show initiative, and he could expect advancement. His mother, Judy Campbell, a well-known stage actress, was a favorite of Noël Coward's, and his father was, in fact, a Royal Navy officer, a legitimate World War II hero who'd navigated high-speed gunboats across the Channel on moonless nights to drop spies on the French coast.

The British class system was as quietly rigid and unquestioningly enforced at Borehamwood as elsewhere. Offspring of the lower classes were expected to aspire to union cards from the trades; they might become sparks (electricians), chippies (carpenters), plasterers, grips, drivers, and the like. Upper-class kids, on the other hand, could jostle for positions in management and leading creative positions: assistant directors, camera assistants, producers in training.

Although inevitably forced to draw from the ranks of this system, Kubrick was the product of a very different society, one with more social mobility. Blind to such things, he ran a meritocracy. It's why Tony Frewin, the son of a studio driver, could become an assistant to the director—a prestigious position requiring a union card from the ACTT, or Association of Cinematograph, Television, and Allied Technicians. He'd jumped above his station, in other words, making it into the otherwise largely upper-class ranks of "Stanley's chaps": a small group of hardworking young men, all in their late teens or early twenties, who had the run of the studio, made their own hours, drew excellent bottles daily from the wine cellar of the MGM restaurant, and in practice, were "inviolate"—as one of them, Ivor Powell, put it with a laugh.

For his first six months at Borehamwood, Birkin was something of a lurker, fascinated by the studio's goings-on but officially prohibited from visiting the sets except at teatime, or ten in the morning and three thirty in the afternoon. This unusual prohibition came about because his commanding officers had quickly grasped that what they had with Birkin was a certifiable film geek in paradise. If sent on errands between tea runs, he invariably became so engrossed in what was being shot that he lingered at the margins, sometimes for hours—a figure half

Andrew Birkin in the summer of 1966.

in the dark. A certain voyeuristic tendency occasionally came into play. One day, for example, he set off to reconnoiter one of the few sound-stages not jammed with *2001* sets—the studio where director Roman Polanski happened to be filming a ravishing Sharon Tate, draped only in bubbles, for the bath scene of *The Fearless Vampire Killers*. Looking as official as possible, an empty film can tucked under his arm, Birkin adopted his best bored expression, climbed the zigzag staircase to the catwalks above, made his way stealthily to a position directly over the brilliantly lit bathtub, and parked himself where he could greedily take in the scene.

Once he'd recovered from this epic sight, the job also inevitably required our tea boy to know the geography of the studio, the names of all the people on the set, their place in the hierarchy, and how many sugars they took in their tea. Apart from this, he absorbed whatever actionable intel he could, even engaging in what amounted to a campaign of espionage on his own behalf. Tasked with copying scripts with a primitive early Xerox machine as prone to belching smoke as pages, he read the whole story, although this had been expressly banned to the lower ranks by a security-conscious Kubrick. Poking around the pod bay set at lunchtime, Birkin managed to lock himself in a space pod and spent a half hour frantically pounding on the front window glass until finally a passing propman sprang him loose—rescuing him from the ignominious fate of being discovered by the returning main unit crew, which could have had serious consequences.

Making his usual morning rounds winding office clocks, he found himself alone in Victor Lyndon's domain one morning and became engrossed in the film's confidential budget documents—grounds for immediate dismissal, had he been caught. When he was in fact discovered, and Lyndon demanded to know what the hell he thought he was doing, Birkin said, "I'm interested in it. I'm curious about it." Softening, the associate producer told him, "It's more secretive than the script. We can't have people knowing what each is being paid." When Birkin swore not to tell anybody, Lyndon let him off with a warning and sent him back to his tea urn.

Throughout these misadventures, Birkin sought a lucky break. Hearing in early March that food kept falling out of Gary Lockwood's tray when he was rotated to an upside-down position while strapped to the centrifuge dining table—hearing, in fact, that after falling more than thirty feet the glutinous mixture had splatted on the white floor, requiring hours of cleanup and stalling production for the day—he went home to his mother and together they experimented with supercharged double-gelatin mixtures. Returning to the studio the next day, his edible antigravity glue-food was gratefully accepted by the Props Department—if not by Lockwood himself, who had to actually eat it—and used thereafter. But though the tea boy had solved their problem, Kubrick never learned of it.

Finally, one evening in the second week of May, Birkin was summoned for a "ghoster" in Stage 3: a night shift. He had mixed feelings about this—he usually went to dinner with his girlfriend, the actress Hayley Mills—but the pay was good. A small group, including production designers, set decorators, and art directors, had gathered in front of an elaborate set, designed by Tony Masters, which filled three-quarters of Borehamwood's largest stage. At almost nineteen thousand square feet, it had previously housed Space Station 5 but now was filled with a large, rolling, boulder-strewn desert. A painted backdrop continued the African landscape to a rugged horizon under a deep-blue sky. While the three-dimensional part was believable enough, the painted backdrop gave it the slightly artificial quality of a Museum of Natural History wildlife diorama. It was vivid, lifelike, and ultimately unreal.

The huge roll-up cargo door spanning much of the studio's side had been left open to the spring air, and at about six thirty, Kubrick drove directly into the building, hopped out of his Mercedes, and announced, "It's a raid, fellows." After bumming a cigarette as usual, his grin faded to thoughtfulness as he surveyed Masters's simulation of South West Africa. Birkin, some distance away, busied himself making tea as various lights were turned on and adjusted. Production manager Clifton Brandon stood nearby. By now, the director's face showed concern, and the two of them were just close enough to overhear him say, "Gee, fellas, this isn't going to work."

"What did he say?" Brandon asked Birkin.

"It's not going to work," Birkin repeated.

"What do you mean, it's not going to work?" Brandon demanded. "It cost fifty-five thousand dollars!"

"I don't know, that's just what he said," replied Birkin, reasonably enough. He was, after all, only the tea boy—though it was his last day on the job.

Rolling his cart closer, he heard Kubrick exclaim, "I just can't believe there isn't a desert somewhere in England. You know, I'm not asking for the Sahara. I'm just asking for some sand dunes!"

To which art director John Hoesli replied, "Well, Stanley, you know, we've looked everywhere, and there isn't one."

At this, Birkin flashed back to a grainy textbook photo of dunes—a vision immediately accompanied by a burst of adrenaline. His moment had arrived.

"I know where there's a desert!" he announced. Turning as one, the high-ranking group in England's second-largest soundstage fixed its collective gaze on the lowest-ranking figure there.

"Who are you?" Kubrick asked.

"I'm the tea boy," said Birkin. He made an instinctive justificatory gesture toward the cart.

"You know where there's a desert?" Kubrick demanded. "Where?"

"I can't remember exactly, but it's in a book I have at home."

"Really? Are you sure?"

"Yeah, I'll bring it in tomorrow."

John Alcott, who was Geoffrey Unsworth's closest assistant and was representing the camera crew at the meeting, observed the exchange with deadpan amusement. Birkin, who'd totaled his Mini in a crash months before, had an arrangement with Alcott, whose mistress lived a block from Birkin. Alcott had a car, but no license. Birkin had a license, but no car. So they carpooled daily. As a result, the camera assistant was well aware of the tea boy's career frustrations. "This could be your chance," he said on their drive home that night.

Birkin rapidly located an ancient school textbook with two black-and-white photographs of Formby Sands—an expanse of coastal dunes facing the Isle of Man just north of Liverpool. Cutting them out, he mounted them on cards and brought them into the studio the next day. Accompanied by Alcott—who'd vouched for him the previous day when Kubrick had asked—he made his way to the director's office.

"Okay, well, go and photograph it," said Kubrick, after inspecting the shots. "Do you know how to use a Polaroid camera?"

"Probably," said Birkin, doubtfully.

"Oh, I can show you to use that," said Alcott. Taking him off to the Camera Department, he outfitted him with equipment, including a compass so he could specify a north-south orientation in the photographs. Advising Birkin to also bring a portable typewriter so he could document everything in words as well, Alcott outlined what Kubrick would want to see and read. His final words were, "Don't let me down."

Birkin hurriedly stopped off at home to pick up his mom's Olivetti, took the tube to Euston station, and hopped a Liverpool train. People he'd been supplying with tea for months had already reserved him a room at the swank Adelphi Hotel and given him an envelope of cash. Realizing on arrival that he had only an hour of light left, Birkin hopped a cab and directed it to the coast, where he was dismayed to discover that although dunes did indeed still exist at Formby Sands, low buildings and pine trees had encroached on the scene. Making the best of it, he chose low angles to mask out the buildings and shot a succession of Polaroids, not forgetting to mark the northern ones. Back at the hotel, he worked until two in the morning pasting the pictures together into several large panoramic mosaics, each shot from a different angle. He topped this off with detailed typed notes, wrapped the whole thing in brown paper, wrote "Stanley Kubrick" on it, and caught the four thirty milk train back to London.

Arriving at Borehamwood an hour or so before call time, he persuaded the guard at the gate to let him into Kubrick's office, where he deposited his research directly onto the Great Man's desk. Then he quickly exited the scene before anybody else had a chance to catch wind of his presence, made his way back to Euston, and jumped a train back to Liverpool. Asked why he'd bothered returning to Liverpool that day, Birkin said, "I wanted it to look like magic. I wanted it to look extraordinary. Because I knew this was my one chance."

Back at the hotel by ten thirty, he was greeted at the front desk with "Oh, Mr. Birkin, they've been trying to get hold of you," and the phone was already ringing when he returned to his room. It was Victor Lyndon. "I don't know what you did," he said, "but Stanley wants to triple your salary and get you a union ticket. Come back as soon as possible."

Proceeding back to Liverpool Central Station yet again, Birkin caught his second London-bound train of the day. This time he wedged himself against the window and slept. He'd spent ten of the previous twenty-four hours on rails.

Back in Borehamwood, he was immediately sent to crash a late-afternoon meeting of the senior art department staff. Kubrick, Lyndon, Masters, Hoesli, and Ernie Archer were all in attendance, and he just had time to glimpse his own panoramic Polaroid dunescapes arranged on the central table when Kubrick said, "Oh, hi, Andrew," as though they'd known each other for years. "You know everybody?"

"Yeah, sure, hi, hi," Birkin said, graciously acknowledging those he'd been supplying with beverages for most of the year.

"Okay, gentlemen," said Kubrick, "I want you to explain one thing. How is it that you spent thousands of pounds and six months scouring Britain for a desert and came up with nothing, and we sent out the tea boy and within twenty-four hours and at a total cost of twenty pounds, he found us a desert?"

■ ■ ■

Apart from a continuing maddening uncertainty regarding location, the question of how to portray our prehistoric ancestors had so far proved intractable, and had been going through multiple conceptual iterations for months. Neither Kubrick nor his makeup man, Stuart Freeborn, was the slightest bit racist, but given human origins in Africa and the inevitable corollary that our earliest ancestors had dark skin, from the beginning their quest was fraught with the possibility of mis-understandings.

Early in 1966 Freeborn had devised a first-draft ape-man body suit with full head mask and had enlisted a diminutive, older Afro-British MGM extra to wear it. With help from Tony Masters, who'd got-ten into the spirit of the problem—and who'd noticed with approval their extra's "wizened" appearance ("We thought, 'Well, that's a good start,'" he said)—Freeborn dressed his subject in the hallway of Build-ing 53, helped him pull the mask over his head, and the two of them brought him in to see Kubrick.

"Oh, jeez, fellows, oh my God," said the director encouragingly. "That's awful. That's a terrible suit. Oh God, we haven't even started yet! This is terrible. Okay, tell the poor guy to take the mask off." After a bit of a struggle, their model succeeded in liberating his face from the tight-fitting foam rubber. Kubrick approached, suddenly intrigued. He inspected the man's features. "Terrific!" he exclaimed. They decided the Dawn of Man could be set in one of prehistory's earlier epochs, enabling them to present more Neanderthal-like protohumans. This would require less overt interventions by Freeborn, and so would pre-sumably look more realistic on-screen. Accordingly, Freeborn issued a casting call for young black teenagers of both sexes, researched Ne-anderthal physiognomy, and set to work augmenting teenaged Afro-British brows, cheeks, lips, and chins. In the end, his results were entirely credible: his subjects looked halfway between modern humans and ape-men.

While this eliminated the man-in-ape-suit problem, it introduced two new drawbacks, only one of which Kubrick took seriously. What-ever its paleoanthropological accuracy, their new approach was poten-tially a PR disaster in the making—as Roger Caras quickly pointed

Stuart Freeborn with an early man-ape suit.

out. At the height of the US civil rights struggle, and with sub-Saharan Africa casting off the yoke of colonial oppression, were they going to risk equating blacks with protohumans? The second drawback was that Neanderthals were far less hairy than Australopiths, which, in turn, inevitably meant that their genitalia and breasts would be visible. This simply wasn't acceptable for a general-release movie. Surveying Freeborn's newly minted savages approvingly, Kubrick acknowledged the problem. He couldn't get away with showing their privates, he said, but, "That's all right—either I'll shoot them from the waist up or go back far enough that you won't be able to see anything." He soon realized that this was untenable, however, so he asked Freeborn to disguise their crotches with a subtle covering—one barely noticeable and yet somehow entirely concealing.

Freeborn, a good-humored, elfin figure who, when breaking into the closed world of film makeup in the 1940s, had successfully transformed himself into an entirely believable Haile Selassie, then had a friend drive him around England in a gleaming open-backed limo— resulting in numerous news reports that the Emperor of Ethiopia had been spotted touring England unannounced—was always up for a challenge. He brought his teenagers up to his top-floor workshop—so positioned in order that toxic fumes from his foam rubber prosthetics didn't pollute the whole building—and subjected them to maximally invasive crotch-castings, with his assistants apologetically daubing on the Vaseline so as not to yank their pubic hair out by the roots when pulling off the plaster.

He then crafted "these little plastic things . . . with some special lace that we use for wigs." They were affixed rather broadly, he explained, "because anything narrow would show, because it would bite into the flesh." The final result was flat, "almost invisible" crotch merkins—a term familiar to dedicated Kubrickologists because of *Dr. Strangelove*'s President Merkin Muffley, named in dubious tribute to these pubic wigs, originally used by prostitutes, that had been eagerly adopted by actors intent on concealing their privates during nude scenes. "We did it pretty well, and you didn't really see them," Freeborn recalled.

> But the problem was that the characters looked completely neutered. The essence of the picture was that aliens came down on Earth to make sure that these creatures would not die off because of famine and whatnot. The aliens knew that these were the only creatures on the planet that might eventually become like themselves . . . Obviously these creatures had to procreate to continue the species, but it was pretty clear this wasn't going to happen with the creatures we came up with. And so the whole thing started again. The only solution was to go back another million years or so—to ape-men rather than man-apes. That way they could be covered entirely with hair.

As a result—and while still aiming to film that summer—Freeborn labored in the spring of 1966 to create more convincing Australopiths. At the time, Kubrick was half-heartedly entertaining the idea of going on location in South West Africa, the deserts of which most accurately corresponded to the period and situation they were trying to convey: namely, four million years ago, a time when our semi-stooped ancestors, only recently descended from the trees, were ostensibly trying to survive in looming drought conditions.

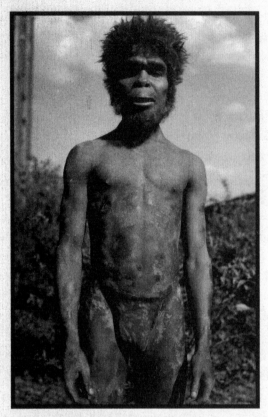

One of Stuart Freeborn's
Afro-British Neanderthals.

Reluctantly letting his neutered Neanderthals go, Kubrick noticed that his photographer Johnny Jay's assistant—an amiable, intelligent eighteen-year-old named Keith Hamshere, who'd already proven himself as a performer by playing Oliver in the original West End stage production—had exactly the right stature to play the drought-famished lead man-ape, Moonwatcher. (At seven shows a week, Keith had, in fact, held down the lead role for thirty-two weeks, so when it came to food, he'd already asked for more of it over two hundred times in front of paying audiences.) Kubrick prevailed upon Jay to lend him Keith, and he sent the kid up to Freeborn's aerie.

There Hamshere was asked to strip for a full-body mold so that "a very vivacious assistant, with red hair" could pour rubber latex gunge all over his body. Next, straws were stuck in his nostrils, and his face was done as well. "So then [Freeborn] turns around and says, 'Well . . . we need the teeth now,' recalled Hamshere. "So then we go for the teeth mold. So that's done, by which time everyone knows my innermost secrets, of course."

> But the worst part of all was having the eyes done. I had to go to this optician up in London. And they had to take an eye mold. And they put this sort of funnel cap in your eye, and you had this funnel coming out, and you're laying back, and they pour this, like, cold cream. And you see it coming down toward your eye, and there's nothing in the world you can do about it. And it suddenly hits your eye, and it makes you feel—you feel *on edge*. And then you have to wait for twenty minutes for *that* to set. And then they try and take it out. And that's quite difficult, to take it out. And there's this huge suction . . . You think, "Great, can I go now?" "No, no, no. The other eye."

Shortly after this ordeal, which had required the type of wire eyelid restraints Kubrick would later make famous in *A Clockwork Orange*, the day came for Hamshere to be outfitted with new teeth, brown contacts, a form-fitting latex mask, and a shaggy hair suit. Five hours or so of painful, invasive work, which ended long after lunch—something he hadn't had. In the latter stages, Kubrick, who'd had lunch, arrived to watch Freeborn, who'd also had lunch, put the finishing touches on Hamshere—getting the teeth to project more prominently, brushing his full-body shag, and so forth.

"It really is beginning to look very, very good," said Kubrick approvingly. Standing back after Freeborn's ministrations, the director considered his freshly made man-ape. A thought occurred to him. "Keith, how are you going to survive, you know, in 120 degrees?"

"Stanley, I have no idea," said Hamshere, startled.

"Keith, I've got this great idea," Kubrick said after further consideration.

"What's that, Stanley?"

"We'll get Geordie to put a bank of 10Ks together," he said, referring to electrician Geordie Walker and 10,000-watt film lights. "And we'll create an area, we'll wait until the temperature gets to about 120, and you'll run up and down in that outfit for as long as you can, okay?"

"Sure, Stanley," said Hamshere. ("As one did," he recalled thirty-five years later.) He still hadn't eaten lunch.

So I had this whole outfit on, and I was running up and down for probably an hour and a half, and then I just collapsed. It was probably less than that. All the rubber started to come undone, you know. In the end, you know, he was saying, "Well, it's just totally impractical. There's no way we can do this. The length of time, the makeup, and to do it all out there." And I don't think Stanley wanted to go out there, because he hated traveling anyway.

.   .   .

Kubrick had asked Clarke not to show the novel to anyone until he'd found time to revise it, and the author had reluctantly agreed—at least to his face. But given Kubrick's continuing prevarications, he clearly hadn't felt particularly obligated, and in fact Scott Meredith had spent the spring shopping it to various publishers. In early June Meredith received a significant offer for both hardback and paperback editions from Donald Fine of Dell Publishing, a pioneer of such "hard-soft" deals. The hardback was to have been issued by Dell and the paperback by its Delacorte imprint. Fine's cable to Meredith identified the book as authored by Clarke and Kubrick, offering a $160,000 advance with a highly favorable percentage on sales. Meredith's message to Clarke stressed the unprecedented nature of the deal, due both to the size of the advance and the royalty. He ended his telegram with "congratulations and wow."

The figure was substantially higher than the agent's previous estimate, which had reached $250,000 only when various subsidiary rights and foreign editions were assumed. Dell's offer promised to exceed that figure when such rights were added. If the deal went through, the financially strapped Clarke would receive almost $100,000 under his sixty-forty agreement with Kubrick—about $720,000 in today's dollars.

Far from solving Clarke's fiscal woes, however, the offer precipitated the single worst crisis between the coauthors of *2001: A Space Odyssey*. Acutely security conscious concerning their story—certainly a major reason why he didn't want the novel to see print before the film's release—Kubrick felt upset and betrayed that Clarke had authorized Meredith to shop it around. The agent responded to this on June 13, writing Clarke that "the blunt truth is that it was a choice of showing the manuscript to *all* possible buyers *now* or forgetting about publication." He stressed that revisions really weren't necessary; the publishers "consider it a masterpiece just as it stands." If Kubrick insisted on them, however, the choice was either producing his suggested changes by July 15 or "abandoning hope on the book."

The nature of the polite, but also politely furious, exchanges that passed between Kubrick and Clarke that summer are only hinted at in Clarke's 1972 book *The Lost Worlds of 2001*, in which he wrote, "During one of my more frantic arguments, he remarked, 'Things are never as bad as they seem,' but I was in no mood to agree." Other clues can be found in his abortive piece for *Life*, in which he wrote of Kubrick being "absolutely inflexible once he has decided on some course of action. Tears, hysterics, sulks, threats of lawsuits will not deflect him one millimeter. I have tried them all." While he may have intended this to be taken as hyperbole, evidence points to his being literal on every point.

For example, Clarke was well aware of Kubrick's hypersensitivity concerning authorial credit, as demonstrated by his very public statements in 1964 diminishing Terry Southern's role in crafting the *Dr. Strangelove* script. Under the circumstances, any implication that the novel might see print without Kubrick's name on its cover was the equivalent of a cannon shot across the director's bow—but Clarke did just that in a letter on June 15. Forwarding the Dell offer, he wrote that Meredith had apologized for showing the manuscript (he had not) and concluded with: "The publishers are so pleased with the book that they would apparently give the same terms if it was brought out under my name only. But I should hate to do this, even if you agreed—I regard your share of the literary content as considerably more than the 40% financial interest you have."

In response, Kubrick—then deep into filming the Brain Room scene—finally found time for an extensive set of suggested revisions, nine pages in all. Dated June 18, 1966, they contained "some very acute, and occasionally acerbic, suggestions," as Clarke recalled. In the novel, the Dawn of Man monolith was described as a "crystal block." Kubrick requested, "[S]ince the book will be coming out before the picture, I don't see why we shouldn't [sic] put something in the book that would be preferable if it were achievable in the film. I wish the block had been crystal-clear, but it was impossible to make. I would like to have the black block in the novel." On reading Clarke's description of the alien object making Moonwatcher jerk like "a puppet controlled by invisible strings," Kubrick's criticism echoed his letter of April: "The literal description of these tests seems completely wrong to me. It takes away all the magic."

Several of his comments telegraphed how he eventually intended to shoot the Dawn of Man sequence. Reading Clarke's description of the block's projecting of tantalizing images into the man-apes' minds—of well-fed versions of themselves, evidently to be taken as a lure, as the beneficial result of using weapons—Kubrick observed:

This scene has always seemed unreal to me and somewhat inconceivable. They will be saved from starvation, but they will never become gorged, sleek, glossy-pelted, and content. This barely happened in 1966. I think that one day the cube should disappear and that Moonwatcher and his boys passing a large elephant skeleton which they have seen many times before on the way to forage are suddenly drawn to these bones and begin moving them and swinging them, and that this whole scene is given some magical enchantment in the writing and then ultimately in the filming, and that from this scene they approach the grazing animals which they usually share fodder with and kill one, etc.

He also rejected a short chapter of exposition Clarke had used to bridge the four million years between the prehistoric scenes and the twenty-first century, commenting, "I think this is a very bad chapter and should not be in the book. It is pedantic, undramatic, and destroys the beautiful transition from man-ape to 2001." Kubrick's comments were indeed acerbic, though they also hinted intriguingly at his approach to filming that leap. In any case, their evidence of directorial engagement did in fact permit Meredith to negotiate a contract with Dell, and by mid-June, Clarke seems to have believed he'd actually achieved his aim. On June 22 he wrote an editor of *Look* magazine, "I have bullied Stanley into acquiescing, I hope, to the release of the book by the end of the month."

He was far too optimistic. Kubrick's notes had taken him only as far as the end of the Dawn of Man chapters, and soon he was back to pleading overwork again. As Clarke himself observed—not without sympathy—in another letter written on June 22, "He is working about 20 hours a day, and practically sleeping at the studio. My publishers are screaming for the book, and if they don't have it in a few weeks, I am liable to lose at least a hundred thousand dollars. Stanley, however, refuses to let me release it, and he hasn't time to look at the MS. He is doing his best, but is really working to exhaustion, and is utterly unapproachable."

Finally, on July 4 Kubrick flatly rejected signing the Dell contract until he had time for further revisions in the fall. His credulity now definitively curdled, Clarke wrote a friend, "I may be leaving in disgust at any moment, and head back to Colombo to lick my wounds." Clear evidence exists that he threatened Kubrick with a lawsuit and actually explored the possibility of filing one. Anxious not to lose the deal, Meredith persuaded Fine that while there might be an unavoidable delay, publication would indeed move forward. As a result, Dell published "an impressive two-page advertisement in *Publishers Weekly*,"

Clarke wrote, and actually set the book in type by August—all the while anticipating Kubrick's forthcoming signature.

In an effort to placate his collaborator, the director—who may have sincerely believed he'd finish the film in time to keep the deal alive—wrote Clarke on July 12 with an offer that he defer "all or part" of his portion of the "worldwide advance for the novel in order to increase your share of the Dell Delacorte advance to a figure as close as possible to what your original share of the . . . advance would have been. I would then recover this deferment from the first royalty income received on the book, which would be paid to me in full until this sum of money was recouped." And he asked Clarke to sign and return the letter if he agreed—which he did.

Clearly the possibility of legal action was on Kubrick's mind even after this, however, and throughout August and September, his letters to Roger Caras, who by then had moved back to New York, warned his VP to stay away from Dell, "discuss nothing with Arthur or anyone else, and just be totally dumb about the situation. There is a potential lawsuit involved here, and anything you say of any nature might be used against us."

That fall, Dell, "fighting back corporate tears"—as Clarke put it—withdrew its offer, though Fine, who obviously really believed in the novel, left the door ajar for a new deal if Kubrick came around. And by mid-September, Clarke had, in fact, retreated to Ceylon and resumed borrowing money from Kubrick.★

■   ■   ■

Having found his first domestic desert, Birkin was tasked by Kubrick with scouring the British Isles for more. Eventually he located an abandoned eighteenth-century copper mine in Wales: a spectacularly ravaged golden-orange landscape of serrated buttes and rockfalls called Parys Mountain. He also discovered an equally suitable dunescape,

---

★ In fact, he was borrowing from Kubrick's Beverly Hills bank, with the interest shared fifty-fifty with the director and the loan guaranteed by Kubrick as well—which amounts to much the same thing.

Newborough Warren on Anglesey, which, if shot in a southeasterly direction, was crowned spectacularly by distant mountains. The latter in particular was deemed excellent by Kubrick, who even sent Tony Masters to take a look at it. By then, however, they were heading into the fall, and the idea of shooting on location was dropped due to the inclement weather.

With the days getting shorter, Kubrick called Birkin home and reassigned him to the film's burgeoning visual effects departments, where he was to keep tabs on things and expand the in-house miniatures workshop. All the live-action interiors now required credible complementary exteriors, and demand for model makers capable of producing 2001's spacecraft was high. With all the sets now struck and production of the Dawn of Man sequence indeterminate, Tony Masters had completed his contractual requirements and was needed on another film. One day while in Kubrick's office, Birkin and visual effects man Brian Johnson witnessed his parting scene.

"You can't leave!" said Kubrick when he recognized the purpose of the visit.

"Well, I'm terribly sorry, Stanley, but I'm afraid that the time has come," Masters replied politely.

"You *can't*," Kubrick insisted.

"Why?"

"We still haven't got a design for the Clavius landing pad."

"Ah, well," said Masters tolerantly, pulling a pen out of his pocket. "Where's a bit of paper?"

Sitting down, he rapidly drew a schematic for a lunar landing pad dome—"something that looked like a fifty-p coin," Birkin recalled. In other words, a heptagon. "There you are," Masters said firmly, handing it to Kubrick and taking his leave. Though drawn quickly, the multiple retractable petals he had depicted contained just enough detail for Johnson and the other model makers to get started. It was his last contribution to *2001: A Space Odyssey*.

Decades later, Doug Trumbull spontaneously brought up the production designer. "I would say that Tony Masters was one of the absolute centerpieces of the success of *2001*, as the brilliant and gifted production designer singularly responsible for the very complicated illusion stuff," he said. "Not just of the production design of the interior of the spacecraft, but of the clever tricks of the rotating cameras and the upside-down sets. All that planning to work out how to create this feeling of running around in the centrifuge was Tony's brainstorm . . . His praises have never been sung loud enough."

As part of his new responsibilities, Birkin became aware of some work that visual effects specialist Bryan Loftus was doing. Kubrick liked to find young people with some knowledge of a field, invest in their training, and give them responsibility. Loftus was among the youngest people in the country who knew how to work with "separation masters": a photographic technique in which each of the three colors of developed color negative film are reprinted on fine-grained black-and-white stock. With digital effects decades away, this method was the only way to manipulate material that had already been shot without losing significant image quality.

Once the three colors—yellow, cyan, and magenta—had been teased apart, much more control could also be exercised over each color. Kubrick located the only optical printer in the world that could do this for 65-millimeter film, flew it in from Los Angeles, and put Loftus in charge of it. Working with the new machine, Loftus discovered that if he made certain fortuitous mistakes—for example, by reprinting the separated colors back onto color negative stock in the wrong order, or using the "wrong" aperture settings—he could produce vividly improbable, psychedelically strange color inversions and image solarizations.

The discovery, which came after Wally Gentleman had independently brought the technique to Kubrick's attention, arrived at about the same time that Con Pederson was painting alien landscapes intended to augment Bowman's space-time trip—the Star Gate sequence, some of which had already been shot in New York in 1965. The problem was that Pederson had already taken the lead in managing 2001's complex visual effects production chain, and had effectively become skipper of the war room: the visual effects nerve center in Building 53, named after Ken Adam's set for *Dr. Strangelove*. As a result, he was quite busy, his paintings weren't coming as fast as they needed to—and, anyway, they looked, unavoidably, a lot like paintings.

Examining Loftus's incandescent prints, Birkin sensed they might present a faster route to extraterrestrial landscapes than Pederson's method, and he proposed that Loftus take his results to Kubrick for evaluation. But Loftus wasn't as unabashedly self-confident with the director as either he or Trumbull was. Seeing his reluctance, Birkin proposed that he do so himself. With Loftus's agreement, he then took several postcard-sized hard copy examples of the effect—soon dubbed "Purple Hearts," after the triangular amphetamine pills then popular with Britain's subculture, the Mods—over to Building 53.

"What are you suggesting?" asked Kubrick, scrutinizing the prints with interest.

"Well, what about if we bolt a camera to the floor of a helicopter and just shoot landscapes in Scotland, maybe, or flying out to sea," said Birkin. "Then Bryan can do his magic."

"Okay, well, do it."★

. . .

One of the few things that Birkin had found interesting at school was cartography. Now he got out maps of Scotland for a preliminary paper reconnaissance, soon augmented by charts containing information about topography and elevation, allowing an evaluation of how the light would fall on various landscapes. Meanwhile, he began testing helicopters and looking for a pilot. He was twenty-one years old and had effectively been appointed producer-director of aeronautical second-unit film production by one of the hottest directors in the world.

They settled on an Alouette, a small, highly maneuverable French helicopter with a lot of glass in front, and a daredevil French pilot named Bernard Mayer. The camera operator they hired, Jack Atcheler, had worked under John Ford and Otto Preminger, and had also shot

---

★ Loftus has a different memory of these events. In his telling, Birkin was not involved, and Kubrick saw the effect in the cutting room and encouraged him to experiment, thus producing the Purple Hearts process.

both Beatles films, *A Hard Day's Night* and *Help!* Gyroscopic mounts didn't yet exist, so they devised a way of fastening the camera onto layers of shock-absorbing rubber sheeting on the cockpit floor. There would be room in the cramped bubble only for the three of them, a camera, and a stack of loaded film magazines.

Production was scheduled for the second half of November. Kubrick wanted to see a shooting plan, and Birkin had planned a series of low-level runs over various locations: the island of Skye; the rugged Outer Hebrides islands; Ben Nevis, the highest mountain in the United Kingdom; and an elongated saltwater fjord, Loch Roag Beag. He presented Kubrick with twenty pages of detailed notes. His initial methodology was simple: "I just thought of all the places I'd ever wanted to go to." There would be targets of opportunity along the way, of course. A fuel truck would follow the chopper around, rendezvousing with it on tiny Highlands airfields.

The first few days they had iffy weather, but Birkin, impatient to get going, insisted on flying anyway. "Having come up with the idea, I felt that I had to deliver," he said. From their first base in the northwest, they proceeded to get a series of low-level shots over the Galloway region of Scotland, with the helicopter frequently shuddering as gusts of wind jerked it around. Conscious that the panoramic wide field of their 65-millimeter film stock required altitudes as low as Mayer could manage, Birkin experimented with terrain-hugging low-level passes in marginal conditions. "If you've been in a helicopter with a little wind like that, I mean, you are in this plaything of the gods," he recalled. After three days of this, they'd arrived in the town of Tarbert, the main settlement on the Outer Hebrides island of Harris, and Atcheler had had enough.

That evening, he invited Birkin for a drink in the hotel bar and demanded to know how many children he had. Hearing the figure he'd expected—zero—Atcheler said, "Well, I have three, and let me tell you something, no film is worth a life."

"Bernard wouldn't have done it if he thought it was unsafe," Birkin protested.

Andrew Birkin and crew with helicopter in Scotland, November 1966.

"Are you kidding? He's as crazy as you are!" Atcheler said. ("And indeed he was," observed Birkin. "He got killed on the next movie, on *Battle of Britain*.") "You're risking my life as well as yours," Atcheler continued. "So good luck to you, but I'm off."

Seeing Birkin's dismay—in Britain's film industry, nonunion personnel weren't supposed to touch a camera, let alone run one, and Birkin's new union ticket was on the directing side—Atcheler said, "Don't worry. My lips are sealed, and your lips are sealed. Do you know how to load a magazine in the dark?" Birkin, who'd loaded 16-millimeter but not 65-millimeter mags, got some instruction from Atcheler, including on aperture and light-meter settings, after which the cameraman took the next ferry, leaving Birkin and Mayer to shoot the footage that would actually appear in *2001: A Space Odyssey*.

The Panaflex couldn't pan or tilt anyway—that was entirely up to Mayer's piloting skills. They hadn't really needed an operator in the first place. "So having felt somewhat deflated, I now felt inflated," Birkin recalled. Flying low, they proceeded over waves, sand, mud, lochs, inlets, islets, valleys, mountains, pastures, rocky coastlines, natural stone archways, and the 4,400-foot summit of Ben Nevis. The latter footage included a brief passing glimpse of an abandoned meteorological station, something still discernable, albeit transformed via Purple Hearts processing into incandescent shades of orange-and-blue during astronaut Dave Bowman's trajectory "beyond the infinite."

. . .

Mike Wilson had been having a great time in London. Financed by his ever-tolerant sugar daddy, he'd imported his glamorous wife, Liz—a Scottish-Sinhalese beauty so stunning that Kubrick had quietly commented to Arthur that she was the most beautiful woman he'd ever seen—and half-heartedly supervised the printing of his Bond parody, which would be released in Ceylon in September. When he wasn't partying with the likes of Rolling Stones lead guitarist Brian Jones, Wilson also accompanied Clarke to Borehamwood, where he looked for opportunities and tried to make himself useful.

One of the people he'd made friends with was New Zealand poet John Esam, who'd followed the beatnik trail to London via Morocco, Greece, and France, staying for a year at the "Beat Hotel" in Paris, where he'd met writer William S. Burroughs and poet Lawrence Ferlinghetti. In 1965 Esam had worked with American mime Dan Richter, the editor of the underground literary magazine *Residu*, to stage a landmark international poetry reading at the Royal Albert Hall—a hugely successful countercultural event that had included readings by Allen Ginsberg and Ferlinghetti. In February 1966 Esam became the first person ever arrested in the UK for dealing LSD. He beat the charge, though, and was acquitted the following year.

Kubrick's mysterious production had caused much comment in London, and Wilson had found considerable cachet in being privy to the inner goings-on at Borehamwood. While sharing a joint with Esam one day, he mentioned in passing that the director was wrestling with a difficult filmmaking problem and had thought a trained mime might possibly help. "I know a mime, Dan Richter, he's a great mime," Esam replied immediately. "He's a friend of mine." He soon sent Wilson over for a visit, and Wilson told Richter that Kubrick wanted to "pick his brains."

A diminutive, mop-headed figure with a large moustache, bright eyes, and an elastic, expressive manner, Richter had followed his own hippie trail to London from the United States. Only a few years before, he'd been the lead performer at the American Mime Theatre in New York, but under the influence of writers Aldous Huxley and Jack Kerouac, had decided to resign, experiment with various forms of consciousness expansion, and tour the world. During a stint in Japan, he'd studied Noh and Kabuki theater techniques and also started a lifelong friendship with conceptual artist Yoko Ono. After a winter in India and a stint in Athens—during which he and his wife, Jill, had launched their literary magazine and acquired heroin habits—he'd made his way to London, where he knew junkies could register with the authorities and become legal.

In contrast with European mime, the American Mime Theatre philosophy was to start with acting values and extend them into purely physical movements. Founded by Paul Curtis—who'd studied under Lee Strasberg, the father of Method acting, and mastered the technique—the theater had produced a hybrid art form. Curtis had recruited Richter when he was still a student and trained him in the unique discipline. Richter had subsequently starred in extended pieces and was adept at holding an audience for as long as twenty-five minutes at a time—a rarity in mime. In one physically demanding performance, *The Pinball Machine*, he'd played four different pinballs, each with its own personality. Kubrick didn't know it yet, but he was about to meet one of the few people in the United Kingdom equipped to deal with the problem he was facing.

For his part, Richter wasn't looking for a job. He'd received a good deal of media attention after the Royal Albert Hall reading, was enjoying a free, unburdened social life among his fellow hipsters in London, and had been teaching private mime classes. But he admired Kubrick's work and was happy to meet him. He arrived at the MGM gates in late October for what he thought would be a consultation, not a job interview. Impressed by the "gorgeous" studio buildings, he was surprised to be directed to a small prefab structure "that looked like a place you'd leave the garbage cans for the janitors." There he met Victor Lyndon and was taken to "a very small office . . . piled floor to ceiling with books and drawings and plans and objects . . . it looked like a hoarder's cave." A few minutes later, he overheard a Bronx accent issuing instructions to somebody in the hallway, and in came "this ruffled guy with a nice smile on his face," who, from the beginning, "was so unpretentious, so open, warm, friendly" that Richter felt completely relaxed.

Kubrick outlined his problem in some detail, concluding with "Dan, they can't look like men in monkey suits." Surprised by his own audacity, Richter immediately said he thought he had the answer. It wasn't something that could be solved with a single solution, but rather by a combination, including "tricking" the audience into a state of belief via something called "conventions." A classic example of a convention, Richter said, was French master mime Jean-Louis Barrault's swimmer, in which his entire body was horizontal apart from one leg, which extended stork-like to the ground while the rest of him appeared to float and swim. That leg, said Richter, somehow made him *more* convincing, not less. If the convention was saying, "You shouldn't be believing," Barrault's movements were so well done that a tension was set up—tipping the audience toward belief, not the reverse.

"Your problem is that no matter how well you design the suits or train the people, you'll still have actors in monkey suits," Richter observed. "And so you've got to get beyond that, and the way to get beyond it is to get the audience involved in the motivations—the feelings—to believe what's going on. If the man-apes are real characters who have real feelings and objectives, you have a chance of getting the audience to buy into it. And once they buy into it, then it's just down to doing it properly, and pacing." He finished his soliloquy by stating flatly that acting values could be used to get around the costume problem.

Richter had been speaking with the self-confidence of somebody well positioned to understand the problem. Despite his successes onstage, it hadn't occurred to him that he might be considered a performer or choreographer himself. He assumed Kubrick would "end up hiring some famous person" to do the job, and he was enjoying himself. In the rapid evaluations one makes in such circumstances, he'd been impressed with Kubrick's candor, and he'd also been struck by where the man chose to work. While the director clearly had the run of one of the largest studio complexes in England, and could easily have set himself up in far grander surroundings, he was working from "this unpretentious little office in a prefab unit." Unlike many who found Kubrick intimidating, Richter, who like the director was of short stature and Jewish descent, found him disarming. "He was stocky, but he wasn't that big. He was like an elf, almost; he was just this humorous little character."

After a half hour or so, Kubrick said, "Well, this all sounds great, but I don't know you. My problem now is that I really like what you're saying, and it sounds like this would be a solution, but how do I know it would work?" He referred to Dan's "not having much of a resume." At this the mime stood up. "I can show you, if you've got twenty minutes," he declared. Smiling, Kubrick said he could have more time if he needed it. "No," said Richter, "I just need twenty minutes, two towels, a leotard, and a stage."

Soon after, Lyndon ushered Richter into the backstage dressing room of a small theater. Dan knew he could imitate a monkey or an ape, but what he really wanted was to show Kubrick a couple different characters. He'd been living for years with one named Joe—a dim-witted, pushy, paranoid type with no material form unless Dan chose to bring him forth—and now he called Joe forward and told him that they were going to pretend to be apes together. Joe thought this "was a dumb thing to do," Richter remembered, but the mime explained that a big director was out there waiting, and proposed that they go out there together anyway. "Joe didn't like it that much, but he agreed to do it," Richter said.

Climbing into a full-body black stretch leotard with the help of a solicitous young assistant, Richter stuffed a towel under his shoulders to bulk them up. Standing in front of a mirror, it dawned on him that what had been an enjoyable discussion had become an audition in front of one of the world's great film directors. Taking several deep breaths, he emptied his mind, called Joe forward, and let him take over. His chin moved forward, his eyebrows descended, his arms extended, and his chest rose. On springy knees, Dan's dimwitted pal Joe walked out

Dan Richter in a Freeborn body
suit for a costume test.

onto the stage in front of Kubrick and Lyndon. Squinting and turning
in the bright lights, Joe's body language transmitted a "What the fuck's
going on?" followed by an "Oh shit, what's that?" as he spotted the
two seated figures. Richter, who'd channeled Joe a lot onstage, gave
his new man-ape version free rein to walk around in the character's
"tough, nosy, tentative way." After a bit of this, he went to the lip of
the stage, still in character, and fixed Kubrick with a resentful, dimly
comprehending simian gaze.

"Oh, that's wonderful!" the director exclaimed. At this, Joe jumped
backward, startled, and retreated wordlessly to the dressing room—
only to reemerge a moment or so later not as Joe but as a nervous,
tentative second character: a shy man-ape fearful of his surroundings
and dazzled by the bright lights. He was now a "delicate, nervous little
thing, which was a complete contrast to Joe, you see." Finally, Richter
moved to the center of the stage, looked again at Kubrick, and came

out of character. "Well, I've seen enough," said Kubrick, smiling. "I think you've made your point. That was very good."

Asked about it years later, Richter said of these transformations, "The character is going to do what the character wants to do. And you let it. It's almost as though you're riding around inside him. And you have this relationship—'Maybe we could go this way; a little bit, maybe. Do you have to be so *nosy*?'—and the character is also a door to magic, from a creative point of view. Things happen that you have no idea are going to happen. You surprise yourself."

At the time of Richter's visit to Borehamwood, tentative plans still called for a shoot only ten weeks later in southeastern Spain. Having survived his own man-ape experience, Keith Hamshere had returned gratefully to photography and was now running Kubrick's bustling photo operation with its multiple darkrooms at the age of nineteen. A few months previously, he'd been sent on a photoreconnaissance of locations in Spain's Tabernas Desert, and now Kubrick flooded Richter with questions. Where could they get a whole tribe of man-apes? If there weren't that many mimes in the world, could Richter teach others to move like that? What about in costume? Would it be possible to shoot in Spain in ten weeks?

Richter answered as best he could. He responded to the latter question with a firm "I can't do that, if you want me to help you with this." Relenting, Kubrick asked him to write a brief proposal as to how he'd tackle the Dawn of Man sequence.

When Richter returned with it the next day, he was hired on the spot.

. . .

Although most of *2001*'s sets had been struck, perhaps the single most extraordinary sight during the film's production could be seen high above Stage 4 on certain days in July, August, and into the fall of 1966. That was where stuntman Bill Weston was doing EVA wirework: extravehicular spacewalk sequences. As with the Emergency Air Lock and Brain Room scenes—only now a good thirty feet above the unforgiving concrete floor, with absolutely no latitude for error—the wirework was handled by Eugene's Flying Ballet and supervised by its leader, Eric Dunning, who'd been trained by Arthur Kirby, the man who'd first demonstrated the technique in a sensational London staging of *Peter Pan* in 1904.

Weston's audacious performances, which unfolded without a safety net, comprise some of the most physically and technically demanding scenes shot during the making of *2001: A Space Odyssey*. Decades before digital effects, they constitute an extraordinary, largely unsung moment in film history. In their convincing simulation of zero gravity, they appropriated and expanded on techniques pioneered by Soviet director Pavel Klushantsev in his 1958 docudrama *Road to the Stars*, which Kubrick almost certainly studied. But they were entirely successful on their own terms. With the camera always positioned directly below him, Weston could rotate and maneuver freely on an x-axis. It was the only way to simulate weightlessness in a gravitationally bound studio environment, and the portrayal later won rave reviews for its realism from those who'd actually done it. ("Now I feel I've been in space twice," said Soviet cosmonaut Alexey Leonov, the first spacewalker, on seeing the film in 1968. Connected only by a thin umbilical cord, Leonov had floated free of his Voskhod 2 capsule in Earth orbit only three years before *2001*'s premiere, on March 18, 1965.)

Dan Richter, fortunate to witness the scene, provided a vivid description in his 2002 book *Moonwatcher's Memoir: A Diary of* 2001: A Space Odyssey. "Passing through the door onto the great stage is like entering a cathedral," he wrote. "All around are vast curtains of black velvet. High above, hanging from invisible piano wires, stuntman Bill Weston, in a space suit, slowly turns like a modern angel in the black abyss . . . For a moment, I have trouble breathing. It is stunning. Stanley is in a huddle around the big 65-millimeter camera with Bryan Loftus, Peter Hannan, and other crew members."

A square-jawed, striking presence who looked not unlike the young Clint Eastwood—and yet with "a beautiful, educated speaking voice, sort of upper class but not one of those plummy, cut-glass, fuck-you type things," as Tony Frewin put it—Weston was over six feet tall and had been brought up in India under British colonialism. Then

Bill Weston being launched from a platform
thirty feet above the studio floor.

twenty-five, he'd become a stuntman after "some freelance soldiering
in Africa." He'd done a number of films prior to *2001*, but nothing
remotely as ambitious as this. Prior to his space walks above Stage 4,
he'd been outfitted with a salt-and-pepper wig and doubled for Keir
Dullea in the emergency air lock scene—specifically, the reverse angle
from behind as Bowman vaults toward the far wall and rebounds back
toward the door. He'd also spent hours hanging at odd angles in the
overheated HAL Brain Room for shots where Dullea's face didn't need
to be visible through his helmet.

In his ceaseless drive for realism, Kubrick had turned down the
suggestion that air holes be punched into the back of Weston's helmet.
He was worried that light might leak through and be visible through
the visor. This refusal came despite the fact that months before, Dullea
had been filmed pressing a button to polarize his visor—ostensibly
shading himself from the sun, and blacking out his face entirely. This
then allowed Weston, whose helmet was also fitted with a polarizing
filter, to take on the more hazardous full-body wire shots without fear

of being recognized as a double. (He played both Bowman and Poole for the spacewalk sequences.) When Weston countered by proposing that the air holes he was suggesting could be screened by black gauze, which should have taken care of any possible light leakage, Kubrick still refused. The director also insisted that Weston wear his Bowman wig in the sweaty, overheated environment of the suit—a directive the stuntman soon evaded by discreetly flipping the thing into a corner of his high launch platform.

Kubrick's intransigence meant Weston's space suit was hermetically sealed. While he did have a small tank of compressed air stashed in his backpack, it contained only ten minutes' worth—and this was unregulated, simply feeding into his helmet via a tube until empty. Given the complexity of the shots, and the amount of time it took simply to remove the platform used to prepare the stuntman's wires and suspend him, ten minutes wasn't enough. And there was another problem: even when the tank *was* feeding air into the suit, there was no place for the carbon dioxide Weston exhaled to go. So it simply built up inside, incrementally causing a heightened heart rate, rapid breathing, fatigue, clumsiness, and eventually, unconsciousness.

One of the first space-walk sequences, shot on July 8, required the stuntman to open the back door of the space pod and exit, with one hand holding *Discovery*'s boxy replacement antenna guidance unit. The pod had been hoisted to the studio ceiling and attached securely to a pipework frame below it, its back facing the stage floor and camera. The frame provided just enough room for Weston to climb on top and squirm inside through the pod's open front window. Although Eric Dunning recommended using two wires for safety, Kubrick had insisted that Weston wear only a single one for the shot. (Each wire was actually comprised of tightly coiled cable strands.) Now truly concerned at the director's uncompromising stance, the stuntman pointed out that since the camera was positioned directly below him, his body would hide the wires anyway. Worried that shadows on the pod would betray the trickery behind Weston's apparent weightlessness, Kubrick was unyielding on this point as well.

Giant carbon arc lamps, mounted on heavy tripods and clustered together to create the illusion of sunlight's single point source, cast a powerful illumination on the pod from below. Backed by seamlessly joined sheets of expensive French black velvet, the spherical vehicle already seemed suspended in interplanetary space. Having been attached by Dunning's crew to only the single wire, Weston crouched inside, ready for the take. As soon as the rickety platform was pulled aside, Pike rolled the camera, Cracknell called "Action!" through his megaphone, and the pod's narrow door opened, revealing Weston's red space helmet. Maneuvering himself through, he descended headfirst toward the camera.

Almost immediately, a sickening *ping* resounded across the sound-stage as one of the wire strands snapped. Whipping 180 degrees forward, the severed strand sliced through a fastener holding Weston's pistol-grip control unit to the space suit's front side. Instinctively righting himself, he reached upward, grabbed the pod door, and yanked himself to safety. Meanwhile, the control unit—a genuine part from a Royal Air Force Vampire fighter jet—spun downward, striking camera assistant Peter Hannan a glancing blow to the head. Kubrick, who'd been hovering beside the Panavision, jumped backward. The rest of the crew scattered as well.

With blood streaming from an open wound in his scalp, Hannan was rushed off to nearby Barnet Hospital, where he required multiple stitches. He returned to work that afternoon, shaken but alive. "If it had caught him straight in the middle, it would have killed him," Weston observed. With Dunning's crew helping to reel him in, the stuntman had managed to climb back into the pod. If he'd fallen, he would have certainly killed himself and likely others as well.

Shaken, Kubrick withdrew any further objections to using dual wires. He also had a cage constructed around the camera crew, who from then on had to wear helmets. Kubrick himself never stood directly below the stuntman again. "One of the great things about Stanley was he had an incredible, tremendous artistic integrity," observed Weston. "I think morally he was a little bit weaker." His words were borne out by another incident.

One of Weston's shots called for him to hang with his back to the camera, his body slowly turning. It would be used in the sequence when Bowman goes to replace *Discovery*'s antenna guidance unit. A bracket mounted on a pivot bolted to the ceiling enabled the rotation. After hours of hanging in a horizontal position, with his arms and legs extended against the pull of gravity, Weston's lower back began throbbing from the constant strain. He asked for a piece of ironing board and a broom handle, which were rushed to the platform, cut up, and inserted under the back of his harness. "That sort of amused me," he recalled. "The low-tech piece of kit inside this high-tech suit."

Though Kubrick and Cracknell could use megaphones to issue orders from below, there was no real communication between the stuntman and the ground. Weston had only those ten minutes of air siphoning into the helmet, with the exhaled carbon dioxide building up the whole time. In view of the perilous circumstances, he'd worked out his own way of mitigating the risks. When Weston started to get groggy from the CO_2, "I was doing the alphabet backward, and as long as I could do it, I figured I was okay." He'd arranged a signaling system with Dunning's wire man: if he extended his arms directly outward in a crucifix shape, he was near his limit and should be brought in very soon. If he did it twice, it was an emergency, and he needed to be recovered immediately.

It took almost five minutes to move his launch tower out of the shot and that much time to bring it back. Added to this, ever since Hannan's injury, Kubrick had checked the camera frame only from off to the side, "because he was still terrified of something dropping on him," remembered Weston. This, in turn, required that after the stuntman's release from his platform, and after the inevitable discussion down below, the camera still had to be moved back into position. All this took yet more time. With just ten minutes of air to mitigate the CO_2 buildup in the first place, they were operating at the outer limits of Weston's endurance.

"The first time I went out, Kubrick was really sort of agitated because it had been explained to him that I had limited time," Weston recalled. With the tank almost empty and the air growing inexorably more poisonous, he recited the alphabet backward until he started to

Bill Weston recovering from oxygen
deprivation at the base of the launch tower.

lose his thread. He gave himself another couple minutes and then with
reality greying into a buzzing haze all around him, he extended his
arms in the crucifix position. Illuminated by a powerful beam of light,
rotating slowly in a black abyss, Richter's modern angel had become
a floating, crucified spaceman. Through the helmet, Weston heard
"someone tacking up to Stanley and saying, 'We've got to get him
back.'" He also heard Kubrick's response: "Damn it, we just started.
Leave him up there! Leave him up there!"

By now, he was using his last ounces of strength to make repeated
cruciform shapes with his arms. Then he blacked out. "They brought
the tower in, and I went looking for Stanley," the ex-mercenary recalled. "I was going to shove MGM right up his . . ." He paused.
"And the thing is, Stanley had left the studio and sent Victor to talk to

me." Kubrick didn't return to the studio for "two or three days," said Weston. "I certainly remember it as two or three days . . . I know he didn't come in the next day, and I'm sure it wasn't the day after. Because I was going to do him."

Lyndon arranged that Weston get "Elizabeth Taylor's dressing room, including a refrigerator with beer and stuff in it." He also got "the biggest coming up for R&R, and Stanley would pay for it"—that is, a large raise. By the time Lyndon judged Kubrick could safely return to the studio, all was forgiven. Even so, Weston commented decades later, "Stanley, if he got involved, it was pretty much an empire of destruction."

· · ·

The single most complex shot Weston had to bring off was when playing the dead astronaut Frank Poole, following his murder by HAL via remote-controlled space pod. Bowman rushes to board another pod and sets off in chase of his deputy. The shot in which he intercepts the spinning, lifeless astronaut is one of *2001*'s kinetic miracles, and perhaps the most effective simulation of weightlessness ever filmed. We see Weston in a yellow space suit, his severed air hose extended, rotating in space as the pod, its servo arms raised, enters from the left. He makes two full spins before slamming into the arms with his helmet, his body reverberating from the impact. As the pod continues moving left to right, his body becomes entangled in its embrace, creating a kind of interplanetary *Pietà*.

Because the pod was mounted statically below the ceiling, the only way to bring off the shot was to have Weston meet it, not vice versa. And in order for it to seem as though the pod were entering the frame from the left, the camera had to travel with the stuntman. To accomplish this, Dunning's crew worked with MGM riggers to mount a track for Weston's wires in the studio ceiling. Below, a dolly track was laid along the same axis for the camera. Both Weston and the camera would move laterally across the studio, with the stuntman released from his launch platform toward the pod, and the dolly matching his movements far below.

As sometimes happens in film production, the spectacle of a shot being filmed equaled or even exceeded the impact of the shot itself. High on his platform, Weston was suspended horizontally from wires invisible from the studio floor. Set at its lowest rotational rate and operating through a gear system, a drill motor connected to his wire pivot set the stuntman spinning slowly. Directly below on the dolly, Kelvin Pike started the camera, which was slightly overcranked—or shooting in slow motion. Safely out of range beside Kubrick, Cracknell called "Action" through a megaphone. The Kirby wire crew started reeling Weston's spinning body across the studio ceiling toward the pod—which faced him, its mechanical arms raised. Simultaneously, the grips who'd laid the dolly track now rolled it forward, matching the stuntman's path across the velvet blackness.

Shortly before his impact, Wally Veevers's crew operated the cables leading to the pod's arms, opening and lowering their paired Ys to intercept the turning body. Weston's helmet hit the right-hand Y with a *bang* that resounded through the studio. Paint chips spun across space. With the Kirby crew still reeling him in, the stuntman used the aluminum ring at the base of his helmet to protect his neck and allowed his body to collapse realistically against the pod's arms. At Cracknell's "Cut," the crew knew they'd nailed it. A platform was hurriedly rolled forward. Liberated from the helmet, Weston drew deep gulps of air to clear his head. He was under no illusions: Kubrick would want another take.

Years later, having performed in eight James Bond films, driven his motorcycle into a lake at high speed in *Raiders of the Lost Ark*, carried out extremely dangerous feats in James Cameron's *Aliens* (after which he commented, "He's going to get someone killed one of these days"), and plowed through the freezing water and explosive pyrotechnics of *Saving Private Ryan*'s D-day invasion, the stuntman was most satisfied with his work for Kubrick. "I was part of a group that actually made history in the purest sense," he said. Ruminating about the "funny juxtaposition" of the director's "absolute integrity on a picture" and his "very funky . . . moral turpitude," Weston said that in the end, "I just found myself brimming with pride."

Not a man normally given to metaphor, he summed up his work on *2001: A Space Odyssey* by referencing a saying that originated with the thirteenth-century Japanese Buddhist priest Nichiren. "It's really the story of the blue tail fly," he explained. "The blue tail fly can fly along at three miles per hour. He hangs on the tail of a galloping horse, which will do thirty miles per hour. I think everybody who worked on it was aware that it was very special. And however exasperating Stanley was, and certainly how demanding he was, the man was a genius."

· · ·

Not quite a fly on a horse, more of a fish on a bicycle, a writer on a film set is frequently an awkward, misplaced creature indeed. Arthur Clarke was a regular presence at Borehamwood. He was fascinated with the production, many of his fondest dreams were being envisioned there with stunning power, and Kubrick wanted him around—though the director was usually busy, and Clarke had plenty of time on his hands. Con Pederson remembered a figure "kind of running by the side of the wagon." Andrew Birkin put it somewhat less charitably: Clarke, he said, was "like a spare prick at a wedding."

Birkin has fond memories of the writer, however. Clarke would sometimes hold forth on his vision of the future at Building 53's equivalent of an office water cooler, namely the coffee machine in the main Hawk Films unit production office. Since the office also had the production's sole Xerox machine, Birkin was there a lot. Unabashed by his own youthful callowness, he remembered the following exchange:

Clarke: "What do you know about calculus?"

Birkin: "Was he before Nero, or after?"

Clarke: "No, that was Claudius."

For his part, Dan Richter vividly remembers his first full day at the studio in early November 1966, following his initial preparatory meetings with makeup artist Stuart Freeborn and Kubrick in October. He'd been handed the keys to a spacious office with "Dan Richter, 'Dawn of Man'" inscribed on the door. Upon entering, he experienced the disorienting sensation of having no script, no instructions, and no real game plan. After sitting quietly for a while in this hushed equivalent of a writer's blank page, he picked up the phone and called his wife. "I'm just sitting in this office," he said plaintively. "I don't know what the fuck is going on."

Soon a knock sounded, and Richter saw a balding man of about fifty at the door. "Here you are, Moonwatcher. You probably need this," said Clarke, handing him the eighteen-page Dawn of Man script. With his usual caginess, Kubrick was keeping the rest of the story under wraps, at least for the moment.

While Richter was pleased to meet the author, whom he had read with appreciation, he told him he wasn't sure that he'd actually play Moonwatcher. His agreement with Kubrick was to choreograph the sequence, not necessarily act in it. They spoke for about forty-five minutes, during which Clarke referenced Raymond Dart, the anthropologist who had discovered the first Australopithecus fossils in 1924. On recognizing that its brain case was too large for a baboon or chimpanzee, Dart had deduced that he'd unearthed remnants of perhaps the first prehuman species to walk upright and use tools. As for Moonwatcher, Clarke said, he was supposed to represent a particularly shrewd representative of the species.

They also discussed Robert Ardrey's book *African Genesis*, which had brought Dart's work to public attention and been so foundational in Kubrick's and Clarke's conception of the scene. Then the author rose, shook Richter's hand, and wished him luck. He was off to Ceylon the next day, and really had to go.

CHAPTER EIGHT

THE DAWN OF MAN

> *No individual exists forever; why should we expect our species*
> *to be immortal? Man, said Nietzsche, is a rope stretched*
> *between the animal and the superhuman—a rope across*
> *the abyss. That will be a noble purpose to have served.*
>
> —ARTHUR C. CLARKE

Stuart Freeborn was facing a problem the likes of which he hadn't encountered in his entire thirty-year career. He'd done many innovative and even extraordinary things in that time. Most recently, he'd helped divide Peter Sellers into three readily distinguishable characters in *Dr. Strangelove*, including an earnestly balding President Muffley and the film's namesake, a megalomaniacal wheelchair-bound Nazi rocket scientist with a bad case of phantom limb disorder. Two decades previously, in a kind of inverse cosmetic surgery, he'd affixed a convincing pot belly on an otherwise slim Roger Livesey, whom he'd also morphed into believable late middle age in Michael Powell and Emeric Pressburger's 1943 film *The Life and Death of Colonel Blimp*—something that would have been relatively easy were it not for the fact that Livesey wore only a bath towel throughout one extended scene.

Most notoriously, he'd created the prosthetic schnozzle that had transformed Alec Guinness into Fagin in David Lean's 1948 version of *Oliver Twist*—a hooked beak so extravagantly anti-Semitic that it wouldn't have looked out of place in *Der Stürmer*, the notorious Nazi tabloid, though it was actually based on George Cruikshank's illustrations from the Charles Dickens book's first edition. (Half Jewish himself, Freeborn had suggested to the director that it might cause offense—but Lean overruled him.)

By the time a skinny, moustachioed kid named Dan Richter showed up in his studio at the end of October 1966, Freeborn had already tried an ape-man suit (which Kubrick had found artificial and wanting), produced a small posse of believable but freakishly neutered Afro-British Neanderthals (which since they really needed to be able to go forth, procreate, and inherit the Earth, also failed to fit the bill), and when Dan knocked on his door on a Monday morning, he'd been working for weeks on an expertly crafted new man-ape costume that he was quite proud of. Freeborn picked up the mask and showed it to Richter—who seemed an intelligent sort, albeit evidently from some kind of countercultural fringe. "What do you think?" he asked.

For his part, Richter had been impressed by Freeborn's studio, with its skylight, makeup chairs, and worktables crammed with various plaster, wire, and urethane forms. Clearly this was a place of transformation. People came in looking one way, and left looking entirely different. As for Freeborn, Dan thought to himself, "He looks like a character out of Beatrix Potter": balding, bespectacled, high strung but not necessarily nervous; an alert, energized creative force with a bit of leprechaun in his veins.

But as he examined the mask, he knew immediately it wouldn't work, though it had clearly been made with a great deal of care. He didn't want to hurt Freeborn's feelings and was acutely conscious that he was in the workshop of a master. And yet the mask was simply too thick. "It's just useless for what I'm trying to do," he thought. "He's a brilliant artist. But it's not what I want to do." He decided to be up front.

"Well, you know, I have a little bit of a problem with this, Stuart," he said. "This is really beautiful. The thing is, wearing a mask like that would be like . . . a bag over my head. I couldn't express myself."

The performer quickly tried to explain that he was hoping to wear as little costume as possible. He would be seeking to convey acting values. For him, masks and costumes were barriers to be overcome. Accepting these thoughts with a nervous laugh, Freeborn showed him the suit that went with the mask. To Dan, it seemed like an inordinately

264 · SPACE ODYSSEY

thick second skin, and much too heavy. "I can't do this," he said. "I just can't do this. Our characters have to show through. Whatever you create for us, it can't just be a surface that covers us. It has to enhance who we are and what we feel."

Although Dan could see him covertly fighting annoyance and disappointment, Freeborn took it well, and as they continued to talk, he seemed to understand that Richter was seeking as thin a layer as possible between performer and camera. Rather than making something apelike and then putting a man inside, he would need to build something up from a man's body. He suggested that Dan come back as soon as possible for a full-body cast.

Remembering that first meeting decades later, Richter commented, "Here's this American kid, obviously a dopey, who comes in to one of the great makeup artists of British film—*The Bridge on the River Kwai*, a major guy—and says, 'None of this will work.' At first, he thought, 'Who the fuck *is* this guy?' He's very polite, very English, he's very polite and everything. And I'm sure he must have gone to Stanley and said, 'What's going on?' And Stanley said, 'No. You guys have to work together; Dan's onto something here.'"

■　■　■

At its best, science fiction takes our post-Enlightenment way of understanding the world and extrapolates, using the findings of science and projections concerning the future of technology and putting them at the service of truths expressible through fiction. By the midsixties, astronomy and astrophysics had radically expanded the universe's dimensions, and the emerging science of paleoanthropology—a discipline rooted in Darwinism, paleontology, and biological anthropology—was starting to revolutionize our understanding of human origins. But rarely had the findings of these broad disciplines been incorporated into artistic expression. Instead, science had effectively been over *here*, and the arts elsewhere.

It was one of Kubrick's and Clarke's great achievements, and a wellspring of *2001*'s lasting power and continuing significance, that they took the complex, sometimes haunting, sometimes magnificent truths revealed by modern science, polished them with all the care given an expensive piece of Zeiss glass, and used them as a window to view the human condition within a staggeringly vast universe. Projecting back into the past, *2001*'s authors examined human origins. They weren't particularly doctrinaire about it. This was fiction, not a peer-reviewed paper in the journal *Nature*. But they always deployed scientific research—and its miraculous offspring, technology—to define their story, refine it, and expand it to its farthest possible limits. It's how they reached the border between the known and unknown—that place science is always probing like a tongue exploring a broken tooth. It's where they wanted to take their audience, because beyond it, something like magic prevails.

As with other players in his production, Kubrick's direction of Richter involved providing him with tools, suggestions, responsibilities, and the freedom to explore. He signaled his serious intent by handing him hard-core research papers by paleoanthropologists, popular science books by Robert Ardrey and Desmond Morris, and invitations to highbrow London seminars with titles such as "Kinship and the Family in Primates and Early Man."

The director also pointed him in the direction of the Regent's Park Zoo monkey house, and informed Richter that, by the way, he now had responsibility for Hawk Films's own small menagerie, which had been either leased or bought outright back in July for *2001*'s prehistoric prelude. Then under the care of Jimmy Chipperfield's Circus, Kubrick's ark included a leopard, two hyenas, two vultures, two peccaries, two large snakes, three zebras, and twelve tapirs—the latter a large South American herbivore with a prehensile trunk destined to serve as Moonwatcher's first kill, and never mind that the species was foreign to Africa.

In retrospect, one of Kubrick's most important moves was simply to give Richter a beautiful 16-millimeter Beaulieu film camera, arrange instruction on how to use it, and provide unlimited stock and processing. Richter also received a projector fresh from the disassembled centrifuge, and with that, the director had provided his incipient man-ape with filmmaking's two most important tools, channeling him to think not only as performer but also as participant in the cinematic process. Concluding a soliloquy on the merits of the Beaulieu, Kubrick said, "Now just go out and do all the research you want. Just collect information, and then we'll make decisions."

One of Richter's later stops was a visit to illustrator Maurice Wilson at London's cathedral-sized Natural History Museum. Wilson's exquisite full-color renditions of wildlife rivaled James Audubon's in their vivid detail and beauty. For several years, he'd been examining the museum's fossil fragments of early man and producing paintings reconstructing what Australopithecus and other protohuman species may have looked like, and how they might have behaved. By the time Richter got to Wilson the mime had been researching for more than a month, and he was growing increasingly frustrated to learn how little we truly knew about our earliest ancestors.

In Wilson, however, he found an artist who'd already gone through exactly the same process, and who was willing to share what he'd gleaned. The Australopithecus specimens that Raymond Dart had uncovered were indicative of small creatures: no more than four and a half feet tall and very slender. The scientific thinking at that time was that they probably weren't yet bipedal, though this has since been revised upward in favor of bipedalism. In Kubrick's and Clarke's conception, 2001's mysteriously powerful alien artifact was supposed to suggest not just tool use but also foster an upright stance. In choosing *Australopithecus africanus* and setting their prelude four million years ago, they had fortuitously arrived at exactly the right species and time frame.

Wilson also provided Richter with a visceral opportunity to convert the dry words of the scientific papers he'd been reading into something far more tangible. Taking him behind the glass-fronted display cases of the public museum into a hidden network of research labs and storage areas, he brought him to a spacious room lined with shelves and cabinets.

He comes to a stop by a large Victorian wooden cabinet. As he opens it I see that it is filled with pieces of bones and skulls as well as casts from other collections. There is a smell of great age in these dusty corridors that seems to add authenticity to the moment. Mr. Wilson lets me hold models and bits of original bones of Australopithecus. I feel a taste in the back of my mouth, and my heart races. I have made contact. Holding a skull that Professor Dart cast from the real fossils of a young boy is stunning. I can feel the ridges and pockets of its surface. There are two holes from the canine teeth of a leopard, which may have been the cause of his death.

Returning after such revelations to Borehamwood, Richter invariably discovered that Kubrick had been working the problem just as hard. "He'd be, 'Hey Dan, you got a cigarette?' Because Christiane wouldn't let him smoke," the performer recalled. "I'd give him a cigarette. He'd say, 'Listen, you know I was seeing this thing Desmond wrote.' 'Wilson wrote this.' Or: 'Hey, have you seen the Jane Goodall footage that Hugo van Lawick shot?' You know, 'Victor called up some people, we got *National Geographic*, they said they can give us some outtakes.' It was like *gold*, this stuff." ★

Apart from the museum, Richter repeatedly visited a mournful, thoughtful London resident, Guy the Gorilla, at the Regent's Park Zoo. Throughout the winter of 1966–67, he spent so much time with the twenty-two-year-old silverback that Guy started to acknowledge his presence "with a calm look. He looks through ordinary zoo visitors, but if I move around in front of his cage, his eyes follow me." Bringing along his Beaulieu, Dan observed that though the gorilla had a limited range of movement due to the small size of his cage ("I can't help feeling that Guy is like an innocent man in jail for something he has not done, with no idea of what he has been charged with"), when he did move, he had a way of shifting his weight directly from the center of his massive body. Richter started experimenting with his own body language.

★ Kubrick was referring to the important midsixties researchers into human origins and primate behavior Desmond Morris, Edward Wilson, and pioneer British primatologist Jane Goodall, whose work was documented on film for *National Geographic* by Dutch wildlife photographer Hugo van Lawick.

What great control! I reach for something, the movement begins in the very center of my body. I stand up, turn, and run—all the movement starts from my center. Moving this way does many things. It immediately removes the humanness from my movements. It creates size—suddenly I'm bigger, weightier. Animals move with their whole bodies. Try moving this way. It creates energy and power. Guy gives me this. He gives Moonwatcher size and dimension . . . Thank you, Guy, old friend.

. . .

Kubrick's alternate plan for shooting the Dawn of Man was to use a technique then fairly new to feature filmmaking called front projection. Throughout Birkin's search for a usable desert landscape in the United Kingdom, front projection was probably more a plan A than a plan B in the director's mind—though the concept was closely held. He simply didn't want to go far from the controlled setting of Borehamwood, to say nothing of the comforts of home, but he wasn't entirely sure that front projection would work. So he enlisted help from the Academy Award–winning head of visual effects at MGM's British studios, Tom Howard, and quietly set about doing camera tests with John Alcott.

Previous to *2001*, many films used *rear* projection as a special effects technique. The classic example is the couple in the car with the road rolling back toward the horizon behind them. Rear projection was also used extensively in *2001* to provide a convincing simulation of high-resolution flat-panel electronic screens in the spacecraft sets. But the problem with rear projection when using bigger screens—screens big enough to put a car, or an entire desert landscape set, in front of—is that the projected image has to fight its way *through* the screen material rather than bounce off of it. This reduces both brightness and sharpness. Although Kubrick had used rear projection extensively in *Lolita* and *Dr. Strangelove*—notably when actor and rodeo star Slim Pickens, as Major T. J. "King" Kong, rode his H-bomb down to a terminal encounter with Mother Russia—he was keenly aware that its inherent artificiality wouldn't work in *2001*.

In 1964 Kubrick had methodically watched all the films produced by the Japanese production company Toho, the studio behind the Godzilla franchise, among other sci-fi offerings of the 1950s and early 1960s. He almost certainly saw its 1963 film *Matango*, about a group of Sunday sailors who lose their yacht in a storm and take refuge on a mysterious island, where they mutate into grotesque shiitake people after eating the local mushrooms. While the horror component wasn't particularly well handled, director Ishirō Honda pioneered the use of front projection for maritime scenes on the yacht. These were several orders of magnitude more realistic than anything achievable with rear projection, and Kubrick would certainly have taken note of this.

Front projection relied on a highly reflective material, Scotchlite, that was invented by the 3M company in 1949 and subsequently used in reflective road signs. The material contained millions of tiny glass beads, which reflected light back at the source with incredible efficiency. A Scotchlite front-projection screen was hundreds of times more efficient at bouncing a projector's light back to a camera's lens than rear projection, and because it reflected from the front, soft focus was less of a problem. There were certain limitations, however. Because all those glass beads reflected light in a narrow beam back to its source, the camera had to be aligned exactly with the projector lens—a seeming impossibility, given that both projector and camera were bulky pieces of equipment and couldn't possibly be in the same place at the same time.

The front-projection process developed by 3M researcher Philip Palmquist solved this problem by placing a two-way mirror in front of the camera lens, angled at 45 degrees. Positioned at 90 degrees from the camera was the projector, which cast its beam on the mirror, which, in turn, reflected it onto the Scotchlite front-projection screen, such that the light bounced back directly at the camera behind the mirror. While the projected image also fell across the actors in the foreground—say, a man-ape with a bone in his hand—it was far too faint to be seen on anything or anyone not draped in Scotchlite. And just as Bill Weston's body hid the cables used to suspend him on Stage 4, producing a near-perfect simulation of weightlessness, in practice the shadows cast by

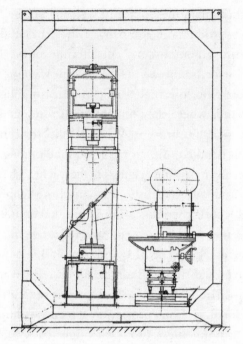

Dawn of Man front-projection
rig, with camera to the right and
projector suspended vertically to
the left above an angled mirror.

performers on the screen from the projector's beam were hidden from the camera by their own bodies.

There were other drawbacks, however. The camera either had to remain fixed or bring the projector and mirror system along with it wherever it went—a cumbersome proposition. And the lighting and color temperature of the foreground set had to be tweaked so as to ensure that the set and its performers merged seamlessly with the projected background image; otherwise the illusion that the actors were playing their roles within the wider frame of the image collapsed. But with Howard and Alcott working the problem, Kubrick was confident these issues could be overcome.

■   ■   ■

Among the locations that Kubrick had sent scouts to photograph was South West Africa—today's Namibia—as well as neighboring Botswana, collectively a vast desert expanse defined to the east by the Kalahari Desert and bordered on the west by the Skeleton Coast, so called because the absence of fresh water for hundreds of miles gave shipwreck survivors the reliable prospect of a long, slow, agonizing death. In 1966 South West Africa was still ruled by South Africa, and thus subject to the strict racial segregation of apartheid.

Of all the locations Kubrick had seen pictures of, the Kalahari and the Namib seemed to provide the most variety and possibility. These ocher deserts also had the advantage of absolute sun-baked legitimacy as a central stage for early man's struggles. Having abandoned the idea of filming his man-apes on location—a decision that compounded his reluctance to leave the studio's controlled environment with the impracticalities of shooting supposedly parched desert exteriors in the iffy British weather—Kubrick decided to pull together a small production team and send it to the region to capture realistic backdrop stills. Because of the resolution of the 65-millimeter film frame, the landscapes would have to be photographed as large-format eight-by-ten-inch fine-grained positive transparencies.

Prior to leaving the production, Tony Masters had suggested that his assistant Ernie Archer, who would be elevated to full production designer upon his departure, should be sent to Africa as well. That way he could ensure the backdrop photographs were framed such that the foreground sets he'd be designing and constructing would match. Masters's idea came during a meeting in which all three were present, and though Kubrick immediately agreed, he suddenly looked concerned. "How will I know what you're looking at, Ernie?" he asked. "I mean, how will I be able to tell if you're shooting the right thing?"

This hadn't occurred to Archer, who replied, "I don't know, Stanley. You can't, really. Just leave it to me."

Never one to accept an easy out, Kubrick was having none of it. "Oh, no, God no," he said. "I'm not going to know what you're doing." Thinking it over, suddenly the director grew enthusiastic. "I'll tell you what. Here's what we do: no matter how far out in the wilds you are, there's always a village with some drums or something, and you can send a message back to the capital, where they have phones. What you do is, you have a piece of glass on the back of your camera with graph lines 'A-B-C' across the top, and '1-2-3' down the side, and you draw the scene you're looking at and call it over to me: A-3, B-9, and so on. I'll be in the office back here in England with my graph, and I'll be drawing it too. Then I can look at it and tell you: 'Yes, that's great. But Ernie . . . three feet to the left.' "

At this, Archer shot an amused glance at Masters. "Stanley, it's never going to work, you know," he said. "You can't do something like that!" And they all laughed. Remembering the conversation, Masters commented, "It was madness—but out of madness came a lot of very good ideas, too." In fact, however impractical it may have been at the time, Kubrick had hit upon much the same method that would be used years later when transmitting digital pictures across great distances—or even simply copying them from one hard drive to another. Minus the drums.

In January Kubrick summoned Birkin and instructed him to gear up for another scouting expedition, this time to landscapes not readily accessible from Euston station. He'd already hired Pierre Boulat, a French photographer who'd worked for *Life* magazine. Boulat would be flying into South Africa with an assistant the second week of February. Kubrick asked Birkin to book a ticket to Johannesburg, proceed to Windhoek, the capital of South West Africa, hire a safari expedition, meet Boulat and Ernie Archer, and make the long desert drive to the Spitzkoppe hills—an ancient granite formation rising abruptly from the otherwise flat central Namib. All of Boulat's shots would have to be taken at dusk or dawn, which should allow Birkin plenty of time to scout other locations by Land Rover or light aircraft. Go arrange the thing with Victor, said Kubrick. Mind your expenses, bring plenty of Polaroid film, watch your back, and report back frequently.

Birkin, who'd recently broken up with Hayley Mills, welcomed the chance to fly into summer. He wasn't prepared to leave his sense of morality behind, however, and his African experience got off to a rocky start. Arriving at Jan Smuts Airport via South African Airways DC-7 on February 1, 1967, he got to a blank in the customs form where he was supposed to specify his race and wrote "human" inside—a nice allusion to the Dawn of Man, possibly, but not particularly amusing to the border police. Compounding this, when they searched his bags, they discovered a contraband copy of *Playboy* as well. Confiscating the magazine ("for their own use," Birkin assumed), they took him into a "sort of locker room," ordered him to drop his trousers, and subjected him to what's known euphemistically as a "full-body cavity search"— though Birkin recalls the incident in more direct terms.

When he'd recovered from this South African welcome, he opened a bank account in his own name—neither Hawk Films nor MGM was allowed to do so because of British apartheid restrictions—extracted a dangerously large amount of cash, stuffed it into his suitcase, and flew on to Windhoek, where he was met a few days later by Archer. Together they purchased a pair of Land Rovers, hired safari organizer Basie Maartens, and arranged for a small plane to serve as a shuttle and location-scouting service. It would fly exposed film directly from the desert to Windhoek, where it could be airfreighted on to London. Then on February 7, "I go to meet the airplane for Pierre Boulat," Birkin recalled. "Somewhat to my dismay, because I thought I had a broken heart at that point, his assistant is wearing a miniskirt. She was about twenty-one, my age. So there was a fairly immediate, *Uh-huh*." Her name was Catherine Gire.

Rumbling up to the Spitzkoppe hills in a multi-vehicle expedition kitted out in the style expected by the Texas millionaires who usually hired it for big-game hunting—fourteen tents, including one large enough for a grand dining room straight out of *Lawrence of Arabia*— they discovered a rugged bronze landscape of jumbled boulders and smoothly serpentine outcroppings. They'd arrived at one of the chief backdrops of *2001*'s Dawn of Man sequence. Returning by light aircraft to Windhoek after a week to get supplies, Birkin wrote his father.

We've been in tents for over a week now, and in spite of the fleas, bugs, mosquitoes, horned-beetles etc, etc that invade one's sleeping bag every night, I love it. The country is as wild and desolate as you can imagine—the hills where we are, Spitz-koppe, consist of huge rounded boulders rising out of the scrub. Every morning at 5 AM we go off and wait for the dawn to creep over the horizon. And again at dusk. After the rains there are the most beautiful sunsets I've ever seen—very strange pinks and indigos throwing shadows for miles into the desert . . . I sent the first batch of photographs off to England this morning. The photographer is shaking with fear that Stanley will ring up and have him recalled.

Boulat, who couldn't see his own results, needn't have worried. His first batch of exposed Ektachromes already contained the evocative sunrise and sunset images that Kubrick would use as his film's opening shots, including the Dawn of Man title card picture. As they moved be-tween locations, however, Boulat's agitation only increased, now due to clear signs that his assistant, who'd evidently been brought along not entirely for her photographic skills, was focused not on him but on Birkin. For his part, when he wasn't chatting up Catherine during the long days, Birkin would drive off in a Land Rover scouting new locations, or sometimes sit in the shade with a typewriter and pound out a draft of a screenplay he was writing based on Thomas Hardy's novel *Jude the Obscure*.

At night, they convened in the dining tent, where they were served by the six local men brought along to do the grunt work, all of whom lived in a single tent. Basie Maartens's safari workers subsisted on grain mash—something akin to animal feed—while the Europeans dined on steaks and drank fine South African Cabernet. "We've all talked and argued about apartheid," Birkin wrote his father. "When I was making up the food list for the safari, Maartens, our guide, said, 'Just get 10 pounds of pump nickel for the n . . . . ers, that'll last them 2 months.'" After dinner, the group gathered around the campfire, where swarms of multicolored butterflies and moths appeared out of the des-ert darkness and hurled themselves into the flames—apparently, Birkin surmised, because they'd never seen fire before. A pair of thrumming gas-powered generators provided electricity to a refrigerator packed

Pierre Boulat and Catherine Gire.

with film, and a record player, which Birkin used to play Shostakovich and the Stones under the stars.

Although the Central Namib usually contained no carnivores big enough to present a serious danger, highly venomous small scorpions emerged from the rocks at night. Catherine Gire "was the unfortu- nate finder of the first scorpion—in the loo!" Birkin wrote his father. "There she was, helpless, screaming into the night air." She hadn't been stung, however, only scared. Not long after, it was Birkin's turn. After washing in one of the shower tents, he unthinkingly sank his face into a towel—and received a sting like a high-intensity bolt of electricity to the nose. Screaming in agony and pawing at himself, he had to be held down by two workers while Maartens shouted that it was for his own good—without restraint, he'd likely do himself serious damage. "The pain is so intense, you try to tear your face off," Birkin recalled.

In the mornings and evenings, Maartens's laborers helped them haul camera equipment to the positions Kubrick had identified in Archer's and Birkin's Polaroids, which had been airlifted to London. A chain gang then passed unexposed film extricated from the fridge up the line from their campground to the boxy Sinar camera, which Boulat operated by sticking his head under an old-fashioned black hood—but not before checking it for scorpions. Freshly exposed film was returned in the same way. After bedtime, Catherine would sneak into Andrew's tent or vice versa. "She told me he'd taken her out as his assistant, and she made it conditional, that was strictly where her occupation ended," said Birkin. "So she could see nothing to bar the two of us having a bit of fun." They kept the arrangement as quiet as possible under the circumstances, until one night Pierre "came into our tent at like five o'clock in the morning, screaming with rage, not so much at me, but at her. Everyone's naked running tangled in the middle of the Kalahari Desert."

One of his early shipments of Boulat's plates and Birkin's own scouting Polaroids contained shots taken in the outback near Swakopmund, a coastal town they'd adopted as a kind of home base, since it was closer to areas of interest than Windhoek. Some featured a distinctively spiny, branching giant aloe: the quiver tree, or *kokerboom* in Afrikaans. Seeing these strange plants with their giant, wrinkled bark and fleshy star-shaped leaves, Kubrick grew excited. They seemed to convey exactly the kind of prehistoric exoticism he was after.

When Birkin called in from Swakopmund after a couple weeks in the field, Kubrick mentioned that he loved the trees. He didn't like their current location, however. Could they be moved northwest to an area they'd been calling the Mountains of the Moon, and positioned there? Birkin said he'd look into it, and discussed the matter with Maartens, a fifth-generation South African who'd pioneered big-game hunting in the region. An old-school Afrikaner, Maartens had been visibly uncomfortable as Birkin fraternized with his black crew, but otherwise they got along reasonably well. He told his client that the *kokerboom* was endangered and protected by law, which was why the largest local concentrations had chain-link fencing around them. Some were over three hundred years old. Plus, they stored a lot of water, were extremely heavy, and would be difficult to transport anyway. Better to think of another plan.

When Birkin reported this to Kubrick, the director responded, "Well, I'm very fond of those trees. I'm sure you can find a way. You

Kokerboom trees.

can just sneak in there and take a few." Birkin thought this over. "What if I get caught?" he asked finally.

"You wouldn't," said Kubrick. "See if you can make it work, because it's very important to me."

Maartens, who told Birkin that his safari license was worth more than the risk of getting involved, nevertheless gave his client the number of a local trucking company, and Birkin rapidly established that while the job could be done, because of the risk it would be expensive. He'd need to rent two large trucks and a team of workers. It would cost about £400 to fell and move the *kokerbooms*—about $10,000 today. "Okay, well, do it," said Kubrick. "Make sure you don't mention MGM. Maybe pretend you're Fox or something."

Birkin hired two trucks and a dozen workers. "I had to do a lot of bribing," he recalled. He was working for 20th Century Fox on a film project, he said, and they'd be driving south. Procuring a pair of wire cutters, he had Maartens start out with Boulat, Gire, and Archer in the direction of the Mountains of the Moon up north, and took his trucks and workers to a large fenced-in preserve that he'd already surveyed. Leaving the road in the late afternoon, they motored across the desert to a section well out of sight of any traffic. There Birkin himself snipped the fence, making an opening large enough to accommodate the trucks. With light beginning to fade, the workers set about sawing down two of the largest, most impressive *kokerbooms*. When they hit the ground, however, they broke into multiple pieces. The weight of their water had shattered their brittle trunks.

Standing amid the draining shards, Birkin was still instructing that the next ones be brought down more gradually with ropes—when an ominous buzzing resounded from the broken trees. A furious swarm of *kokerboom* hornets arose from the trunks. Yelling in agony from the stings, the workers scattered. Birkin, the only white man present, withdrew unscathed into a truck cab. After the insects had dispersed, his crew continued working, now under headlights. For every tree successfully felled, four lay in pieces. They loaded six reasonably intact trees onto the trucks, braced them with cork-stuffed bags, and set off across the desert, avoiding roads in case of pursuit and heading northeast.

Navigating with a compass and a flashlight, Birkin rode in the lead truck. Because the Namib is mostly flat and bare north of Swakopmund, they made reasonably good time—until they encountered a manifestly improbable sight: an ephemeral river, snaking through the gloom, evidently the result of a rare flash rainstorm upstream. As they drove along the bank to find an area shallow enough to cross, a laborer in the second truck lit a cigarette and unthinkingly flicked the glowing match amid the cork-stuffed bags. Within seconds, a brilliant yellow sheet of flame had erupted in Birkin's rearview mirror. With one truck blazing from the back, the small convoy screeched to a halt, and everybody spilled in confusion out onto the desert floor.

Realizing that he'd better think fast, Birkin ordered that the flaming truck be backed up and its precious contents dumped into the river. Half their cargo spiraled into the night, surrounded by gradually subsiding flames from the cork bags. Restarting the trucks, they pursued the *kokerbooms* downstream. Eventually the trees rammed into a sandbar shallow enough for the workers to recover them. Being congenitally tough and heat resistant, they were surprisingly undamaged.

When they finally reached their location near the Maartens safari encampment late the following day, the workers helped set up the trees in several places by propping boulders around their trunks. Then leaving Birkin behind, they drove off across the baking desert—apparitional vehicles gradually vanishing in shimmering ripples of heat. And at dusk, Boulat set to work documenting a landscape suddenly festooned with transplanted *kokerbooms*.

Most of the pictures taken in Africa were framed with an emphasis on background and middle-ground elements. Foreground distractions had to be as few as possible, because the sets being constructed at Borehamwood would provide the foreground. As a result, only a couple of the trees that Birkin had transported so laboriously across the desert can be glimpsed, far in the background, in *2001*'s man-ape prelude.

Still intrigued by the *kokerbooms*, however, Kubrick asked the MGM art department to fabricate some new ones, and several are prominently visible at the Dawn of Man. They were made in England.

■   ■   ■

Back in London, Dan Richter had recruited three others to assist him, including another refugee from the American Mime Theatre, Ray Steiner, a diminutive dancer named Roy Simpson, and Adrian Haggard, an untrained amateur with an "uncanny physical freedom that has him bouncing off the walls as an ape-man," as Richter put it. Their work was predicated on the understanding that they'd be core performers in the prelude. With Stuart Freeborn busy in the background devising new methods to construct light, flexible costumes, they worked together under Richter to establish behaviors and body movements capable of transforming themselves into credible representatives of a long-extinct prehuman species.

Richter had rapidly established a kind of upper-body vocabulary based partly on chimpanzees, and partly on Guy the Gorilla's chest-first way of moving. "They have a longer torso relative to the lower body, so by raising the shoulders and straightening the back, that movement translated," Richter observed. As a result, he, Ray, Roy, and Adrian could transform readily into "chimps with a touch of gorilla"—at least above the belt. But the lower body was a different story. "The problem was these big human legs . . . they just didn't work, they looked wrong," he said.

Richter had gotten used to taking his Beaulieu to film the primates at the Regent's Park Zoo, and one day, during a visit with his collaborators, he felt that his close observation of Guy and the chimps was yielding diminishing returns. He decided to pay a call on the gibbons—an Asian ape species characterized by fast, agile movements as they swing from branch to branch. Dan noticed that they periodically descended to the floor and walked rapidly across it, their arms raised for balance.

The walk is interesting because the gibbons are not knuckle walkers; they have long legs. Their walk has a swaying rhythm that seems to be an extension of their swinging from branch to branch. I point the Beaulieu at them and begin filming. Something is still wrong . . . They are moving too fast to be models for the man-apes. I sense that the movement is what I am looking for, but the speed makes the dynamic wrong. Suddenly I have an idea that really excites me. I switch the Beaulieu to half speed, forty-eight frames a second, and take more footage of them. Having exposed quite a bit of film, I put the camera down and move alongside of the cage with them, imitating their walk. One male gibbon gives me the queerest look. A crowd is starting to form and watch me.

The next day, the team gathered in Richter's office to project the footage. "There on the wall, in black and white, is the solution," Richter wrote later. "It's uncanny. A gibbon walking in slow motion was the control. I could do it, describe it, and teach it. The initial stage of my choreography is finally complete." He now had motion templates for both the upper and lower body. Kubrick's intuition to provide Richter with a filmmaker's tools had paid off.

Throughout his long working days at MGM, Richter was shooting a potent speedball: a blend of pharmaceutical-grade heroin and cocaine. As a legal addict, he was under the care and supervision of Dr. Isabella Frankau, "an aristocratic lady in tweed suits with a gold lorgnette hanging from a black ribbon, which she raises with deft aplomb whenever she needs to read something or write the coveted prescriptions." The mixture, which he injected up to seven times a day, was designed not to get him high but to provide a highly medicated form of stability, with the cocaine countering the heroin to produce a simulacrum of normalcy. "Lady Frankau is convinced that addicts need to be stabilized with a constant and controlled supply of heroin and cocaine so that they do not go through the up and down cycle that most constantly experience with withdrawal symptom followed by being high again," he wrote. Whenever Dan's regular blend didn't do the trick and the coke wasn't keeping him "bright and alert," he also always had some state-supplied methamphetamine on hand—crystal meth.

Although Dr. Frankau's prescriptions allowed him to work more or less normally, Richter had a very large habit and estimates he was injecting thirty to forty times the quantities a street addict would use. So far, he'd managed to keep it under wraps, and he intended for it to remain that way. One day, however, he forgot to lock his office door while shooting up, and Roy Simpson entered without knocking. Dan assured his visibly shocked collaborator that he was registered and under government medical supervision, and he asked Simpson not to talk about it.

With Birkin scouting locations and sending Polaroids and large-format Ektachromes from the desert, Richter attacked the tricky problem of casting the Dawn of Man. At first, he'd resisted Kubrick's assumption that he'd play Moonwatcher—he had more than enough work on his hands just casting and choreographing the sequence, he told the director. "It's part of my nature, I guess; I really want to be wanted," Richter said. "I think it was probably passive-aggressive behavior on my part. I wanted him to say, 'We really *need* you.'" He laughed. "Because I think I knew that nobody else could do it."

Meanwhile, Freeborn had been using his body cast of the performer

Freeborn's extraordinary final mask.

to build a lightweight suit, and was working on a thinner, more so-phisticated mask design as well, one permitting Richter to transmit a variety of facial expressions. In March Kubrick announced, "It's you. We built the costume around you, you know how to do it. It's you, and we'll make everybody smaller than you." This meant that Steiner and Haggard had to go—leaving only Simpson, who was shorter than Richter, as a potential performer. He could play a female, and Kubrick was planning to show babies as well—though how Freeborn would bring *that* off hadn't been determined yet.

The new height requirement further complicated Richter's difficult casting process. They'd been advertising among jockeys, long-distance runners, and high school athletes, but even so, casting calls that winter had yielded only about six candidates with the stature required, out of hundreds of respondents. As a result, Kubrick had agreed reluctantly to reduce his tribe from sixty to about twenty—still a lot, given the weave of their net.

They seemed to have met an impasse, and meanwhile time was

passing. Finally, the director came into the studio one spring day with a gleam in his eye. With three young daughters to entertain, the Kubricks' TV was frequently tuned to hokey family fare, and a kids' variety show featuring a dance troupe called "the Young Generation" had appeared on their screen the previous evening. The dancers were indeed young, and evidently chosen for their small stature so they would appear even younger. This could be the solution to their problem, Kubrick told Richter excitedly.

Richter immediately arranged that the Young Generation performers come by the dance studio in Covent Garden he'd been using for tryouts. "As I walk in . . . I have to work hard to hold back my excitement," he wrote. "Here are enough performers to complete my tribe of man-apes. They are small, slight, and, best of all, while they look like kids on TV, they are all sixteen or older professional dancers—they can move!"

．　．　．

After photographing the Mountains of the Moon with their new, slightly singed trees, Birkin received help from Maartens's safari workers to chop them up and dump them in a ravine. The *kokerboom* massacre was complete. A week or so later, he went on one of his periodic aerial sweeps, searching for vistas potentially of interest while the rest of the group traveled with Maartens to a new location. When he landed back in Windhoek, he received an urgent message: there had been an accident. The Land Rover carrying Boulat and Gire had slalomed off a dirt track and slammed into a rock outcropping, smashing the front end and breaking both of Boulat's legs. He'd been rushed to the Windhoek hospital. Catherine was shaken but unhurt.

Having received dual plaster casts, a pair of crutches, and an envelope of cash from Birkin, Boulat flew back to Paris several days later. Catherine elected to remain, at least for a few weeks. "Stanley stopped his pay the minute that the crash occurred," Birkin remembered. "Just like what happened to the crew members on board the *Titanic* when the last bit of water went over the top." He gave a hollow laugh. "Then an insurance claim was filed."

Kubrick rented new equipment and hired a well-known London fashion photographer, John Cowan, to replace Boulat. Famous for his energetic pictures of supermodels suspended in the air, an effect achieved with a trampoline positioned in front of various landmarks, Cowan was the basis for the photographer in Michelangelo Antonioni's 1966 film *Blow-Up*. His actual studio and darkroom had been central to the film—which also happened to include a brief but frisky appearance by Birkin's sister Jane. Cowan showed up in late March in crisp new safari khakis, minus the pith helmet. He'd never shot in large format before but had brought a more experienced assistant who had.

He proved temperamentally unsuited for the job. In London, Kubrick had explained the narrow strictures of his role: Cowan was expected to photograph exactly the views that had already been chosen, at precisely the time of day specified, so as to achieve the required lighting. That way foreground sets could be constructed accordingly. Instead, he kept on saying, "I think *this* is more interesting," and choosing his own angles—much to Ernie Archer's exasperation. He also insisted on sending plates to London with conspicuous foreground elements to them, although it had been explained that they would be completely useless to the project.

After much bickering, Birkin's compromise with the photographer was that he could take the pictures *he* wanted, as long as he *also* took what Kubrick needed. The director, however, soon became "pissed off with all these extra shots that were of no use to him," Birkin recalled. When Kubrick's complaints were passed on, Cowan, in turn, grew irritated that his artistic prerogatives were being disrespected. After a couple weeks of this, Kubrick decided to replace Cowan with Keith Hamshere, and on April 27 Birkin summarized the situation in a letter to his father: "We now have another photographer here—the fella from 'Oliver!' was hurriedly sent out following a hasty return of Photo No. 2. (Who got upset with Kubrick's criticisms)—what a *pain* he was—always going on about The Shrimp & Twiggy★ . . . It would have been rather different if he'd brought them along!"

★ Sixties supermodels Jean Shrimpton and Lesley Lawson.

When he returned to London, Cowan discovered that Kubrick was suing him for a week's safari costs plus other expenses.

. . .

Stuart Freeborn had been recruited in 1965 with a phone call and letter from Kubrick promising an "interesting makeup film. Schedule: five months, maybe six." He would be on the production for more than two years. This was almost entirely due to the Dawn of Man sequence. It was by far the most difficult job he'd ever had.

After his ape-man suit with a rubber pull-on mask had been deemed too thick and expressionless, Freeborn had followed it with his Neanderthals. Their augmented facial features had been built up, in a labor-intensive process, from foam rubber prosthetic pieces, and their crotches covered by light, flexible wigs. In the spring of 1967, his emerging solution to the challenge Dan Richter had posed—make a costume thin enough to allow a gestural, actorly expressivity—was a fusion of the two approaches. It had the great virtue of not requiring hours of work in the chair per subject to look right—something he couldn't afford, even with a tribe of "only" twenty.

The removable full-head man-ape masks Freeborn devised were made of multiple highly flexible pieces of form-fitting foam rubber, or polyurethane, which fitted over an underskull of more rigid material tailored to the individual. The latter didn't encase the performer's head but served as faceplates attached by straps, and could be removed and replaced in seconds. The foam pieces were highly variegated, with different degrees of flexibility depending on their function. When layered properly, they gave the convincing effect of muscle and sinew at play. Even in the simplest masks made for secondary performers, Freeborn had devised a wonderfully economical hidden internal system of threads, such that when the actor opened his jaws inside, the top lip outside raised three fourths of an inch, and the bottom went down a half inch. Depending on the mood at hand, the effect was of either a highly realistic warning or a greeting. Certainly the exposure of fanged teeth—the only Australopith defense, at least until weapons came around—was an entirely believable threat display. With this innovation as a baseline for all the masks, Freeborn had already tran-scended the dreaded man-in-ape-mask cliché.

But this was only the beginning. From there, he worked on refin-ing a way for Richter and the other principal performers to frown and use their tongues as well. Kubrick wanted his man-apes to be able to lick their lips weakly, in a sign of famine. In response, Freeborn had devised an internal mechanism permitting this action, which required the construction of believable mouths with dense foam rubber tongues that could extend and turn upward and downward to lick their lips. Each performer outfitted with this kind of mask had to have his tongue cast—not the most comfortable procedure. "I made a little acrylic cup that by suction fitted over the end of their tongue," he recalled. "They could force their tongue in by biting on the cup, pushing their tongue in it, and then they'd release their own jaws. Then they could push their tongue out and operate it, and it worked perfectly." The jaws themselves had small rubber bands inside to keep them closed most of the time. Opening them, however, required real muscle power.

A very small part of each performer's face was visible through each mask: the area immediately around the eyes. This was made up to match the dark skin tone surrounding it, and all the performers wore dark brown contact lenses as well—which tended to vacuum up the dust of the set, fortuitously producing the wretchedly bloodshot effect Kubrick was after. The masks' polyurethane layering thinned around the eyes and was affixed to the eyelid area by a type of sticky glue that remained viscous throughout, so that the performers could easily re-move and replace the masks between takes.

Freeborn's approach to the body involved creating what amounted to a body wig closely tailored to each performer: a thin, stretchy knit-ted wool with hair woven into it, augmented at the shoulders and at the small of the back with form-fitting urethane pads underneath. "It was lightweight, airy, comfortable to wear, and fitted perfectly," he said proudly. He'd discovered that by weaving in hair of different con-sistencies and colors, he could create a more believable pelt, and after much experimentation, came up with a second skin for each performer that allowed air to circulate and was more flexible than even the leotard

Richter had worn at his impromptu audition. Each ensemble contained human hair, yak hair, and horsehair, with the latter "down the spine, where you get longer, shiny, crispier hair," Freeborn recalled. Velcro strips allowed the top and the bottom pieces to join, with hair combed over the seam.

If art is a form of fanaticism, one soliloquy Freeborn made to visual effects magazine *Cinefex* about a decade after *2001*'s release provides a clear measure of how single-minded this unique artist's pursuit of absolute realism was. It concerned the intricacies of his man-ape's jaws, and specifically how to get them, and their lips, to close in a realistic manner.

For the performer to open the jaw, he would have to stretch the foam rubber on the sides of the face, which was difficult enough. In addition, he would need to pull cords and toggles to open the lips, top and bottom. So the performer's jaw muscles really had to work to open the mouth of the mask. Then the mouth had to come back and close—which it did not want to do. We found that it would hang open a bit—both the jaw itself and the foam rubber mask on top of it—which looked silly. The problem was that once the rubber stretched and loosened up, no matter what we did, the mouth would never snap shut again with just the tension of the foam rubber cheeks. So we had two separate problems: making the jaw snap shut and making the foam rubber over it snap shut. Springs were out because they make such a noise, and they're awkward to use. Elastic bands worked great for the first part of closing the jaw—but at about the halfway mark, they would get very weak and couldn't close all the way. Finally, I tried using deep field magnets. I had seven magnets hidden in the teeth, so that when the elastic band began weakening, losing its power to withdraw back, the magnets would take over and shut the jaw the rest of the way. At first, I found that if the magnets were positioned exactly flat, facing each other, it was a helluva job for the performers to break that magnetic field and open the jaw. I solved that problem by tip-tilting them slightly. The magnets still had their power, but by putting them at a slight angle, there was enough give that the performers could open the jaw quite easily. It made all the difference.

I had seven magnets hidden in the teeth. Who other than Stuart Freeborn could have produced such a phrase?

Throughout this wizardry, Kubrick arrived periodically to evaluate the work and up his demands. "I saw him look at it," Freeborn recalled in conversation with Richter in 1999, "and he went around, [going] 'Hmmm, hmmm, *hmmm.*' He never said, 'Oh, that's fine.' Never, ever said that. 'Hmmm, *hmmm.*' And went out. But I knew that in his mind, he's thinking, 'The bugger's done that, and if he can do that, he can do something else. I'll think of something else he can do.'

"And damn me, the phone went, and it was him. 'Stuart?' he says, 'I just got an idea. I'm writing in a scene now where I want to see them with their jaws closed, and I want to see them snarl with their teeth closed.'" He laughed. "There's no reason for it. It was just that he knew that it was almost impossible for me to do that, because all the other mechanics was depending on the opening of the jaws that made everything work. And now he wants to see the lips still working, without the jaws opening. So they could snarl. He said, 'I want to see them snarling.' He was just pushing me, you know. I knew that."

The British Empire was built on the uncomplaining execution of orders from the higher-ups. Entire subcontinents had been subjugated on that basis. It took a stiff upper lip to redesign his man-ape's lips after so much effort had been put in already, but Freeborn set to work on the problem. He devised a single tongue-operated acrylic toggle that functioned as an internal lever. When the tongue was pushed against it, it pulled the internal thread system connected to the foam rubber lips, exposing the teeth with the jaws still clenched—producing a dangerous-looking snarl. Underneath lay an extraordinary accumulation of hard-won experience.

No matter how extreme the request, Freeborn didn't ever say no. "He never yelled at Stanley or anything like that, but he would just grin and bear it and suffer," Richter recalled. "He showed a lot of stress, and he was a perfectionist just like Stanley was. He'd work and work for days to get something, and Stanley'd say, 'No, that's not right. Can you change it?' And, of course, changing it meant you had to work day and night for a couple days to make a change, which he'd bring, and Stanley'd say, 'Well, that's closer but still not right. Could you do *this*?' I think other people would have just said, 'Fuck this shit. I can go work on this other picture.'" Freeborn never even considered doing that. Throughout, he worked such long hours that he regularly slept exhaustedly in the car as his wife, Kathleen, who was working just as hard assisting him, drove him home in the early hours of the morning or returned him to Borehamwood at dawn.

Upon seeing the man-ape snarl that Freeborn had managed to produce against all odds—and contrary to all his prior designs—did Kubrick bare his own teeth in a smile? Actually, he remained unsatisfied. He wanted ever more nuance. He didn't want a simulation of life, he seemingly wanted life itself. "So Stuart, in a surge of inventiveness, separates the toggle into two sections, each affecting a different half of the mouth," remembered Richter.

It works! By balling my tongue and pushing both toggles simultaneously, I produce a very effective snarl. If I do it and keep my jaws wide, I produce the look of Moonwatcher roaring. Push the right one, and the right side pulls back. Push the left one, and it does the same on the left side. Rolling my tongue across them both produces a great effect. He has made the mask's tongue hollow so I can put my tongue inside it. I can lick my lips!

Of course, all that slavering tongue action within the sealed confines of all the masks under the high heat of the film lights presented a new kind of ordeal for the performers inside. "Believe me, it was revolting," said Richter. "And the only thing that kept me from vomiting was the realization that if I vomited inside that mask, I'd probably suffocate."

On top of all these innovations, Freeborn remained keenly aware that Richter's man-apes constituted different characters; they were individuals. "What I had was the positive of the artist's head, plus the urethane rigid mask on top of it," he recalled of constructing the masks on head molds in his studio before fitting them onto the actual subjects. "Well, now I have to build, on that, the patterns and indents as well as the outer shape of the foam rubber mask."

> The two things I had to think of were the age and the character of each particular ape-man. They're all separate. I tried to fit the personality, the shape, the expression, and the age of the face I was modeling to the particular performer—I knew all of them fairly well at that point. I knew their personalities, which ones moved quite fast, which ones were slower in their movements, and so on. Of course, the slower ones I made the older ones. And so forth.

Throughout, he and Richter compared notes daily. It wasn't just Kubrick who came to him with new demands; Dan was choreographing the sequence, and much of Freeborn's work was in response to his imperatives as well. "Stuart and I are working together in what has become a very subtle manner," Richter wrote. "He cannot build a costume with urethane from plaster molds alone, and I cannot create the illusion of a man-ape with movements alone. During the first few months, we slowly forged a way of working together that was very symbiotic. As a result, we have become very close. As Stuart's frustration grows, our friendship and respect for each other grows as well."

Asked about it a half century later, Richter said, "I think that both of us understood after a week or two that we were both onto something. I was working with this incredible artist who had changed film history and would go on changing film history, and he saw that he was working with somebody who actually knew how to make this thing work if he would work with me."

The artist Richter was referring to was Freeborn, not Kubrick.

. . .

By late June, Freeborn had established a costume production line with the help of another leader in the field, Charlie Parker, who'd done makeup for films such as *Ben-Hur* and *Lawrence of Arabia*. As spring turned into summer, the stresses of whipping his band of twenty man-apes into shape—which came on top of preparing himself for the lead role—was starting to hit Richter hard. He'd been taking them through uncompromising exercises designed to build leg strength and purge any remaining vestiges of trained dancers' movements, replacing them with a specific kind of wild, primitive energy—not random by any means but one adhering to the vocabulary of upper- and lower-body movements that he'd established and was now seeking to choreograph.

His goal from the beginning was to get his performers to establish kinship groups—with one of them led by One Ear, the rival gang leader to Moonwatcher. They'd watched films of Jane Goodall with her chimpanzees, and learned to exhibit various behaviors visible in them, including foraging, grooming, aggression, and submission. Incarnating an *Australopithecus africanus* required an intense physicality particularly hard on the lower body. The only way to disguise those long human legs was to crouch and weave, gibbon-style, with the knees turned outward most of the time. Accordingly, Dan had established a brutal physical training regimen—boot camp for man-apes. Every day, they met in the fields behind the studio, ran laps, and performed calisthenics.

Film productions, particularly extended ones staffed with unusually creative people, sometimes produce their own strange subcultural behavioral characteristics. Even without overtly simian conduct, they can be anthropologically interesting, and one of the ways that Doug Trumbull and his animators blew off steam was to go into the studio back lot, where he'd stashed a large rented trampoline, and take turns using it. In doing so, they had a fine view of rolling green meadows. The MGM back lot merged imperceptibly into pastureland and woods. Sometimes Doug went alone, and he distinctly remembers seeing Richter and his trainees shrieking at one another as he bounced up and down, "playing monkey amongst the trees and ravines and stuff. It was the weirdest stuff." Asked if he'd sometimes wear his cowboy hat when he did so, he said, "A lot of the time. It was my signature as the young California kid."

Picture, for a moment, the bouncing cowboy on his trampoline, hand on Stetson to keep it from flying off, observing from a distance the shaggy man-apes prowling among the back-lot trees, thinking they looked weird. As one snapshot among many, it could do worse in representing the multifarious backstage goings-on of *2001: A Space Odyssey* during its last year of production.

. . .

Despite his hipster background and drug habit, Richter was something of a disciplinarian. He didn't fraternize with his men, and he carried a three-foot wooden dowel—something his teacher Paul Curtis had called a "mime stick"—which he used to poke errant parts of performers who weren't performing as well as the rest of them. "You were quite strict. You were a bit scary," one of his tribe, David Charkham, told him in 1999. "You were the leader, you were definitely the boss. You were kind of on another planet at times; I didn't quite understand what it was . . . You were slightly not all there."

Apart from the incident with Roy, Richter had so far succeeded in hiding his addiction. When Freeborn had subjected him to full-body casts, he'd hidden his tracks with adhesive bandages. If Stuart had suspected anything, he was too diplomatic to say. As production neared and pressure grew, however, the drugs imposed an "awful burden," Richter recalled. The hardest part was measuring the dosages so that he wouldn't be sick from using too little, but also not groggy from using too much. "The only benefit I get out of it is that it keeps me very skinny, so I won't look like the Michelin Man in my costume," he wrote.

The strain of what he knew would be one of the most important creative commitments in his life had also affected Richter's self-confidence. Like Kubrick, he sought to hide this from his subordinates. Increasingly aloof, he kept upping an already high drugs dosage. His daily interactions with the director and his regular attendance at screening sessions, in which the visual effects team projected spectacular footage of wheeling space stations in Earth orbit and atomic-powered spacecraft arriving at Jupiter, only drove home how high the stakes were. The Dawn of Man sequence would open this remarkable film, and its success relied significantly on him. "It was a pressure cooker," he remembered. "You're surrounded by the brightest minds of our generation, and you had to deliver. And you were in the spotlight. You couldn't hide."

One day as Richter was putting his monkey men through their paces, he started to feel a constriction in his chest and shortness of breath. Worried he was having a heart attack, he asked Simpson to take over and drove into town seeking medical attention. Scrutinizing Richter's EKG readings, his doctor deduced that it was almost certainly stress and said he'd need to develop coping mechanisms. "They also recommended that I take less cocaine," Richter recalled with a laugh.

The sole holdover from his original group, Roy Simpson, hadn't been able to match the raw physicality of the young dancers then in their final preparations. Simpson had been kept on with the idea that his short stature and slight build would allow him to play a female, but Kubrick had decided that the females would all need to be significantly smaller than Moonwatcher, and with Freeborn now in the process of making individual costumes, they decided that Roy, who was about Dan's height, would have to go.

It fell on Richter to inform him of this. Simpson, who'd been helping supervise rehearsals and training for months, didn't take it well, and within an hour of his departure, Dan received a call from Kubrick's secretary requesting that he come immediately. The tone of the summons worried him, and when he arrived, the director looked as somber as he'd ever seen him. "Dan, we've had a very serious complaint from Roy Simpson," Kubrick said. "He said that you're a drug addict, and you were trying to hold him down and make him take drugs."

Richter was stunned. "It's bullshit, Stanley!" he exclaimed. "Roy is hurt because we let him go, and he's striking out." Suddenly all of his work was in danger. Realizing that he was "scared shitless," he thought, "That's it, I'm finished." In the stress of the moment, it didn't occur to him that Kubrick needed him every bit as much as he himself wanted to finish what they'd started.

"He wouldn't say something like that if there wasn't a basis for it," Kubrick said, making it sound more like a question than a statement. He scrutinized Richter solemnly with his dark eyes.

Dan realized he'd better confess immediately, and damn the consequences. Their whole project was clearly at risk, but he owed this man who'd invested confidence in him the truth. "Well, yes, I'm a drug addict," Richter admitted. "But no, I didn't force him, nor was I doing it around him. He did walk in on me once. I'm a registered addict. It's all legal. And if you want, I'll give you my resignation, and I'll do whatever I can to help. I just want to make sure the project does as well as it can."

Kubrick assessed the situation. "You're legal?" he asked, intrigued.

"Yes, I'm legal, and I'm registered with the Home Office," Richter confirmed, referring to the British governmental department in charge of immigration and law enforcement. "I'm not breaking any laws." He told him about Lady Frankau and said that Kubrick could call her or his Home Office caseworker for confirmation.

"Well, if you haven't done anything wrong or illegal, I want you to stay on," Kubrick said. "I've got a lot invested in you, and I need you." He admitted that Simpson's story about being forced to shoot drugs had been hard to believe. With the tension ebbing, his curiosity kicked in and the questions came thick and fast: "How does it feel? What's it like? How do you shoot up?" He was already converting the situation into another opportunity to absorb potentially valuable information about the world and its strange inhabitants.

Each performer would have a studio dresser assigned to him when filming started, and later Kubrick would arrange that Richter have a large dressing room with a private bathroom, so he could do what he had to do without his dresser knowing. "Suddenly we have an intimacy in our relationship that hasn't existed before," Richter observed.

. . .

To an untrained eye, the purposeful preparations under way in Stage 3 for the Dawn of Man may have looked like any other tableau of busy film workers in action. In fact, they were laying the groundwork for one of the most ambitious, technically complex shooting situations ever attempted. Everything about the front-projection technique pioneered at Borehamwood from August 2 to October 9, 1967, was new and untested. Front projection had never been done at such a scale.

After much trial and error, a giant, sixty-foot-wide backing screen had been covered with innumerable irregularly shaped pieces of 3M's Scotchlite reflective material. The manufacturing process for it had produced an inconsistent reflectivity, so that if it had simply been hung in strips on the backing screen as received, the image reflected back at the camera lens from the projector would have displayed visible horizontal brightness and color variations—something like a bad TV signal. Effects supervisor Tom Howard set an army of MGM scenographers to work with scissors and glue. Their stochastic patchwork of Scotchlite fragments randomized the problem, thereby cheating the eye, which subliminally discounted the variations as a natural part of the background scenery and sky. This solution was only partial, however, and in the shots with minimal cloud cover, a kind of stipple effect can still be seen in the sky—particularly if you're looking for it.

All the shots taken in South West Africa had been exposed at dawn or dusk. This allowed Kubrick and Alcott to light their sets in a specific way. In practice, they were meant to be largely in shadow, with the backgrounds brighter. The dawn, in other words, was a perpetual, controlled environment during the Dawn of Man. But those shadowed areas weren't actually in shadow. In fact, they were well lit from above, and that light had to be totally even, as secondhand light from the sky usually is. Every effort had to be made to avoid multiple shadows being cast by the performers—a dead giveaway that a studio lighting situation was in effect.

All this was extraordinarily hard to manage, particularly at the brightness levels they required, and Richter recalled Alcott and Kubrick in "very collegial, very friendly, usually softly spoken" conversations as they grappled with complex lighting issues. "You know if you'd listen to him," said Richter, referring to Kubrick, "he'd be saying, 'Oh, John, I wonder if we should . . . I think we should do *this*; I'd like to.' And if you listened to those words carefully, you'd say, 'He just asked him to do something that's going to be incredibly difficult!' And John would

Chief lighting technician Bill Jeffrey, one of
the unsung creative talents behind *2001*.

say, 'Okay, Stanley'"—here Dan muttered indecipherable words, imitating a conversation conducted in low tones. "It was all very quiet, but
you're flying at such an altitude!"

The requirement that they avoid the illusion-shattering effect of
multiple shadows enforced a particularly high-altitude innovation. "He
said to me, 'Well, how are we going to do it?'" Alcott recalled of
Kubrick. "And I said, 'The only way to do it is to make the entire ceiling of the stage a big white sky.'" Working with a particularly innovative and creative lighting gaffer, Bill Jeffrey—who'd lit the entire film
and was one of *2001*'s unsung creative resources throughout—Alcott
also devised an insanely precise studio lighting control system.

So I hung mushroom reflective-type globes from the ceiling—
500-watt photo floodlights—and I covered the complete ceiling
area of the stage with those, which worked out to about two
to three thousand bulbs. So Kubrick said to me, "Well, that's
fine, but then we've got these mountains, and then it's going
to be all hot [overexposed]," which was true . . . And he said,
"We'll have to turn bulbs off over the mountains." I said, "Well,
the only way to do that is to have every bulb on an individual

switch." So he said, "Okay, put every bulb on an individual switch." Normally, you'd never get anyone to go along with that. If you said that to a producer now, he'd say, "You're bloody crazy!" And it *was* ridiculous. But Kubrick wanted the lighting to be true and real; so we had a switch for every light, and I was able to control every bulb on that set. I could light the ground level fully and then have perhaps only two or three bulbs lighting the higher areas. The heads of the departments hated the idea because it made a lot of work for them; but they made a switch for each light, which required something like eight miles of cable all together.

Watching this process with fascination, Richard Woods, who played Moonwatcher's rival One Ear, had the presence of mind to take notes. He recorded thirty-seven rectangular crates filled with bulbs being hauled up to the ceiling, each the size of a bed frame and containing fifty 500-watt lights. Each light was controllable from below, making a total of 1,850 switches. This allowed for an extremely precise reduction of any hot spots that might emerge as the set's higher elements were wheeled into place. With Kubrick examining Polaroids throughout the process, overexposed areas could simply be made to disappear by incrementally switching off individual bulbs.

With all that firepower on, however, the total per crate was 25,000 watts. With thirty-seven crates suspended high overhead, the "sky's" total power was 925,000 watts. And that didn't count all the brute lights that had been brought in on the sides, creating better definition and giving a sense of sunlight being cast across the landscape from near the horizon. For the upcoming battle scenes alone, nine brutes would simulate sunshine from the side. With each of *them* cranking out 25,000 watts, the sunlight alone was 225,000 watts. Added to the sky, Woods calculated that the Dawn of Man sets were lit by 1.5 million watts.

Unsurprisingly, temperatures in Studio 4 soon spiked well above 100 degrees—as hot as the Namib Desert in summer. By now, Freeborn's costumes were far more permeable than the suit that Keith Hamshere had worn when he collapsed in similar conditions months before. Still, Woods remembered losing seven pounds during the first week of shooting alone—and like the other performers, he'd been hired specifically because he was already skinny. Concerned that the grueling conditions would impact his performers' energy levels, Kubrick brought in a large refrigerator packed with sodas.

He also tried another technique, soon abandoned. "They brought in a big wind machine, which was on very low, because they didn't want to create a dust storm," Woods recalled. "They fed it with some enormous amount of dry ice, putting it in the intake and blowing cooler air into the studio." But the temperature differential soon caused film lights to explode, and with glass raining down from the African sky, Kubrick ordered that the device be removed. Richter's man-apes would have to endure actual desert conditions.

• • •

Apart from the heat, as Richter and his men began filming scenes that August, one of the problems that quickly cropped up was near asphyxiation. Freeborn's highly realistic mouths with their prehensile tongues meant that the performers were almost as sealed inside as Bill Weston had been in his space helmet—all while performing highly strenuous movements. As a result, the carbon dioxide buildup was immediate. "You start to die," Richter recalled. "It's like being in the death zone on Everest; the minute you put the mask on and the lights go on, you're starting to die. You only really have seconds before you can't function anymore."

Ranks of nurses were on standby at the peripheries in case the performers passed out. Each man-ape had to pop salt tablets twice a day. Compressed air tanks were used to flush out the costumes and masks between takes, with negligible results. The performers weren't supposed to wear their head masks for more than two minutes at a time, and though Freeborn had created masks that were easily removable, in practice this was rarely possible during actual filming. Lengths of stiff hose were cut into sections and used to pry their magnetized, elasticized mouths open between takes, permitting some air to flow inside.

Finally, Freeborn devised a "specialized breathing apparatus . . . by making what I call a schnoggle, a special cap—a seashell cap that fitted over their own nostrils, but just short of the shape of their nose—it had two little tubes going up their nostrils." This allowed waste air to be expelled, partially ameliorating the problem. But the innovation came fairly late, and for most of the Dawn of Man, Richter's troupe was bounding about in near agony as carbon dioxide built up and their red blood cells carried more and more of it—and less and less oxygen—to their straining muscles.

Man-ape David Charkham vividly recalled shooting the nocturnal scene when Moonwatcher's tribe awoke at the mysterious appearance of the monolith and scampered out to investigate. "We had to wake up, run or jump around, and then I had to go up to it and calm down, and slow down, and I was breathing so hard, and my head was pounding. To then reach up and touch it, you know, it was so hard to do. It was so hard to not just collapse at that point. And to get down to those small, delicate movements."

Charkham's touching of the monolith came only after Richter was directed to do so. Because the Dawn of Man was shot entirely MOS—without sound—Kubrick and Richter could communicate throughout the filming, with Richter's voice emanating from behind his mask, something he could achieve without actually moving Moonwatcher's face. Kubrick hadn't told Richter in advance that he'd asked a space-suited William Sylvester to reach out and touch the monolith with his gloved hand on the Tycho Magnetic Anomaly set nine months before.

"I had to be in a squat, and you get very tired," Richter remembered. "And I had to have control of my body so my legs aren't shaking . . . And I reached up, and I knew the camera was *here*, so I reached up, and I didn't actually touch it; it looked like I'm touching it, but I didn't actually touch it because my hands were covered in dirt, and I knew there were going to be a lot of takes, and I didn't want to start getting it dirty."

Seeing this, Kubrick said immediately, "No, no, I need you to touch it."

Still in character, Richter's muffled voice issued from behind his mask: "I'm afraid I'll make it dirty."

"That's okay," said Kubrick. "They'll clean it up." Afterward, he showed Richter his lunar monolith footage with Sylvester and told him he'd been trying to match that seemingly instinctive human gesture.

Animator Colin Cantwell, who'd arrived as a late hire from Graphic Films in Los Angeles, remembered watching the monolith scene being shot. "He was directing Dan," Cantwell said of Kubrick. "Just seeing the stage and him calling out occasionally, you know, 'Now back away from it . . . *Yeah!*' It was so extraordinary as it was unfolding. Of course, Dan's interpretation was extraordinary acting in its highest level . . . All of it without a spoken word. It was beyond goose bumps." He laughed. "The intensity of that is really something to live."

■   ■   ■

No shooting day was more fraught with anxiety than the leopard attack scene, set in a dry riverbed and filmed on September 18. Because the Dawn of Man sequence kept being deferred, animal trainer Terry Duggan had been wrestling with the leopard for almost a year, and had established a relaxed friendship with the animal. Whenever he worked with it, however, he removed his wedding ring. The leopard didn't have its claws clipped, and he wanted to avoid one of them catching on the ring—a potentially dangerous situation. He also always wore a thick army jacket, so that any inadvertent scratches would maul the material and not his skin.

Although the leopard knew Duggan well, it had never encountered him in a man-ape costume before. Nor did it know Richter, who'd agreed to participate in the scene—not without a good deal of trepidation. The big cat also had no experience of film studios, with their brilliantly illuminated sets, nervous crews, electric atmosphere, and shouted commands. The animal was positioned on top of a craggy cliff face, with a newly hirsute Duggan foraging in the dirt below. Behind him squatted Richter, also nervously foraging. In the film as released, a couple other man-apes filled out the scene. But because they'd refused

Terry Duggan working with the leopard.

to be in the actual shot with the leopard, they were filmed separately and added later via a painstaking rotoscope animation technique.

It has been recalled, perhaps conveniently, that the entire film crew was behind a safety barrier while filming the leopard attack. Stuart Freeborn and his wife, Kathleen, were both present, however, and when discussing the scene with Dan Richter in 1999, a somewhat different memory emerged.

"I was just wondering if you had any remembrances of that scene," Dan asked.

"Yes," said Kathleen. "Well, I remember the leopard being on the set, in fact."

"Yeah," said Richter. "Did we take any special precautions, do you remember?"

"No," said Kathleen. "Only for Stanley. He had that cage thing."

At this, Stuart joined the conversation. "He was in a cage, yes," he affirmed. "Funny that, wasn't it?"

"He was the only one in the cage," Kathleen continued. "Nobody else had anything."

"What did the cage look like?" asked Richter.

"Just like a lion's cage, you know," said Kathleen.

"It was sort of rounded on the top and came down," Stuart recalled. "He was directing from inside the cage."

As for Richter, he distinctly remembers that "everybody was very nervous because they figured Stanley's in a fucking cage; *he* doesn't have a problem." Before they attempted a take, Duggan wrestled with the animal a bit so that it knew who was in the costume. He also reassured an edgy Dan that if anything happened, he'd handle it.

"I remember being very nervous, because I had to be in the middle ground," said Richter—or halfway between the far bank of the dry riverbed and Duggan. After a false start when the leopard, confused by the film lights, failed to budge, they tried again. At "Action," this time the leopard dropped from its cliff onto Duggan as planned—but then it immediately noticed an interesting new element: a second monkey-man. Forgetting Duggan, it started toward Richter. "If you've ever had a leopard walking toward you with a look in its eye, I tell you it's very disconcerting," said Richter. "I knew, 'Don't run.' I trusted Terry. Terry said, 'Don't worry, I'll take care of it.' He jumped up and dove and grabbed the leopard and rolled around with him and whatnot, and I got out of there." Asked if he did, in fact, run now that the animal was otherwise occupied, Richter demurred: he'd simply conducted a kind of midtempo hustle to safety.

The shot was unusable, and now the leopard knew that Richter was part of the scene. They tried again, adrenaline pumping through Richter's system. While he wasn't alone in this, he had the most reason to be afraid. Duggan's assistants brought the animal back up to the ledge, and they rolled the camera. This time the leopard dropped onto Duggan as planned, spun him over in the dusty riverbed, crouched as his familiar sparring partner attempted to back away, sprang at him, and turned him over again. It was as close to perfect as could be expected in the circumstances. Kubrick called, "Cut!" Then "Okay, that's it." They were done.

Surprised, Richter—now fully into the spirit of the scene—left the stage and approached Kubrick's cage. He was fretting that Duggan's reactions hadn't been believable enough. "I was worried because Terry didn't look like an ape," he recalled. "Maybe we could work on his movement a little more." Kubrick was unequivocal: "Look, I got what I needed, so it's okay."

Asked to evaluate the director's mood at that moment, Richter meditated for a moment, then said, "I think Stanley was probably afraid. I think most of the other people on the set were nervous . . . I think he was afraid of the fact that it was a dangerous situation for other people, something might happen that could shut the film down—somebody gets killed or maimed on the set. The film had enough problems. He wanted to get his shot and put an end to it."

The next day, everyone hurried to Theater 3 to watch the shot. Unbeknownst to the crew the bright light of Tom Howard's front-projection system had bounced off the tapetum of the leopard's eyes, reflecting brilliantly into the camera lens during filming. Seeing this, Kubrick exclaimed, "What the hell's that!" and jumped to his feet. They ran the shot again. The cat's radiant eyes effectively doubled its menace as it closed in for the kill.

Kubrick couldn't contain his glee. "He was bouncing all over the place with delight at this lucky accident," remembered Richard Woods.

■ ■ ■

One of the more humorous attempts at producing a thoroughly credible prehistoric scene followed Kubrick's directive that Freeborn produce female man-apes capable of nursing babies on camera. The mothers would be played by Richter's smaller males, albeit outfitted with convincingly simian urethane breasts.

The babies, of course, would have to look equally authentic. Freeborn's first thought was to produce foam rubber baby Australopiths, with wires inside to make them simulate nursing behavior. But he suspected Kubrick would reject this solution, because "it was cheating—and Stanley didn't like cheating—so I knew I had to have something else up my sleeve. I thought maybe I could use baby chimpanzees: cover their enormous ears with more human-looking ones and then do makeup over their very pink faces," he recalled. "I wasn't about to bring that up first, but if Stanley balked at the doll idea, I would tell him the only alternative was to make up baby chimps, supposing we could get a baby chimp anywhere near us to do this."

Sure enough, having nailed his monolith scene, Kubrick called Freeborn. "Oh, by the way," he said, "I hope you haven't forgotten, but in two or three weeks' time, we will be getting to the family sequences, where we'll want to see the babies suckling from their mothers' breasts. How are you going to do that?" Freeborn outlined his foam rubber dolls plan. "But that's no good," Kubrick said. "I want to see them suckling at the breast."

"Well, I have another idea," said Freeborn, unfazed. "It could have problems, I don't know. But it's a question of if I can make up some real baby chimps, you see."

"That's good," said Kubrick. "I like that idea. How are you going to make them suckle from the breasts?"

"Well, I'll put a little bit of honey on the foam rubber teats and let them suck that off," Freeborn said.

"No, that's no good," said Kubrick. "They'll suck that off too quickly. I want to see them really suckling."

"Oh, I see," said Freeborn. "You want practical, working breasts." He wasn't surprised.

"Yes," said Kubrick, satisfied that he'd once again provided a challenge commensurate with Freeborn's abilities.

"So here we were, getting deeper and deeper," recalled the makeup man. But as with every other thing the director asked of him, he proceeded to make it so. "I went through my reference books, and I couldn't find any female chimpanzee that was 'in milk'—that was in the act of feeding babies. And I knew if we were going to do this, I was going to have to fool those little devils."

The breast would have to look and feel just right; the mechanical action and flow of the milk would have to be just right; the taste and smell and consistency of the milk would have to be just right. On top of that, the milk couldn't come out when we didn't want it coming out—it had to come out only when the babies suckled at the breast. But what the hell did I know about the mechanics of a real female chimp's nipples? And I thought, My God, how can we reproduce all this with foam rubber or plastics or something, as well as the milk itself would be a problem. Then I've got to reload it. That's another problem.

Freeborn called up Mary Chipperfield of Chipperfield's Circus, the same outfit that had supplied Terry Duggan and his frisky leopard. He shared his problem with her and in short order had arranged to borrow a couple baby chimps and two trainers. He dressed the latter in his new female ape suits with milk-filled artificial breasts and, after a couple weeks, the babies were actually drinking companionably from their new wet-nurse's nipples. Kubrick's profoundly male movie had suddenly become touchingly maternal. Behind the scenes, at least.

The baby chimps were, however, still in their native pink face, plus their ears were too big. In between nursing sessions, Freeborn and his wife carried the little chimps around his studio, and they'd all become quite trusting of one another. So when the time came for them to have blackface makeup put on, they sat "very still," Freeborn remembered, and also endured the attachment of flexible foam ears without complaint. But when released, "they went straight up to each other and began licking the makeup off," he said. "They loved the taste of it. So we had to sort of smack their bottoms every time they did that, take them back, remake them up. Then immediately they started going back to each other and licking it off. Smack their bottoms again. Eventually, though, they learned not to lick the makeup off."

So far, so good. They had well-trained, friendly, thirsty, blackfaced baby chimps with small ears ready for their close-ups, and Freeborn had also worked out an easy way to refill the breasts of his lactating females; deploying "one of these syringes that they use for inoculating cows—a big one—with an ultrathin needle." Then came the day when the breastfeeding scene was to be shot. Stuart injected the milk, tickled his chimps, and accompanied his mothers (now played by the two smallest of Dan Richter's troupe) and their babies to Stage 3. To his dismay, he saw that Kubrick had decided to shoot something else that morning.

Instead of feeding the babies, he had the performers jumping and rolling around—which churned up the milk in their breasts, you see. On top of that, it was getting very, very hot, and the milk was separating into butter and watery liquid. And after a while, every time the apes jumped, it would squirt out of their nipples. Stanley, of course, didn't see the funny side of that at all. He said, "Oh God, this is no good. Get them out of here." So I had to take them back to my lab and clean them up and then fix them up again with a new load of milk. Then we went down, and Stanley shot on them, and it was fine—but, of course, it was all cut. You *do* see one going up to the breast, but that's about it. In the end, there was no need for *any* of that work at all. But we're lucky if we get a third of the work we do up. We recognize that that's the way it is—that's our job. We get a third of it . . . But that's a true story.

Apart from a brief view in the darkness of their cave of one baby chimp turning to nuzzle its mother's breast, the chimps did, in fact, merit their own dedicated shot in *2001: A Space Odyssey.* There they were, examining a small bone with great curiosity as their elders shoveled back the raw meat that, under Moonwatcher's leadership, had suddenly become available in large quantities. No milk necessary.

. . .

One day in late September 1967, Dan Richter and his wife, Jill, were invited to the Kubricks' home at Abbots Mead for dinner and a screening of Roman Polanski's new horror comedy, *The Fearless Vampire Killers*—the one with the bubble bath scene that Birkin had snooped on from the studio rafters the year before. Polanski was there, but Tate, soon to become his wife, had remained behind in Los Angeles. Richter observed that the diminutive Polish director was nervous about what Kubrick would think of his latest effort. Apparently Polanski was "a bit in awe of" his host. "He doesn't need to be, the picture is great," he wrote. "We are all laughing at the endless jokes."

Afterward, they went to the kitchen for coffee, and Christiane put multicolored sugar cubes on the table. At this, Polanski and Richter started riffing on how sugar might look like that if they were tripping on acid, and how each cube, in fact, looked like a little LSD delivery device. Seeing that their host was refraining from comment, Polanski asked him if he'd ever taken drugs. Kubrick answered that he never had and never would—not because he had a problem with getting high, but because he didn't know the source of his own creative gift, and was afraid that he might lose contact with it and never get it back again. Polanski said that, on the contrary, it actually boosted creativity. He urged Kubrick to try it sometime. During this exchange, Richter covertly observed the director. At previous social occasions, he'd noticed that the director sometimes seemed a bit ill at ease, as though uncomfortable when he wasn't in charge.

> I am struck by the loneliness of such a great artist—not only to have a precious gift that one might lose, but also the burden one must have to use it well. Stanley knows how important his work is, and I know he derives great joy and satisfaction from being a good steward and doing his work very, very well.

In a set of preparatory notes, prepared eight weeks before shooting the Dawn of Man, Kubrick had meditated on the core message of the sequence. "Man emerged from the anthropoid background for one reason only: because he was a killer," he wrote. "Long ago, perhaps many millions of years ago, a line of killer apes branched off from the non-aggressive primate background . . . We learned to stand erect in the first place as the necessity of the hunting life . . . And lacking fighting teeth or claws, we took recourse by necessity to the weapon."

The location for Moonwatcher's realization that the heavy thigh-bone of a zebra skeleton could be weaponized was the same dry river-bed set where the leopard attack had been filmed a week before. The bulky, overheated front projector had been loaded by two men wearing surgical masks and gloves so as to minimize any chance of fingerprints or condensation appearing on the eight-by-ten-inch transparency of distant hills and an overcast Namib sky.

Richter knew that one reason why Kubrick shot so many takes was because he was discovering the meaning of a scene and learning something new each time. He compared it to a painter working on a painting: "Well, I'll add a little red here." He understood that the reason they would sometimes spend the entire day on a single wide shot and not even get a close-up was "because he was going to be learning and crafting and building something that a lesser director would have stopped along the way someplace, or settled for something less."

Months before, Richter felt he'd made contact with mankind's ancient ancestors at the London Natural History Museum when Maurice Wilson had allowed him to hold casts and genuine fossils of *Australopithecus africanus* bones. Now he would have to transmit another form of epiphany, also while holding a bone. He was facing one of the most difficult challenges of the sequence: that of conveying Moonwatcher's emerging recognition that the skeletal remains he was foraging among could have a previously unsuspected use. Looking at the camera, he saw through the angled glass of the front-projection system that Kubrick had elected to use an 85-millimeter lens. It meant he was framed in a medium shot. And as usual, his every movement would be magnified by the huge Cinerama screen.

It was the single biggest acting problem Richter had to solve, and he could have handled it in multiple different ways. Aware of the lens, he elected to play the first part as an enfolding series of small realizations, each made visible by slight shifts in the angle of his head as he contemplated the skeleton. As a subtle, unmistakable transmission of realization dawning, it drew from all his prior experience as a performer. "I was trying to do as little as possible and still make it happen," he recalled. "The moment when I first get the idea in my head, I have a lot of things to deal with there. I can't just pick up a bone and break something, because I don't even know what a weapon is. How do you hold it? What does it feel like? What does it smell like? I have no context for this. I don't want to just go there right away."

Keir Dullea has called Richter's performance during this scene his favorite moment in the entire film. Arthur Clarke commented, in his introduction to Richter's book, "That frozen moment at the beginning of history—when Moonwatcher, foreshadowing Cain, first picks up the bone and studies it thoughtfully, before waving it to and fro with mounting excitement—never fails to bring tears to my eyes."

Picking up the bone, Richter smelled it a little, and then started meditatively thumping it down on the skeleton fragments. Lined up along the reflected beam of the front-projection plate, both the camera and Kubrick were fairly close, and Dan could communicate with the director as he did so. In an early take, Dan banged the bone down, and a rib spun up into the air. "Oh, sorry," he said from behind his mask. "No, no," said Kubrick. "Use it, it looks good. Keep doing it. Keep doing it." And so Dan continued, smashing down on the smaller bones around him in an escalating frenzy of liberated violence. Finally, he rose on both legs and brought his bone club down on the center of a large skull—all in the fixed framing that the front-projection technique required, with the dry banks of the dead river behind him.

■ ■ ■

Every morning, Richter's band would come to Theater 3 to watch its work from the previous day. The man-apes were already wearing their body suits, dark rings of makeup around their eyes, masks off. Kubrick always sat in front, holding the small hand controller the projectionist had rigged for him after weeks of constant interruptions by the director as he demanded the focus be changed. As their work unfolded on the screen, the man-apes stayed in character, greeting any critical comments from Kubrick with rasping croaks of disapproval and positive ones with high-pitched cheeping cries of ecstasy.

Upon watching Richter's dry riverbed scene, however, everyone fell silent. Dan had played it perfectly, all the way to the smashing of the skull, which functioned as a euphoric visual crescendo. Although he knew it was good, Kubrick discovered he was unsatisfied. He instinctively wanted to build on what had been achieved. This moment was the absolute crux of the Dawn of Man. He wanted to get in closer and shoot from a low angle. He wanted to see Dan's arm with its weapon from below, in slow motion with the sky above—two things impossible to achieve with their front-projection technique, which required that the camera remain rigidly in line with the cumbersome projector.

In any case, even if they had the luxury of changing to a low angle in the studio, slow motion required more frames per second, which, in turn, required more light per frame—and they couldn't increase the brightness any further without melting the African backdrop transparency. Their filmic research project had reached the very limits of what was possible with the front-projection technique, and later that day, Kubrick asked production designer Ernie Archer to prepare a raised platform high enough so that a shot of Richter from below would be possible. It should be big enough to strew sandy gravel and warthog bones across the foreground. They'd wait until the London skies looked reasonably similar to the partly cloudy day visible in the background transparency, and film outside.

Now that Moonwatcher had discovered that the bone was usable as a weapon, the script called for him and his tribe to walk upright and eat meat for the first time. Once again, Kubrick came to his makeup wizard. "Well, that's all right, I'll make some phony meat," said Freeborn, not without trepidation. "You can't have real meat because the fat will destroy the foam rubber."

"No, I want real meat," the director responded predictably enough.

The new requirement forced Freeborn to design a new face, one easily detachable from the mask. When Richter's performers chomped down on the raw meat, its fat caused the faces to swell almost immediately. (It also sometimes found openings and squirted in, disgustingly, on the performers.) Each new mask required about eight hours of work, and so Freeborn was forced to mobilize three shifts working twenty-four hours a day exclusively to accommodate the carnivore scene—even though the false meat he'd produced in anticipation had looked completely convincing.

Even without meat, the masks' foam rubber absorbed oil from the sweaty, slavering performers, and they became "quite rank by the next morning," Freeborn remembered. After some trial and error, he devised an ozone cleaning method—a dedicated space with a machine producing ozone gas to deodorize and disinfect. "We could peel the rubber off, turn it inside out, and we had all these inside-out faces of all the artists—all these monkey faces, completely inside out—hanging on hooks in rows in a special room," he said. Fans on the windows drew the gas through the masks, "which all night killed any germs and cured it completely so they were quite fresh the next day."

The only scenes requiring the full man-ape contingent were water-hole confrontations between the two ragged bands. These took place within the single largest set built for the Dawn of Man, one matching the rugged backdrop of the Spitzkoppe hills and framing battle scenes replete with twenty gibbering, capering protagonists. Although not really evident unless you look closely, during the second confrontation—the one following Moonwatcher's bone-weapon realization—Richter and some of his band walk upright for the first time, while their weapons-ignorant antagonists remain on all fours.

By now, Woods—whose character, One Ear, would be pummeled to death by Moonwatcher's bone—was suffering from "housemaid's knee," or bursitis. His legs had become painfully swollen and puffy due to all the bouncing and squatting during grueling weeks of shooting. He would now demonstrably be put out of his misery with a blow to

Dan Richter as Moonwatcher.

the neck, which, in turn, required a bone large and straight enough to look lethal but light enough not to actually "do me in," as he recalled, "because they thought it may not go down well."

After much experimenting, Freeborn finally devised a bone built around a stiff piece of bamboo both light and rigid enough to do the job. "But I tell you something, it damned well hurt," said Woods to his old antagonist, Richter, three and a half decades later. "And I recall, I think, thirty-two retakes. You know, we did front and back and side."

At the end of that sadistic production day, after Woods had been wearily resurrected from his death crouch on the dusty ground for one last time, and he and Richter's tribe had limped, disheveled and exhausted, back to their dressing rooms, Kubrick asked his victorious, murderous alpha man-ape to stay behind for one more shot. Framing it far more tightly than the battle scenes he'd just captured, he asked Richter to throw his bone triumphantly upward into the million-watt African sky.

■   ■   ■

With the Dawn of Man in the can, Derek Cracknell took all the man-apes out into the countryside, as far away as possible from engine noises and church bells, and recorded them shrieking, howling, gibbering, and muttering. On noticing with irritation that another animal sound—the cawing of a flock of country crows—was intruding on his session, Cracknell borrowed a shotgun from a local farmer and blasted away at the sky in an effort to scare them away. After a couple deafening discharges, the crows retreated and they got what they'd come for, and all of the sounds that Moonwatcher, One Ear, and their bounding, squalling subordinates generate in *2001: A Space Odyssey* were made by the performers themselves—an auditory confirmation of the profoundly intimate genetic linkage between Homo sapiens and its distant ancestors.

Then everyone was let go except Richter. In part, this was because Kubrick was waiting for the perfect puffy clouds for his low-angle shot of Moonwatcher smashing bones, and in part because he simply liked having him around, and suspected he'd find something useful for Dan to do.

Although he'd referred to "the beautiful transition from man-ape to 2001" in his notes to Clarke a year before, it's not clear if Kubrick already had the idea of matching a shot of Moonwatcher's flung bone with an orbiting nuclear bomb—an epic match cut spanning four million years. But Richter strongly suspects that one of the most extraordinary cinematic transitions ever presented on-screen originated in that downward thrust he'd made in Stage 3, in which he'd mistakenly caused a warthog rib to spin upward, back on the day they'd filmed his riverbed epiphany. "The key point in that—and the point that gives you an insight into Stanley's genius—is that one little thing that I thought at first was a mistake, or an unfortunate little thing that happened, he took and turned into one of the great moments of film history."

On September 20, 1967, Dan Richter, Stanley Kubrick, and a small team went out onto the MGM back lot under puffy cumulus clouds. "On a field . . . the carpenters build a small rectangular platform perched five or six feet off the ground," Richter wrote. "It is covered with a small strip of the desert a few feet wide, elevated off a green English field a few hundred feet from the road. In the background, buses and cars and folks on bicycles are moving by . . . I feel strangely naked as I walk up the steps for the first time. I am alone on this little pedestal." Knowing Kubrick's penchant for multiple takes, the prop people had acquired a seemingly endless quantity of bones and horse skulls from the local knacker's yard.

In the first takes, Richter was somewhat tentative, put off by the narrowness of the platform. "Throw the bones around, Dan!" Kubrick yelled. "Cut! Can't you give me more energy?" Richter responded that he was just warming up and was worried about falling off. He could never see as well as he liked with his mask on, and the brown contacts didn't help. He selected reference points on the ground to help orient himself and gradually found his way into the scene. Soon ribs were spinning off in every direction as he pounded the bony gravel. "We shoot and shoot. Take after take," Richter remembered. "I smash an awful lot of skulls today. It's clear that Stanley is trying to get into the moment. He is trying to penetrate it and get every bit of energy and then somehow find a key to its door and pass through it into a greater space . . . Clearly everything we have shot until now is going to end on this platform, in this field."

Back from Ceylon and bearing new drafts of the film's narration, Arthur Clarke was visiting Borehamwood that day, and would be a regular presence for the next few weeks. He watched with fascination as the vision he shared with Kubrick materialized into live action and the scene unfolded. It was the only part of the Dawn of Man that he saw being shot, and the only sequence in the whole production filmed outside rather than within the studio's safe confines. Occasionally they took a break to allow an airplane to pass overhead, but otherwise the sky stayed in character—a fair match for that partly cloudy morning in South West Africa that Pierre Boulat had photographed months previously. Exactly a week later, Kubrick would shoot multiple takes of a dead tapir slamming into the platform's sandy ground, which he would later insert into the scene in a rapid series of cuts underlining the consequences of Moonwatcher's orgiastic rampage.*

After dozens of horse skulls had been smashed to pieces, "eventually Stanley was satisfied, and as we walked back to the studio, he began to throw bones up in the air," Clarke wrote. "At first, I thought this was sheer joie de vivre, but then he started to film them with a handheld camera—no easy task. Once or twice, one of the large, swiftly descending bones nearly impacted Stanley as he peered through the viewfinder."

Evidently those shots weren't satisfactory, however, and after filming the dead tapir on September 27, Kubrick asked Kelvin Pike to overcrank the camera and capture more shots of bones spinning up into the air. An old friend and advisor, cameraman Bob Gaffney, was visiting Borehamwood that day and witnessed the scene. Along with

---

* One of the tapirs had died in a panicky stampede off the edge of a high stage earlier in the production, and Kubrick, sensing an opportunity, had asked that it be frozen to preserve it for the shot.

Peter Sellers, Gaffney had brought Terry Southern to Kubrick's attention while *Dr. Strangelove* was still in the script stage. Prior to that, he'd driven around New England with the director, filming B-roll material for *Lolita*. That afternoon he observed the proceedings with a cinematographer's keen eye.

> Arthur C. Clarke and I were standing there talking. They were shooting the bone going up in the air, and the operator couldn't get it. It's a very difficult shot. They were shooting at high speed. When you throw something up, you don't know how high it's going to go, because the guy was throwing the bone differently every time. They threw it to go end over end. So you'd follow it up and then follow it down. The operator missed it three or four times. Stanley got behind the camera and did it the first time, and that's the shot that's in the picture. Stanley operated the camera.

Half of that landmark cinematic transition linking *2001*'s prehistoric and futuristic parts was thus bagged by Kubrick himself. Reflecting on the path to that Elstree platform, Richter said, "What was happening with him is he was getting to that cut. His problem is he's got to move forward three or four million years, you see. And if you think about it, that's a big problem. You can just do a quick cut, and everyone goes, 'Whoa, what just happened?'"

Instead, somewhere in the time between that rib spinning upward in the studio, and then Kubrick's asking Richter to fling his bone up at Boulat's projected African sky—and then still later, those exterior shots on that platform under puffy English clouds—Kubrick had formed a plan. "And I found myself on this journey with this amazing man," said Richter, "and going through this incredible creative process where he kept taking something further and deeper and richer and *more*—and then suddenly we were free of the three million years and just blossomed into the future."

.   .   .

Reflecting, a decade later, on his experience working with Kubrick on *2001: A Space Odyssey*, Stuart Freeborn was unequivocal in his respect for the director. In gearing up for the Dawn of Man, he'd worked seven days a week, "twelve to sixteen, sometimes eighteen hours a day," with one Sunday off every four weeks. Kubrick was working no less hard, however, and Freeborn had noticed, as the dual pressures of money and time escalated, that the director's lighthearted, jokey side had all but vanished and been replaced by something darker and more grimly determined. Conscious of why this was so, he made allowances.

> We worked long hours, but that I didn't mind, because I knew I'd gotten an opportunity here to do things that I would never ever in my life have that opportunity [to do] again. I'd never be able to do those things again. And he was the man who was going to get every ounce out of me, like nobody else ever could. *I* couldn't myself. I needed somebody like that. And I knew that once I'd come out of that, I would be in a position where I had so much know-how that I'd be way ahead of anything I'd ever imagined. And with that one vision in my mind, it didn't matter what Stanley did or said, I was going all the way with him.

Still, because Kubrick had insisted on absolute secrecy, Freeborn was under no illusions that the Academy of Motion Picture Arts and Sciences would recognize his work. And indeed, it wasn't even nominated when Oscar time rolled around. Adding considerable insult to that injury, in 1969 a special award for makeup was given to John Chambers for his work on *Planet of the Apes*—a 20th Century Fox production that had conducted an espionage operation into Freeborn's man-apes at Borehamwood, and had then quietly tried to hire him instead of Chambers to make its man-sized gorillas and apes.

Their headhunt had been unsuccessful because Kubrick's extended shooting schedule conflicted with *Planet of the Apes* star Charlton Heston's—and Freeborn wouldn't even consider leaving *2001* before it was completed. In any case, following Chambers's Oscar, Clarke and

Freeborn with friends of his own making.

others were vocal in their suspicions that Freeborn had been a victim of his own success—that his collaboration with Richter had produced man-apes so convincing, it was simply assumed they were real.

Despite his own experience with the novel, he may not have recognized the extent to which Kubrick's penchant for secrecy had also contributed. In any case, sometime after the film's release Freeborn received a letter in the mail with the director's return address on it—an extremely unusual occurrence. On opening it he read: "I believe the mutual frustration we both felt may have prevented me from adequately expressing my admiration for what you accomplished in *2001*. I think you did a number of things that have never been done and may never be equaled. You have my gratitude and appreciation. Sincerely, Stanley Kubrick."

Freeborn was too busy working on other films to see *2001* until years after its release. Reflecting on how any Oscar aspirations had been "killed stone dead," he thought to himself, "Okay, nobody's going to know about this. But I know what it took—and there's no reward."

But the reward came to me after I'd done another picture, and we went to New York, and I hadn't seen *2001* at that point, and it was showing at Times Square. And I thought, "I'll sneak in there all on my own and see it." After all, I was trying to get it out of my system . . . And I went in there, and up came the Dawn of Man sequence and the monkeys, and I thought, "Oh, it doesn't look too bad, you know." And then there was this American family behind me, and Ma says to Dad, "Are those real monkeys?" And, of course, my ears pricked up, and Dad says, "Hmm. Yes, dear." "How do they get them to act like that?" "Well, you see, dear, they're specially trained." And that for me, that one moment of hearing this going on behind me, that was worth anything else. That made it for me. I thought, "Well, that's it! I did it. That's great." So that was reward enough.

CHAPTER NINE

END GAME

FALL 1966–WINTER 1967–68

> *Man, in a very technical age, must attain*
> *more discipline and control of himself, and thus*
> *become more like a machine. Inversely, the machine,*
> *in order to communicate with man and enlarge his*
> *horizon, must become more human. So it goes.*
>
> —STANLEY KUBRICK

Postproduction, usually just "post," is generally a time of consolidation. The gains made in production have to be retained and augmented, whatever they may be. The crew slims down. Their professions change as well, from set designer to editor, from sound recordist to sound mixer. The director usually puts down his conductor's baton and collapses gratefully into an editing room chair. All the freshly shot film fragments need to be pieced together, as does the sound, and new music is usually composed. An entirely different, less energy-intensive creative ferment begins.

That's in a normal production. *2001* was far from normal. The film had more than two hundred visual effects scenes—unprecedented in the predigital era. Kubrick and Con Pederson sat down one day and figured out that each would require about ten major steps to complete. By their definition, a "major step" meant another technician or department had to do something significant to the shot. Ten steps times two hundred scenes equaled two thousand steps. Not unreasonable for a big film, until one considers that in this analog production, a chutes-and-ladders situation prevailed—the very game that gave us the phrase "back to square one." * Any mistake in any of those two hundred effects shots meant a redo of the entire thing. This Sisyphean procedure soon spawned a mission-specific Borehamwood vernacular. The term "redon't," for example, gained wide currency among Kubrick's visual effects grunts. It meant: redo the damned thing, but don't make the same mistake again.

A mistake is only a mistake, of course, if the correct way to do something has already been established, and all this has to be seasoned with the fact that in the grand research and development project that was *2001: A Space Odyssey*, Kubrick and his tenacious minions were inventing almost everything as they went along—including entirely new visual methodologies and late-breaking critical plot elements. It should have been no surprise that the film's delivery date kept slipping. With Kubrick's uncompromising perfectionism governing the process, most of those two hundred effects sequences had to be repeated upward of eight or nine times. The true total, then, was closer to sixteen thousand steps—an estimate coming from Kubrick himself.

Underpinning all this was that extraordinary bravura element, the "held take"—precious camera-original negative kept on ice until various additional layers of contents could be added. Even with the fine-grained color separation masters as a backup, the held takes would eventually have to be committed. "It was so bold," marveled incoming animator Colin Cantwell, who arrived just six months before the end of the production but made a significant contribution to *2001*. "Anything could blow up along the way, and you couldn't go back and reshoot, largely because the setups were gone."

In response to the overwhelming logistical problems inherent in such complexity, Pederson had largely ceded hands-on image making to Doug Trumbull and others and had taken a management role. His war room, the Building 53 nerve center where the whole process was overseen, was covered with swing-open bulletin boards onto which had been tacked the full production history of each sequence. En-

---

* Originally an ancient Indian (and later, British) board game called Snakes and Ladders.

Con Pederson, photo most
likely by Stanley Kubrick.

larged Xeroxes of film frames, or crude drawings by Kubrick himself, topped each "Sasco" chart, the better to ID the sequence at a glance. (Kubrick's reliance on the planners and other products of this British office supply company became so complete during production that Tony Frewin took to calling him "the Sasco kid.") People came and went, phones rang ceaselessly, and shots inched their way incrementally to completion on the wall. "Every day, the amount of information that kept coming in was just extraordinary," Cantwell recalled.

Meanwhile, a half dozen cameras were in operation throughout 2001's Borehamwood facilities at any given time, many on twenty-four-hour shifts, altogether producing multiple visual effects sequences. The proliferation of such sequences produced by these image-factories were given names as well as numbers in the war room, the better to keep track of them all. Football terms were in common usage. Prompted by

Kubrick, Ivor Powell could ask Con Pederson what the status was on "Kickoff," and Pederson would know exactly what he meant.

The end of live-action shooting, then, was actually production by another name. Such normal postproduction activities as editing and sound mixing would be tacked on at the end—whenever that might come. And, of course, even without actors, most of the effects shots had one or more feet in analog reality, and frequently required studio setups as elaborate as the live-action sequences. Stage 3, for example, now housed the largest "miniature" ever made for a film, the fifty-five-foot *Discovery* spacecraft, which was easily as ambitious as *2001*'s sets. It had required "a four-hundred-feet railroad, impaling the studio from end to end, as a camera track," as Clarke put it. Far too big to budge itself, *Discovery*'s stately progress through space would be achieved with camera movement.

Roger Caras went with Kubrick to witness Stage 3 being prepared for *Discovery*, and remembered the scene vividly.

> He had a [shot] where the spaceship *Discovery*, which is roughly the size of the Bronx, comes by or the camera pans along, dollies along the length of this thing. And any kind of a movement, any tremor, would show tremendously on that big screen. So he had to have it without tremor. And so he had them tear out the floor of the studio—not a small job. That's heavy stuff, the floor. You can push tanks and battleships across it. Tear out the floor and sink enormous concrete pilings. And the concrete pilings came up above the floor, and they are what held this model, which was—oh Jesus—I guess a hundred feet long. [*sic*] And the camera was on dolly tracks that were also braced up. It was a huge undertaking to get that one shot.

As for *Discovery* herself, the spacecraft's design had gone through the same evolutionary process characteristic of everything else in the production. An early NASA-funded concept for the manned exploration of the solar system relied on the controlled detonation of a series of nuclear bombs, with each explosion accelerating the spacecraft to a higher speed. This resulted in a short-lived design in which *Discovery* would have been "farting its way across the universe"—as Kubrick put it to Gentleman, who found the idea ludicrous and quickly put an end to it. A less laughable propulsion system was selected, but Kubrick's and Clarke's commitment to technical accuracy led Harry Lange to depict a skinny object fitted with dragonfly-like wings—actually radiators, designed to disperse excess heat from the craft's nuclear reactor. And the earliest model, which had been made by an outside provider, reflected this faithfully. But a consensus soon emerged among the design and visual effects staff that the thing simply didn't look right. They anticipated the inevitable audience question: Why does something that flies only in space need wings?

In short, nobody was happy with *Discovery*. From there, however, an interesting process unfolded. Con Pederson complained to Kubrick that the design "looks like shit." Kubrick didn't dispute it and invited him to come up with something better. Pederson and Trumbull sat down together and produced drawings with substantial modifications. Masters became involved as well, and so did Lange. Ultimately a kind of extended collective improvisation caused the spacecraft to materialize in its final form. "We all designed it," Masters recalled. "We spent months and months on that. Then it gradually began to get built, and added to and changed and altered. And so whose design it was in the end, goodness only knows. It was sort of . . . it was everybody's."

By spring 1967, the final incarnation of *2001*'s Jupiter-bound giant had been produced. When done and detailed, with hundreds of plastic model kit pieces added to its surface—giving it a formidably machined, flight-ready appearance—Lange had it sprayed eggshell white. Then he dirtied it up a bit with a judicious mixture of turpentine and black paint, giving *Discovery* the slightly used appearance of real hardware. It was an immensely impressive sight, all fifty-five feet surrounded by curtains of flawless black velvet and mounted on velvet-covered poles embedded in rock-solid concrete.

After admiring it alone for a spell, Lange went to fetch Kubrick. The director walked appraisingly around the enormous model. Spher-

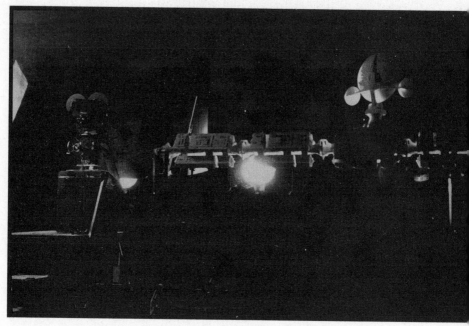

A rare surviving photo of the fifty-five-foot *Discovery*.
The camera on the left gives a sense of scale.

ical in front, it had the segmented spine of a dinosaur, at the center of which was a tripartite communications dish and behind which was a rectangular propulsion area terminating in six howitzer-type engine bells. Seen from a distance, it had a strangely sperm-like shape. From up close, it seemed capable of taking humans to Jupiter right away. Stopping, Kubrick stroked his beard meditatively.

"Harry, I don't like it," he said finally.

Lange was shocked. "What do you mean, 'don't like it'?" he demanded. The director had seen them making the smaller version in the model shop; it couldn't have been a surprise.

"I don't know what it is," Kubrick replied, frowning. "I just don't like it. Do something." And he left.

Lange sat on a chair for a long while, contemplating *Discovery*. "What the hell is wrong with it?" he asked himself. Finally, he covered the majestic spacecraft in plastic sheeting and returned to his office. He

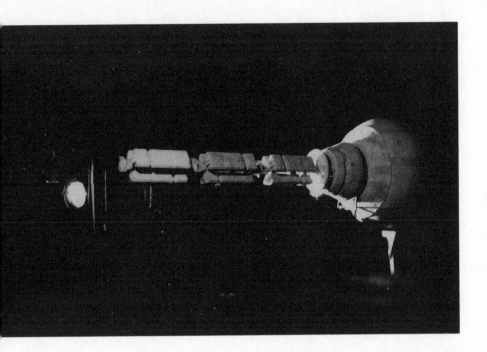

wanted to satisfy Kubrick, but he couldn't think of a thing to change. So he didn't change anything.

Three days later, the director called. "I want to see *Discovery*," he said. Lange took him back to the studio and removed the plastic sheeting. "That's it!" said Kubrick. "Perfect!"

. . .

The smaller spacecraft models had all been completed by late fall of 1966, including the film's orbiting retinue of nuclear bombs, which at the time seemed a likely outcome of the Cold War. These featured inconspicuous national markings, with the French Air Force roundel, the German iron cross, and the Chinese red star all visible in the final sequence if you knew to look for them. They would be the first twenty-first-century technologies visible after Moonwatcher's ecstatic bone throw into the sky—a weapon-to-weapon match cut.

Complicated worm-gearing rigs governed by "selsyn" motors—a portmanteau of "self-synchronous"—had been set up, allowing shots to be repeated with frame-by-frame accuracy. This was necessary when adding tiny people moving around within the space station, for example, or the underground Moon base air lock, or showing Bowman on *Discovery*'s bridge, and interior details of its open pod bay.

In such cases, the model was lit with its window or pod bay areas blacked out and filmed under John Alcott's supervision with a given camera move. Then the camera was recalled down the worm-geared, studio-length track and the shot repeated, now with the model dark and only the window areas alive with front-projected little figures or exquisitely detailed pod bay set photographs. Seen live in the studio, the production of such shots had all the excitement of watching a minute hand creep across the face of a clock, because in order to get maximum depth of field, each frame typically had to be exposed for a second or more. It was an excruciatingly painstaking procedure, and it didn't respond well to human error.

Andrew Birkin remembered one such sequence, in which the camera passed directly through Space Station 5 as it rotated in Earth orbit. The nine-foot-wide space station model had been mounted on a spindle, and the camera was able to move laterally on a cog-wheeled rod across the studio, its incremental progress as it approached the turning station imperceptible to the naked eye. As part of his pre-Namib duties, Birkin was monitoring visual effects for Kubrick, and had scheduled completion of the shot for the July 30 weekend. Accordingly, he'd asked the effects unit to work overtime—but had received a blanket refusal: England was playing West Germany in the 1966 World Cup final on Saturday, and no way were they going to miss that game.

He soon came up with a counterproposal: a TV could be put in the studio. At this, all resistance faded away. Suddenly overtime pay looked good, and that Saturday, the fifteen effects-unit personnel sat around watching the game with the space station shot clicking away incrementally behind them. In principle, Birkin's solution was elegant: at thirty-two million viewers, the match still holds the UK record for

Space Station 5 model.

most-watched television event, and there was no way he could have successfully fought its immense gravity field. As the camera inched its way across the studio toward the station, however, the game ended in a draw and went into overtime—during which Britain scored not once but twice. And at each goal, the unit jumped to their feet in glee. "Such was the joy of the unit that the floor rocked," Birkin recalled.

That Monday, a beautiful shot of *2001*'s stately space station pirouetted on the screen, with Kubrick tweaking the focus appreciatively using his hand controller as the camera closed in on the white wheel. Suddenly a squall seemed to blow across space, and a tremor shook the station. The shock wave from the epic goal sealing England's 1966 World Cup victory had reached Earth orbit.

The shot was redon'ted.

. . .

New forms of creativity specific to *2001*'s unique circumstances arose throughout. Engrossed in producing his Purple Hearts psychedelic planetary landscapes from Birkin's kamikaze helicopter runs, Bryan Loftus discovered that an innate bias toward certain color combinations was tilting his results toward a given palette, when what Kubrick really wanted was truly random combinations. Looking at the results, which were excellent but biased toward Loftus's predilections, the director said, "Bryan, I want you to produce stuff that you don't *know* you can do. If you know it, it's no use to us. But if you *don't* know it and find something, *that's* what we want." It was the governing principle of the entire production.

In response, Loftus took three Kodak film boxes, put roulette wheel–style diagrams on them, and mounted a spinner on each: one for color, one for aperture, and one for which separation master to use as the basis for a new shot. (The latter already came in many varieties: negative, positive, high-contrast, low-con, and so forth.) Like William S. Burroughs's Dada-inspired literary cut-up technique in his experimental novel *Naked Lunch*, Loftus's system seemed very much of its time, and it succeeded in producing genuine chance juxtapositions. "We've taken the human element out of it!" said Kubrick gleefully.

The director's management style can be summarized by two stories. One concerns Tony Frewin, who early in his tenure as an assistant was asked by Kubrick to read and summarize a stack of books and scripts. Frewin, who hadn't finished high school, thought that only "posh people on the radio" were capable of such a thing. "I can't do that," he said.

"Well, listen, you can read, can't you?" asked Kubrick.

"Of course I can read."

"If you read a book, and I say to you, 'What's the book about?' you could tell me, couldn't you?" asked Kubrick.

"Yeah, I guess so," Frewin replied.

"Thirdly, you would presumably have an opinion," said Kubrick—who'd barely managed to finish high school himself. "You would think it's credible, it's silly, it's stupid, it's crazy?"

"Yeah."

"Well, what's the problem? Now go off and do it."

Having been liberated from the perception that such things were well beyond his ken, Frewin eventually went on to write three well-received novels, and he edited a number of other books as well. "I think at that time, he had a lot more confidence in me than I had in myself," he observed. "He had a lot more confidence in a *lot* of people than they initially had in themselves."

The other story concerns how Kubrick kept his talented stable on its toes. "He had a list of ten people—and he showed me this—a list of ten people in the crew," Birkin remembered. "And you're on top of the list for a week. Next week you're number two on the list, and somebody else's moved from the bottom, has gone to number one. And the third week, you're number three; fourth week, you're number four; by the fifth or sixth week, you're going, 'What did I do wrong?' And Stanley is saying, 'Nothing, what are you talking about?'" Birkin paused for effect and then mimicked the plaintive supplicant: "*'I seem to have lost your affection!'*"

After which, said Birkin, Kubrick would say it was all a misunderstanding and bring the unfortunate one back to the top of the list, where he was showered with attention again.

■   ■   ■

By the summer of 1966, Roger Caras had moved back to New York to revive the Polaris Productions office, intending to work with MGM on publicity strategies for *2001*. Meanwhile, Clarke had returned to Ceylon to "lick his wounds" following the Dell Delacorte fiasco.

Caras's move inevitably resulted in various expenses as he negotiated rents, hired a secretary, made deals with ad designers, and began taking influential people out to lunch. This, in turn, set up ongoing low-level kvetching from Kubrick, whose eye tended to be fixed on his bottom line when it wasn't looking through a viewfinder. Among other things, the director questioned the need to meet people for meals, and advised Caras to meet people *between* them if possible. In a letter written in mid-August, he also observed that many of Caras's meetings were with people that he and Polaris Productions had already established "fundamentally sound and mutual-interest relationships with." Therefore, he wrote, "I don't think bribery or any subtle form of it really enters." It's a measure of Kubrick's trust in his lieutenant, however, that he left the door open to just that: "If, in your opinion, it really does, then you're certainly free to use your judgment and take them anywhere you want."

To avoid misunderstanding, Kubrick praised Caras's results and said he was concerned mostly with how it looked to MGM, which was footing the bills. After four more months of expenses, however, the director felt he needed to compose another note to his VP. Written in longhand on personal stationery and labeled "Private" at the top in capital letters, it contained a brief summary of Stanley Kubrick's philosophy of the deal.

"You have to learn to always be shocked by the money people ask for," he wrote. "When they say the price, you have to turn pale and say incredulously, 'For *what*?' The $250 fee you agreed to pay for the ad roughs, and which I am weaseling on, is absurdly high."

> You have to learn to bargain in a respectable way. Various outs: I am not authorized to spend more than _____. I have to get OK on that price. Then—price not OK. Can offer _____. The best negotiating out in the existence is that your alternatives are zero. You are not authorized to go beyond _____. Etc. Etc. Then it doesn't make things unfriendly, the vague implication being if it were *you*, you would, but *they* won't. You have, I think, a Diner's Card, expense account, Monopoly money–oriented view of business expenses.

He concluded with "I know you look forward to these lectures and keep a notebook called *The Sacred Words of SK*, so I don't feel so bad about giving them. So better cost-effectiveness drive starts now—right? OK. OK. Spend it like it's your own. Thriftily, Stanley."

Following his return to Colombo Clarke's health went into steep decline, and by mid-December, he'd acquired a bad case of dysentery, which led in turn to dengue fever—a particularly debilitating tropical disease—and his collapse into bed. There he remained for the rest of the month, later describing the illness to Kubrick as "like flu raised to the power of 10." Before that, however, he'd produced the first three of a series of promotional magazine articles intended to foster awareness of *2001*. They would be paid for by MGM and were intended for publication just prior to the film's release. These unsigned puff pieces were a way Kubrick could both channel money in Clarke's direction and keep him on a leash.

As of early 1967, Scott Meredith's deal with Dell was still effectively undead and subject to resuscitation at any time, since the publisher had indicated a continuing interest, albeit with an ever-diminishing advance. In an effort to keep things that way, Kubrick affirmed periodically that he was working on manuscript revisions whenever possible. In addition to arranging payments for Clarke's articles, he also took care to raise the author's loan guarantees whenever asked, and he'd already tried to placate his collaborator by ceding whatever percentage of his advance covered the difference between what Clarke would have originally made and the money he would actually receive when the director eventually okayed the novel's publication.

In early January Kubrick sat down to respond to several letters from Clarke that had accumulated over the preceding weeks. After praising the article drafts and inquiring about his collaborator's health, he said he'd been overwhelmed by production issues and problems, and gave a progress report: "We are getting magnificent shots, but everything is like a 106 move chess game with two adjournments." He also shared in confidence that the premiere date would likely slip to October "at the earliest," writing, "the delays which have caused this have also drained enough of my time to cause further slippage on the novel."

I know you are greatly concerned about this, but this is one of those situations where there is no choice. I am certain it will all come out right in the end. I have no options now. I don't want the novel published in a form I consider unfinished. I am continuing my work on it, a little every night, though there are a couple nights a week I get home so exhausted I cannot think straight.

He went into detail concerning further loans, affirming he'd guarantee them even if he had to take money he'd invested himself and use it as collateral. "Don't forget that I am much more extended on these loans than you are because all you have to do is not be able to pay if they want the money, and I automatically must," he observed. And his mail concluded with the pointed observation that in his view, MGM had been more understanding concerning his delays than Clarke had.

As a closing thought: whatever unreliability there has been about the date of completion of the novel, it has been only a proportionate echo of the delay in completing the film. As you can imagine, there is a considerable amount of money involved in the film, too, and as many good reasons for people wanting it finished. The only difference has been that instead of continual pressure and oblique recriminations, there has been an objective understanding of the problem, something that would be greatly appreciated regarding the novel.

Despite oblique slaps offered for oblique recriminations, Clarke's loyalty to Kubrick was unshakable. A few weeks later, he received a letter from documentary director Thomas Craven, who'd been hired by MGM to make short promo films about *2001*. "I'm very distressed to learn of your financial headaches arising from Stanley's unwillingness to permit the *2001* publication," Craven wrote. He referenced the radiation belts encircling Earth: "I fear it is this kind of insensitivity to the needs of others that is responsible for the Van Allen belt of suspicion which surrounds Kubrick in the film world."

At this, Clarke wrote a decisive "No" in the margins. A week later, he answered Craven: "Actually, I do not agree with you that Stanley is insensitive to the needs of others—he is very sensitive, but his artistic integrity won't allow him to compromise. I have to admire this attitude even when it causes me great inconvenience!"

A few days later, he fed paper into his typewriter carriage again, this time addressing Kubrick. "Thought you'd be amused at reactions to *Jamis Bandu*," he wrote, referring to Wilson's Bond parody. "It seems a smash hit—special midnight shows being arranged, and *18* prints being screened at *19* cinemas (neat trick)."

For all the anxiety and financial strain it had caused, Clarke's investment in saving Wilson's film had seemingly paid off.

■   ■   ■

Throughout the production of *2001*, Kubrick's penchant for secrecy and his unwillingness to share footage—or even production stills—with MGM's publicity and promotions department raised the collective blood pressure of studio executives. As the director's New York representative, Caras inevitably received the brunt of the criticisms emanating from the studio's corporate headquarters on Sixth Avenue.

As early as December 1965, before shooting even began, Mort Segal, the assistant to MGM's top advertising and publicity man, Dan Terrell, was already scolding Caras about Kubrick's failure to send him various types of materials used to foster media awareness of forthcoming films. These included the script itself—which the director absolutely disallowed distribution of beyond a small circle, even as he revised it daily. By February 1966, with filming well under way, Terrell himself weighed in, stating that the studio was "pretty disturbed" by "Stanley's reticence"—in this case, his refusal to allow production stills to leave Borehamwood.

Kubrick's motivations were clear. He didn't know how long it would take before he finished his picture, and he didn't want other filmmakers stealing his set and spacecraft designs and putting out movies or TV shows based on them before *2001*'s release. He also wasn't entirely sure yet what his final story line would be, so he didn't want to convey clues as to its identity before it was fully formed. More generally, his instinctive position was simply to keep the project under wraps until released—something Clarke had certainly discovered.

MGM was pouring a considerable percentage of its total production budget into the project, however, and although studio president Robert O'Brien's default tendency was to indulge the impulses of directors he believed in, in principle, the studio wasn't used to working in this way. On October 10, 1966, matters came to their initial boiling point when Terrell heard, apparently spuriously, that a partner of Keir Dullea's agent had seen more than an hour of *2001* footage—material allegedly withheld from the studio's own top executives—and had found it to be merely "okay." Summoning Caras, he demanded more information and said that it would be nice if O'Brien would be allowed the same courtesy.

At this, Caras "played dumb"—as he reported in a cryptic thirty-seven-word telex briefly outlining Terrell's charge and starting with: "Confidential some noses out of joint at MGM." This triggered an immediate alarmed response from Kubrick. Clearly exasperated, he cited a catty Hollywood gossip columnist whose specialty was descriptions of film-world machinations.

> Please don't send me Mike Connolly pieces concerning important information like [the] partner of Keir's agent. Something [that] obviously required detail. Who told you this story. Where did they find out. What is the name of Keir's agent. What is the name of his partner. What do you mean you're playing dumb. Did anyone at MGM speak to you about this. Whose noses are supposed to be out of joint. For your info no one has seen a larger segment than O'Brien. When did this person supposedly see the segment. What other facts about this do you know. If you don't know all the answers but have a way to find out I want you to do this. I don't want something untrue to be a problem.

Caras soon sent a longer message. "I told him I knew nothing and very much doubt possibility of this," he wrote, asking if Kubrick wanted him to make some calls to try to get to the bottom of the story. The director responded with, "Please write me today full details and maximum length of your meeting with Terrell."

Choose your adjectives carefully . . . I don't like the theme that is developing of them complaining to you. Please detail exactly what your replies were . . . I think we are going to have to think of how to handle this type of play, but first I want a full report. Please be completely factual. If you were caught off guard by their hostility and may have seemed guilty, please say so, if it's true. Try to write a play of the meeting.

That night, Kubrick phoned O'Brien—who he always referred to affectionately as "Big Boy Bob," though not to his face—to explain that Terrell had received erroneous information, and the episode soon blew over. Clearly, however, the incident exposed an underlying tension, one that effectively blew up in Caras's face again less than a month later. Requested to stop by Terrell's office on November 3, he arrived

MGM's president and CEO.

to discover that Clark Ramsey, the head of the studio's Los Angeles division, was sitting in on the meeting. With the Dawn of Man sequence not yet shot at the time, 2001's delivery date had been pushed back to the second half of 1967—though Kubrick knew it would probably be even later. After a brief preamble concerning the sliding release date—which Caras deflected by saying the matter was best left to discussions between Kubrick and O'Brien—Terrell turned to his casus belli.

Loews Theatres, he said, had just been "highly critical of MGM for not taking the simplest of steps toward exploiting their tremendous investment" by having promotional materials for 2001 available for their lobbies. It was "incomprehensible" that they'd been denied drawings, paintings, or stills. Caras replied that the problem lay in Kubrick's concern over releasing the film's unique design elements so far in advance, which would immediately make them vulnerable to exploitation by competitors.

Writing later that day, he told Kubrick, "It was obvious that this was to be a session where pent-up grievances were to be aired." He played it cool, though, joking about being outnumbered but willing to do battle. At this, Terrell said he didn't want Caras to think he was being held personally responsible, but there were unresolved situations that needed to be aired. A jesting tone partially ameliorated the tension in the room. "Throughout the entire conversation, the humor was good, and no temper was displayed by anyone," Caras wrote. "However, it was apparent they are irked by some things."

Terrell said the reason Caras was being subjected to their complaints was because it had been made clear to them that Kubrick was too busy making his film to be disturbed or upset—something Caras understood immediately to mean that O'Brien had warned them to leave the director alone. This, then, was their workaround. By venting to Kubrick's VP, they were adhering to the letter but not the spirit of O'Brien's command, assuming their message would still be delivered.

Dropping any pretense at jokiness, Terrell mentioned that in 1965 David Lean had invited O'Brien, Ramsey, and himself to come look at Doctor Zhivago, to see for themselves what was being done with MGM's investment—which had resulted, he said, in their being so inspired that they'd launched one of the most successful ad campaigns in recent film history. "I and the others involved are professional enough and intelligent enough to look at uncut, unscored film and appreciate the filmmakers' art," said Terrell. "I'm certain from everything that has transpired that Stanley personally thinks my department stinks, and that I personally am not bright enough to work on his film and appreciate what he's doing."

At this, any remaining vestiges of banter left the room—though everyone's tone remained calm. "Stanley apparently doesn't want to leave any of my prerogatives to me, and wants everything done by people other than me," Terrell continued.

> I appreciate the fact that he wants to be advertising manager on his own film, but I don't see how he can find time to do that while in production, and if he waits until the end of production to take up his duties, it will be too late. I feel total frustration at being unable to make decisions on stills, stories, and displays. My department is as good as any in the industry, if not the best, and we're being denied the tools we need to function . . . I find it impossible to believe that there are no stills we can have. This is one more example of MGM being denied the opportunity to protect seven million dollars.

At this, Ramsey jumped in. "The whole thing is being treated like a six hundred thousand dollar art film," he observed. "Stanley's being given complete latitude in absolutely everything. That's fine when he's riding on a budget that can't hurt us, but when he's going all out with a budget like he has, MGM should have an opportunity to protect itself."

Terrell concluded with remarks about how the Columbia Pictures ad campaign for *Dr. Strangelove* had been "terrible" due to Kubrick's insistence on various ill-advised points concerning style. "I don't think he's always realistic in advertising matters," he said. "He's making decisions I should be making . . . I'm more frustrated than I've ever been in my professional life. But I feel constrained not to disturb Stanley with endless arguments and disagreements when he still has a film to make. But this is resulting in definite harm to the film's potential."

Near the end of his seven-page report to Kubrick—among the longest in his three years working on *2001*—Caras commented that Terrell's purpose had clearly been to have his views conveyed to the director, albeit indirectly. "Without a doubt, it was O'Brien who warned Dan about tangling with you while you were still in production," he wrote. "The unfortunate thing, Stanley, is that their position is not totally unrealistic or unreasonable. Many of the things they say are very difficult to argue against, and so I was in the position of keeping my mouth shut."

> One last thing, they were not unpleasant to me, and I cannot have any complaint about my treatment. I don't want this to sound [like] they were abusive or even unpleasant. They were neither of these, but they were firm and unequivocating.

Despite this placatory sign-off, on November 17 Kubrick sent a thunderous riposte directly to Terrell. "These remarks are so inaccurate and unfriendly and so surprising that I scarcely know how to respond to them," he wrote. "I am at a considerable loss to imagine why you should be so impatient to adopt the traditional advertising-department-versus-the-producer stance. Certainly nothing I have ever said or done could warrant such hostility."

He proceeded to answer the executive's complaints point by point. It was a measure of the seriousness with which he viewed Terrell's charges that, like Caras's letter, his response spanned seven single-spaced pages—exceedingly rare for the director, who usually conveyed his views in a page or less.

Just five weeks before *2001* was ultimately released in April 1968, *Variety* ran a story about the film's late-breaking ad campaign, which it implied could damage its prospects. The ads, the trade magazine observed, featured no stills—only illustrations. After going through MGM's explanations for why this was supposed to be a good strategy, the article floated a theory, directly below a paragraph quoting Dan Terrell: "Widely rumored, but denied by Metro, is the additional possibility that stills weren't used (and the campaign itself was delayed beyond the usual starting point for a film of comparable size) because of the necessity of getting approvals for everything from director Stanley Kubrick, who was busy putting the finishing touches on his picture."

Although the rumors' source wasn't given, it sounded strikingly similar to the comments Terrell had made more than a year before.

•   ■   •

During *2001*'s grueling two-year end game, Doug Trumbull's importance to the film increased inexorably. He'd arrived knowing how to use an airbrush but with little knowledge of photography. He would leave as one of the handful of truly innovative practitioners of optics-based visual effects in the world. Apart from being collaborative, filmmaking is largely about solving problems. Like Tony Masters, Trumbull had a flair for solutions, which he devised one after the other in rapid succession. He was intuitive, he was precise, and when the results weren't good, they were excellent, and when they weren't excellent, they were outstanding.

As with most of his key creatives, Kubrick deployed a similar originality-extraction technique on Trumbull that he'd used on Stuart Freeborn. When he saw something he liked, he reserved the praise and upped the game. And though Trumbull may have sometimes muttered under his breath, he always seemed to come up with something better. In response to *2001*'s unique strictures—and to Kubrick's ultimately self-serving yet fundamentally benign provision of resources and time—Trumbull's motto seemed to be "Do it right, then do it better, then do it all over again." In many respects, he became Kubrick's go-to effects guy.

By early 1967, the Star Gate sequence consisted of the weirdly compelling paint-thinner starscapes Kubrick had shot in Manhattan in 1965, and the fluorescent Purple Hearts landscapes that Birkin and Loftus had produced via helicopter and color separation masters. But the trajectory between Dave Bowman's arrival at Jupiter and his checking into Tony Masters's eerie, doorless hotel suite was like an extended space between refrains in jazz improvisation, with the intention being to evoke whatever abstract, nonrepresentational, space-time astonishments might be lurking Beyond the Infinite. Each of *2001*'s image instrumentalists could try stepping in to take turns creating what in effect were melodic solo lines, with ideas building upon ideas as augmented chords, extended chords, and multiple keys unfolded in quick succession—always visually speaking, of course. Like John Coltrane leaning into the mike after Miles Davis was done, Trumbull figured he'd take his turn.

While still at Graphic Films, he had become aware of a pioneering visual effects technique invented by Los Angeles animator John Whitney, in which automated motion control of cameras was combined with layers of rotating, image-bearing tables. Whitney's process created complex, variegated animated shorts. Best known for his title sequence for Alfred Hitchcock's *Vertigo*, made in collaboration with Hollywood graphic designer Saul Bass, Whitney had experimented with leaving the camera shutter open as the tables moved smoothly beneath the lens, sometimes in combination with strobe lights—a technique that produced kinetic, abstract, sometimes pointillist effects, usually on a flat plane. His system linked the camera and artwork mechanically via time-exposure control. He called his invention "slit scan." Interesting as historical artifacts, Whitney's work hasn't dated particularly well.

Although Kubrick was pleased with his Manhattan material and also the Purple Hearts footage, he didn't feel yet that the incipient Star Gate sequence was sufficiently variegated. So he called a meeting and challenged his effects team to come up with other ways of illustrating Dave Bowman's wild ride. As a result, various techniques had been tried, including the deployment of spinning mirrors and other tricks intended to create something interesting—but in Trumbull's opinion, "It was just all terrible." The director also revived his banana-oil-and-paint visual effects production technique at Borehamwood—though according to Con Pederson, the new footage mysteriously lacked the magical quality of the material shot in 1965, and very little was used.

Then one day Trumbull had a "kind of epiphany." He'd "remembered John Whitney and the streak photography thing." If augmented by forward camera movement and focus control, he realized, Whitney's technique could be adapted to produce a sensation of kinetic motion through space. Using the Polaroid bracket on his animation stand, he did a quick-and-dirty test of his idea, cutting a ten-inch slit from a piece of black card and mounting a piece of graphic art behind it. Leaving its shutter open, he slowly moved the Polaroid camera forward toward the slit, simultaneously keeping the image in tight focus.

The result was an abstract plane morphed "at a weird angle. And it was a corridor of light," he said. "One side of a corridor of light." The image confirmed his intuition that if he repeated the movement with slight camera offsets, he could mirror the effect in other parts of the frame and use it to create quasi-architectural forms. The next step would be to get the whole thing to move. If he accomplished *that*— and if the mirrored images were made to flow in opposite directions from the center of the frame—the final effect could be of glowing, two-sided passageways seemingly coming right at the viewer. Done right, it just might produce the kind of cosmic roller coaster effect they were seeking.

Trumbull swabbed the test Polaroid with fixative and speed walked down the hall, and Kubrick examined it while it was still drying as he explained his idea. "I think we can make the Star Gate this way," he concluded.

"Okay, I get it. What do you need?" Kubrick asked. He usually had difficulty envisioning an idea unless he saw a photograph—exactly what Doug had provided.

"He understood completely that it could be patterns, colors, lights, or anything," Trumbull remembered. "And it could be doubled for both sides . . . he got it instantaneously. I said, 'Well, I need Wally to build me this track, and I need to get these giant sheets of glass, and we're going to make this work.' And he said, 'Whatever you need. Whatever you need. Do it. It's approved.'"

Trumbull's breakthrough allowed him to build a substantial edifice—a city, really—on Whitney's groundwork. He devised new methodologies permitting an unprecedented filmic sensation: that of

An early Polaroid test of the slit scan concept.

vaulting forward through seemingly real and three-dimensional, yet entirely synthetic and abstract, space. With help from Veevers, he mounted a camera on a track and governed its motion with a worm screw—essentially a fifteen-foot bolt with no nut—attached to a motor governed by a selsyn controller. This was mechanically coupled to another controller governing the camera lens, which kept focus constant as the camera moved along a y-axis toward the photographic subject during each exposure.

From here, Trumbull worked with artist Roy Naisbitt of the animation department, who assembled various image sources into visually complex ten-foot-high translucent composite color images. About four times wider than they were tall, these collages were also mounted on a track, with their motion also governed by a worm screw—another long bolt, this one on an x-axis, oriented at 90 degrees to the camera's track. The result was a highly controlled, incremental way of filming flat, transparent artwork. When backlit by colored gels and pho-

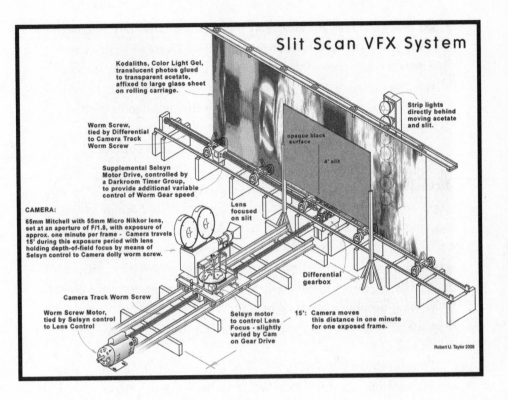

Slit Scan VFX System

Kodaliths, Color Light Gel, translucent photos glued to transparent acetate, affixed to large glass sheet on rolling carriage.

Strip lights directly behind moving acetate and slit.

Worm Screw, tied by Differential to Camera Track Worm Screw

opaque black surface

4' slit

Supplemental Selsyn Motor Drive, controlled by a Darkroom Timer Group, to provide additional variable control of Worm Gear speed

Lens focused on slit

CAMERA:

65mm Mitchell with 55mm Micro Nikkor lens, set at an aperture of F/1.8, with exposure of approx. one minute per frame - Camera travels 15' during this exposure period with lens holding depth-of-field focus by means of Selsyn control to Camera dolly worm screw.

Differential gearbox

Camera Track Worm Screw

Worm Screw Motor, tied by Selsyn control to Lens Control

Selsyn motor to control Lens Focus - slightly varied by Cam on Gear Drive

15': Camera moves this distance in one minute for one exposed frame.

Robert U. Taylor 2008

tographed frame by frame—each time with only a narrow slit of the artwork exposed—the images morphed into highly kinetic, vividly abstract sheets of light. Complex, multiplane structures could be created with additional camera passes, some of which lasted for days at a time. In 1976 Trumbull described the process to Don Shay and Jody Duncan of *Cinefex* magazine.

As a simple explanation of how slit scan works, I think everyone, at one time or another, has seen a time-exposure photograph of city streets at night, where you see nothing except red and white streaks of light caused by the fact that cars are moving while the shutter is open. You do not see the cars at any one point; instead, you see an accumulated exposure of cars. If, during that same photograph, all the cars were blinking their lights on and off rapidly, instead of having them continuously on, you would get a series of dots and dashes. If you expanded that and took not just a point of light but a bar of light—a fluorescent tube, for instance—and moved that toward the camera while the shutter was open, you would create a *plane* of exposure rather than a line. Then, by modulating that light—turning it on or off or changing patterns in front of it—you would create an accumulated exposure that could be fairly complicated in content.

Depending on the nature of the backing artwork, those changing patterns could manifest kinetic striations (for example, *Scientific American* diagrams that had been enlarged and colorized), or morphing, organic undulations (colorful reproductions of op art paintings), or spidery, strobing, jagged-edged abstractions (electron microscope images of plant stems or insect mandibles). Some image categories were chosen by Kubrick himself, as attested by a surviving set of index cards dating to April 1967 and filled with his scribbled notes ("Doug—also use op-art and electron art").

Trumbull's core innovation was to introduce that hurtling y-axis camera motion—an idea that Whitney hadn't explored. It gave his slit scan footage its visceral sense of transport; the viewer was seemingly vaulted down scintillating corridors of cosmic space and time. It's what "elevated the effect from a two-dimensional novelty to a three-dimensional jaw-dropper," as Shay and Duncan put it.

■ ■ ■

Trumbull still marvels at what he and his colleagues managed to achieve in Borehamwood from late 1966 through early 1968. "Every day that I was working on that movie, I felt that I was working on some very extraordinary, unusual event that I was contributing to and that was really important in some way," he said. "It was like going to church every day." He noted that with his slit scan process, Bowman's trajectory through space and time was itself the result of a breakthrough in understanding the relationship between those two elements within the context of film.

One observation he made concerning Kubrick's evolving thinking resonated in his own work for the next five decades. As he and Bryan Loftus continued presenting material to the director, Bowman's trip sequence gradually changed its personal pronoun from the cinematic equivalent of third person to first person. Their original strategy had been to show Bowman's pod during its journey through the Star Gate, with reverse angles revealing the astronaut's reactions along the way. "And that was part of the evolutionary transformation of Kubrick's directing," Trumbull observed. "A reverse shot or over-the-shoulder shot was not in keeping with allowing the audience to *become* that character . . . That was third person. He wanted first person."

In the end, Kubrick retained Bowman's reactions by deploying a handful of brief freeze-frames of him reacting with awe and horror to the extraordinary sights flashing by outside his window, and he eliminated establishing shots of the pod entirely. Dullea's face appears on the screen so rapidly it's almost subliminal, however, and it doesn't interfere with the audience's first-person sense of being along for the ride. We also see macro shots of his blinking eye, its cornea and pupil transformed by Loftus into shades of green, orange, and yellow, making it as abstract as the sights he was ostensibly viewing. This evolution of directorial emphasis, says Trumbull, turned the Star Gate into "seventeen minutes of uninterrupted experiential material."

It's hard to overestimate how important the Star Gate is to *2001*'s larger narrative arc. After two hours of perfectly realized photographic realism, it launched the film into a new realm of purely subjective audiovisual experience, much of it entirely abstract and nonrepresentational. Even today the sequence doesn't seem to have dated. For all their power, contemporary computer-generated images haven't really supplanted or superseded Trumbull's slit scan striations, Kubrick's Manhattan Project cosmological footage, and Loftus's Purple Hearts planetary landscapes. Viewed even with full knowledge of the visual effects breakthroughs of the past five decades, the Star Gate is as spectacular today as when *2001* was released in 1968. No small achievement.

Ruminating, in the year 2001, about a certain lysergic character to Bowman's cosmic trip, Arthur C. Clarke—an abstemious teetotaler content for a lifetime to enjoy the projections of his own imagination—

said he suspected some of its authors may have indulged themselves with certain mind-expanding substances while working on it. Asked about this, Trumbull was unequivocal.

"Were you guys ever, I mean, even rolling a reefer?"

"Never."

"You were too busy."

"Never, no, not, too busy."

"You were straight."

"Straight as an arrow, and so was Kubrick. And we talked about it."

Never, no, not, too busy.

Asked further about that conversation, Trumbull recalled a fragment, reminiscent of Kubrick's exchange with Polanski:

Doug: "Stanley, I've seen friends of mine lose their perspective. They've gone out on this LSD trip, and they come back, and the effects are completely worn off, you would think. But their perspective on reality has shifted. I've had conversations with people who say, 'Wow, this is such a trip,' because they're looking at reflections in their pepper shaker. There's something that gets skewed. I guess I'm really square, but I don't ever want to get screwed up like that. I don't want to lose my perspective. This is hard fucking work."

Stanley: "I agree with you completely. I will never touch the stuff because I will lose my ability as a filmmaker to stay focused. It'll go off, and I'm afraid."

I'm afraid.

Some have pointed to *2001*'s Star Gate as a kind of highbrow culmination of the psychedelia then coming to the fore in popular culture, and there may be something to that. But other comparisons could be to the abstract art produced by the avant-garde movements of the early and mid-twentieth century—and also to the exceptionally detailed, vividly colorful visions of cosmic phenomena that NASA's game-changing Hubble Space Telescope started beaming to Earth in the early 1990s.

Trumbull's slit scan abstractions took the static nonrepresentational art of the Russian avant-garde of the 1910s (Kazimir Malevich, Wassily Kandinsky) and the abstract expressionism of the forties and fifties (Mark Rothko, Jackson Pollock) and transposed it to the kinetic space-time of the moving picture. The result was a new form of visual experience. And Kubrick's morphing nebulae and expanding star clusters—his majestic vistas of stars being born and galaxies expanding within the infinity of deep space, achieved during weeks of labor hunched over malodorous tanks of paint thinner—provided a vivid cinematic preview of Hubble's otherwise unprecedented windows onto the cosmos.

It's a perennial chicken-and-egg story, the history of science and art. In this case, art got there first.

.　.　.

As pressure mounted on Kubrick to deliver his manifestly overdue and overbudget film, his fascination with methods of efficiency improvement broadened and deepened. Just as he'd asked his effects people to go beyond what they already knew and discover something they didn't, now he interrogated his own process for ways of improving *2001*'s workflow. Years later, Trumbull still found this "fascinating, because Stanley was, I think, such a genius-level person. He was just generally frustrated with the ineptitude of mortal men. And every weekend he would spend struggling to try to make things better or find some better way to organize things, or better systems, better communication or whatever."

The enforced weekend production breaks of England's unionized film industry were part of his problem—though they may have prevented him from killing himself, and possibly others, through sheer overwork. In any case, after stewing for many a weekend, the director would return on a Monday with a New Plan. One such new direction was announced during a particularly pressurized period in 1967 when Kubrick simultaneously juggled the requirements of his multicamera, multimodal, multimedia visual effects effort and the demands of the Dawn of Man—about as fecund a creative period as has ever occurred in film history. Standing in the screening room in front of his inner junta of image makers and incipient man-apes, Kubrick complained that people kept coming to him with wastefully lengthy explanations that really could be handled with fewer words. Adopting an erudite British accent, he launched into an effete set painter's soliloquy: "Stanley, you remember when we were on the set of the *bridge*, and we had this column that was kind of an off-purple kind of *color*, and you said something to the effect of, 'Really, it should be another *tone?*'"

Here he switched to exaggeratedly outer-borough New Yorkese: "Ya want blue paint?" And then back to the posh Brit: "Well, yes, Stanley, that's what I'm trying to say. I want blue paint." Kubrick reverted to his normal voice: "Okay, I don't want to *hear* about all the other crap. I just want to hear the blue paint part. And everybody on this production's going to start talking the same way, and we're going to use pidgin English. Don't talk to me unless you can say what you want to say in three or four words. It's not"—back into a British accent—"'Stanley can we put in a requisition for nails tomorrow because we're going to need them for . . . ' I don't want to *hear* all that shit! Do you need nails, or not? That's all I want to hear."

And from that day forward—at least for a week or two—everybody on the production conformed to this directive. If he needed more widgets to make his gadget grow up into the gizmo he knew it could, Trumbull would stop off at Kubrick's office and say, "Need more widgets." And so did everybody else on the production. Asked if they did so with a straight face, Trumbull responded, "Yeah, absolutely deadly serious. Absolutely deadly serious. It was okay. We loved Stanley. So it was not like a big issue. And we knew what his problem was. And he was frustrated, and we would try our best to oblige." The effort soon petered out, of course. "He wanted to streamline human communication and get rid of error, ambiguity, and misunderstandings," Trumbull observed. "People can't do that."

Another efficiency brainstorm stemmed from Kubrick's acquisition of a then hot new technology, the microcassette recorder, which he started using as an on-the-fly dictation device. "He would constantly be mumbling orders, or changes, or ideas into it," Trumbull remembered. "He became very enamored of this thing." When Trumbull came up to him, for example, and enunciated "Need new widgets" in perfect pidgin, out came Kubrick's trusty Norelco, and he'd say, "New widgets for Doug." After which, his secretary would transcribe a day's worth of pidgin from his peeps, yielding a slate of purchase orders. The result: order and productivity incarnate.

From there, though, Kubrick insisted everybody do the same—and provided a crateful of recorders to prime the pump. There was no secretarial staff, however, to do all the transcriptions. Everybody was supposed to do that himself, at the end of the day. This, too, devolved entropically within several weeks from ostensible efficiency enforcement into a slothful putting off of transcriptions—but not before certain screening room hijinks in front of their hawk-eyed, perfectionist director. "One day in dailies—you just knew you were going to go through this litany of rejection in dailies," Trumbull recalled. "So, three of us got together, and we had prerecorded statements in our recorders that we could play. The first one would be, 'I don't like it.' The other would be, 'I don't like it, either.' The third guy would say, 'Yes, I think we should reshoot it right away, don't you?'" He laughed. "We all knew he could handle it. We all knew we loved him, and we all knew we were doing a good job, and we were not failing, and we weren't embarrassed or anything. But there's a kind of camaraderie that came out of that."

Still, as tension continued to build and Kubrick's insistence on absolute perfection took its toll, some in the effects crew began anticipating the dailies with dread. In particular, it became a serious faux pas for anyone other than Kubrick himself to spot a flaw, thereby sending a colleague back to square one. In any case, the director always seemed to catch everything, and one day Trumbull and animation cameraman Jim Dixon had had enough. "On the one hand, there was pride for the quality of the work, of the few things that got through Kubrick's net," Trumbull said. "On the other was all of the strain and agony, toil, and work of doing it over, and over, and over again."

Tired of the "redon'ts," they bought a starting pistol from a local sports store with an extremely loud, stadium-ready blank loaded in its chamber. As the projector started rolling the next morning and the results of their labors were subject once more to the director's hypercritical gaze, they waited for the inevitable. When Kubrick commanded that one of Trumbull's effects sequences be repeated yet again, Dixon rose from his seat with a menacing air. "Who fucked up the shot this time?" he demanded. Trumbull stood. "You're embarrassing me," he said coolly. At this, Dixon pulled out the gun, aimed for Trumbull's chest, and fired. A deafening bang echoed in the enclosed space.

Trumbull spilled onto the floor of the darkened theater. Everyone leapt to their feet in horror. He clutched his chest. He stirred, groaning. He writhed, moaning. Finally, he sat up.

Kubrick started to laugh.

. . .

Much has been written about *2001*'s stately pace, with its spacecraft waltzing across the screen in dignified, measured orbital cadences and *Discovery*'s solemn progress to Jupiter unfolding with the unhurried regal tempo of an eighteenth-century monarchal procession leading inevitably to that Louis XIV hotel suite. It turns out, however, that the film's internal speedometer was limited by an oddly prosaic factor: the flicker factor of stars when filmed at twenty-four frames per second.

Attempts to represent *2001*'s starry backgrounds went through several phases, with initial tests involving drilling hundreds of holes into black metal sheeting and positioning lights behind them. This technique was soon judged wanting because as the lens tracked by, the stars either became elliptical due to camera motion, or their brilliance changed, or they twinkled—the latter a big no-no in the airless environment of space.

With drilled starscapes judged inadequate, another method was tried in which dots of paint were speckled across glass, with the glass then angled and lit from below. But glass being two-sided, each star ended up with a reflected double—another no-no. Finally, a decision was made to film all the stars on the animation stand, and Trumbull was recruited to airbrush random white specks of paint onto black backgrounds. These were then lit from above.

While this produced perfect star fields, it's also when the speed limit of *2001*'s filmic universe became apparent. Most of the shots looked best with those stars moving in one direction and velocity, while the foregrounded spacecraft, filmed separately, moved in another, usually contrary direction and speed. Having grasped this, Con Pederson worked to optimize the best stellar drift directions and speeds, and the same for the spacecraft. He determined, for example, that if *Discovery* was moving from the left to the right of the frame, and the stars were drifting from the top to the bottom, the result was a kind of floating sensation. After much trial and error, a kind of vocabulary of eight optimal drift directions and speed combinations was established.

It was at this stage that the realization sank in among *2001*'s effects crew that at the standard rate of twenty-four frames per second—which they couldn't evade, since all film projectors functioned at that speed—fast camera moves were going to be a problem. The issue was that in analog film projectors, each frame actually flickers twice, producing forty-eight flashes of light per twenty-four frames of image. And because of the inherent persistence of human vision, bright white objects such as stars seem to double. At still-faster star motion rates, they triple—an even worse strobing effect. "And so very early in the production, we hit a speed limit that we introduced," Trumbull recalled. "And that was the law of the movie, to never move faster than that speed so you didn't ever get double stars." The stately pulse of *2001*'s space sequences, then, was governed by the substrobe rate of their moving star fields.

Asked if this meant that *2001*'s waltz rhythm—its courtly choreography of machinery wheeling in approximation of three-quarters time—came in response to their self-imposed stellar speed limit, rather than the film's pace being edited to match such tracks as "The Blue Danube," Trumbull said, "Exactly. That's the story. The waltz rhythm followed the star speed." (This is not to suggest the tempo of Strauss's composition was altered—only that Kubrick's music choices sometimes came in response to a visual rhythm that he'd already determined.)

With this having been established, there remained the question of how to combine those slow-moving star fields with *2001*'s spacecraft miniatures. Kubrick and Veevers had improvised a screening room technique in which two projectors were used to ascertain which speeds and directions worked best together: one to project the star fields, the other the spacecraft. But with this kind of live mix, stars flowed visibly across *Discovery*'s exterior. Obviously, this couldn't be used in the final film, where it would instantly obliterate the audience's willing suspension of disbelief, and so an animation technique was devised to mask the stars. It required the production of rotoscope mattes, permitting the frame-by-frame removal of background stars wherever the spacecraft was present in the frame.

This was all hugely labor intensive, however, and a team of young people was soon brought in to what became known as the "blobbing room." There they spent innumerable hours meticulously tracing spacecraft outlines onto animation cels (the transparent sheets on which shapes or objects are drawn in analog animation), then blacking them in with paint. Their laborious outlining, and equally monotonous removal of individual stars, took the tedium of traditional handmade animation to a new level. Colin Cantwell used to drop by and check up on the "blobbers," who, he noticed, had "gotten pretty strange" under the relentless pressure of their star-killing duties.

The rotoscopers were in there in the dark all the time—just them and some work benches and rows of still projectors—and the most exciting thing they ever got to do was paint blobs of black paint. This was an effort of months and years, and so they were gradually going kind of bananas. Anytime the rest of us got bored or started to think *our* jobs were mundane, a perfect cure was to go and talk to the blobbers. They were always so glad to see somebody—*anybody*. All of them became great friends during the production—I'm sure as a means of survival . . . Most of them went from the finish of *2001* over to working on *Yellow Submarine*—and they were so happy not only to work in the light and look out the window but also to paint in colors. It was a reward for all they'd been through.

. . .

2001: A Space Odyssey was made during a period when President Lyndon Johnson was decisively stepping up American involvement in the Vietnam War, and Doug Trumbull had been allowed to leave the country in 1966 only after receiving permission from his draft board—which required, however, that he leave an address in case his service was required. One day he received a bundle of mail forwarded from home, with an ominously official-looking envelope bearing a blunt request: he was to show up for a physical in Los Angeles, after which he would be inducted. The date of the physical, however, was long past.

Shocked, Trumbull wrote back to inform the board that he hadn't gone AWOL but was simply working in England and thus receiving mail with a delay. He soon received an answer: in that case, he needed to go to the West Ruislip Air Force Base, just a short drive from Borehamwood, for his physical.

Trumbull had arrived in England with his wife, and they'd only recently had a baby girl. He was so deeply immersed in his work that he'd barely had time to register what was going on in the news, though the Vietnam War was almost as well covered on British television as it was in the United States. In fact, 1967 was the second-bloodiest year for American forces during the whole conflict, with more than eleven thousand US servicemen killed in action. They needed reinforcements. Clearly, Trumbull was to be one of them.

He considered his options. He was using a technology he'd built himself to produce what amounted to an entirely new cinematic vocabulary. It was the most creative and exciting period of his young life by far. There was no way he was going to interrupt his work if he could help it. He didn't tell Kubrick about the letter, but he did look into the techniques others had used to avoid service and soon discovered that his air base visit would consist of the physical, a written questionnaire and aptitude test, and an interview with a US Air Force doctor.

In the mid-1960s, homosexuality was a criminal offense in both the United States and the United Kingdom. It had driven one of the biggest heroes of the Second World War, British code breaker Alan Turing, to suicide following his conviction for "gross indecency" in 1952. Although Clarke spoke about it only to close friends, this criminalization was what had led him to settle in Ceylon, where such activity was tolerated. Certainly being gay was grounds for dismissal from the militaries of both countries—and also rejection at induction. "And so I decided I'm going to purport that I'm extremely bright, but I'm gay—bright and gay and married with kids," Trumbull recalled. "And an artist. I said, 'I think this might fly.' Because it was so confusing. That was my strategy."

On presenting himself at Ruislip Air Force Base, he got to the part in the questionnaire asking him to specify his sexual orientation, and wrote "Homosexual." Ushered into his doctor's interview, he sat down, leaned forward, and stared at the floor, avoiding all eye contact. "I'm being the most depressed person on Earth," he recalled. "Like I was depressed out of my skull. Like I might go for his throat."

"Are you gay?" the doctor asked.

"Yeah," Trumbull affirmed.

"Well, but you're married," the doctor observed.

"Yeah," said Trumbull, frowning at the linoleum.

"And you've got a daughter."

"Yeah."

"And how's your life?" the doctor inquired.

"It's okay," said Trumbull grimly.

A couple weeks later, he received his third official letter, this one informing him that he'd received a 4-F classification and was not acceptable for military service under established physical, mental, or moral standards.

He was so busy he barely had time to register the news.

■   ■   ■

Multiple competing memories of events in 1966 and 1967 by collaborators and witnesses record the origins of Kubrick's pathway to the unprecedented use of so-called drop needle music—previously recorded tracks, all classical—in *2001: A Space Odyssey*. No single story reflects the whole truth, but a kind of aggregate accuracy can be established.

The use of original musical scores was standard operating procedure in big-budget Hollywood productions, and in 1966 Kubrick signed composer Frank Cordell to score the film. Cordell, who'd had a series of UK hits with his own orchestra in the 1950s, had been doing film music since 1952. His most recent score was for the 1966 United Artists Cinerama production *Khartoum*, starring Charlton Heston and Laurence Olivier (and written, interestingly enough, by Robert Ardrey). After hiring Cordell, however, Kubrick inexplicably resisted showing him any material, instead insisting that he work from verbal descriptions alone. And when he wasn't avoiding meeting his composer altogether, he had difficulty articulating what he wanted—though he did find Gustav Mahler's Third Symphony, with its vocal section based on Friedrich Nietzsche's philosophical novel *Also sprach Zarathustra*, of great interest. Accordingly, Cordell spent a year writing variations on Mahler's Third.

He'd never tried to work in such conditions before, however, and eventually it got to him. "I know poor Frank had a nervous breakdown afterward, that I can tell you," said David de Wilde, first assistant film editor on *2001* and a close collaborator with Kubrick since *Dr. Strangelove*. "Because he was like this, you know"—de Wilde shook like a leaf. "Poor man. He came to see me . . . He was asking me, 'What's going on? When will I get to see the film? When will he talk to me?'"

> Then he did a [recording] session, but I would be careful what you say about this . . . and it was rather difficult. It was a difficult time for Frank, and his nerves were shot, and Stanley had that effect on people. I mean, how many people had nervous breakdowns on the film? Have you worked that out? A lot of people. Stanley used to put the fear of God into people, without doing anything.

De Wilde may have been referring to associate producer Victor Lyndon, who, after soldiering through more than two years at Borehamwood, left *2001* on his doctor's advice in February 1967, reportedly due to stress-induced anxiety brought about by the production's unrelenting demands. Wally Gentleman and Bob Cartwright also departed the production early in response to the director's methods. Whatever their ostensible reasons, Kubrick didn't forgive such bailouts. None of them received credit in the film.

Cordell's tenuous position can be explained in part by Jeremy Bernstein's November 1966 profile of Kubrick in the *New Yorker*, which revealed that although the director was still trying to decide what style of music to use, he'd settled at least momentarily on German composer Carl Orff "as a model for what he was after." Clarke also cited Mahler and Vaughan Williams as composers of interest to Kubrick—something confirmed by his playing Williams's Symphony no. 7 (*Antarctic Symphony*) while filming Dullea in close-up as he reacted to his wild ride through the Star Gate.

In the spring of 1966, Kubrick had contacted Orff directly to see if he'd agree to compose the score. Then seventy-one, Orff declined to take on a project of that magnitude, and Kubrick subsequently hired Cordell as a kind of also-ran. According to Bernstein, the director had "ruled out most of the ultramodern composers, like the practitioners of musique concrète, and electronic music in general (although he admires some of their work), because the music is so weird that it would dominate whatever scene it was played against." In the piece, Kubrick was described observing that "in general . . . film composition tends to lack originality, and a film about the future might be an ideal place for a really original score by a major composer."

Following Orff—and with Frank Cordell still struggling to understand what was required of him—Kubrick approached film composer Bernard Herrmann, who'd provided outstanding scores for *Citizen Kane* and many of Hitchcock's best-known films, including *Psycho* and *Vertigo*. But Herrmann, who'd refused to take on *Dr. Strangelove*, turned down this one as well—though not before offering to do the job for twice his normal fee. This provocation was probably a response to Kubrick's insistence that several existing tracks share the bill, something he'd also done with *Strangelove*, much to Herrmann's displeasure. The director also reportedly consulted composer Gerard Schurmann and conductor Philip Martell on music choices for *2001*, though neither was offered the score.

Kubrick's ultimate arrival at the use of prerecorded music, then, followed a highly circuitous path. It may have started on the last Friday in January 1966—specifically the twenty-eighth. With live-action production in full swing, the director was bracing himself for an incoming group of MGM VIPs, including studio president Robert O'Brien. Knowing he'd have to show them newly shot material the following week, he summoned Tony Frewin. "Listen, get a couple hundred pounds from the accounts department, petty cash, grab one of the unit car drivers, go to town, and get me a good selection of classical music, and also modern classical," he instructed. "None of that musique concrète shit."

Two hundred pounds in 1966 is worth £3,500 today—or nearly $5,000. Hardly believing his assignment, Frewin hastened to one of London's best record stores, Discurio, and essentially bought the store. LPs ran about £1 each those days, with even the most expensive Deutsche Grammophon offerings just over £2. As he drove back from Mayfair, he worried the police would stop them: his station wagon was so jammed with vinyl it sagged visibly.

Directed by Kubrick to haul the records to his private writing retreat off Stage 6, Frewin was then asked to come in the next morning, and they proceeded as follows: Kubrick, standing by the turntable, instructed Frewin to remove the discs from their sleeves and pass them to him—"and I don't want to waste time doing this; you've got to be quick"—and then he lowered the needle rapid-fire into innumerable grooves as they careened through hundreds of tracks. "We spent most of the morning, and then he had to do something else, and then came back midafternoon until the evening," said Frewin. "And then we did more on Monday and Tuesday."

That Friday, Kubrick and the top MGM brass repaired to one of the studio's "wonderful preview theaters," Frewin recalled—not the usual scruffy dailies room—where the director gave the projectionist several records. Clarke was also present for the screening, which took place on February 4. "He'd used Mendelssohn's *Midsummer Night's Dream* for the weightless scenes," he wrote—referring to the scherzo, over footage of the *Aries* and *Orion* spacecraft interiors—"and Vaughan Williams's *Antarctic Symphony* for the lunar sequences and Star Gate special effects, with stunning results." As David de Wilde recalls it, Mahler's Third was also played: "Mahler's music with that show reel was absolute bliss."

With a great deal of first-rate classical music now divided among Kubrick's private writing lair, his office in Building 53, and MGM's various screening facilities, the groundwork had been laid for the fortuitous introduction into the proceedings of Johann Strauss's 1866 waltz "The Blue Danube" (actually titled "By the Beautiful Blue Danube" in German). Andrew Birkin remembers distinctly an incident that August when reshoots of the tremulous space station were under way, and effects crew member Colin Brewer had trouble staying awake during the dailies. The man would snore, the snore would build, he'd get an elbow in the ribs, he'd stop—and then the sequence would repeat, ad infinitum.

Discerning that Brewer's cyclical snore-fest was having an adverse impact on morale, Birkin had the idea of playing some of the classical LPs in the projection booth as they watched the station turn—a legacy of Frewin's shopping spree months before. "And we're looking at the wheel, and on comes 'The Blue Danube,' and when the lights went up at the end, he turned to us and he said, 'Do you think it would be an act of genius or an act of folly to have that?'" Birkin remembered of Kubrick. "A rhetorical question. He was asking himself. And this pure piece of serendipity, as I would call it, happens then."

At that stage, it was still a full year before editing commenced, and by the time it did, Kubrick's position had solidified. "The Blue Danube" was then considered a kitschy, musty, nationalistic composition—the Austrian anthem in all but name—and in the fall of 1967, Kubrick played it over the space station scene and asked de Wilde what he thought. Used to his beloved Mahler, de Wilde said, "I don't like it."

"Well, you're overruled," said Kubrick, grinning.

Asked in 2017 about the Strauss composition, Kubrick's informal music advisor, his brother-in-law, Jan Harlan, responded:

Ray Lovejoy, the editor, told me the story how the editing crew were quite in despair when he played "The Blue Danube" again and again, and they realized that he was thinking of using this "old-fashioned" Viennese waltz. He decided to recut the whole sequence to the music, including the introduction—everything turning and turning. No cut, no fading out, the lot. I suppose Christiane supported him in his decision, while most people, including film critics, later thought he had lost his marbles. By now, Johann Strauss is "space music." He would have been pleasantly surprised.

Because the cutting rooms in Building 53 had thin walls, it soon became common knowledge among the Hawk Films contingent which music was being considered, and Colin Cantwell verified a kind of systemwide consternation at Kubrick's flirtation with "The Blue Danube." "Around the lot, there was a strange reaction when people were hearing that cut in as he was editing," Cantwell recalled, laughing. "There was this, 'Has Stanley gone round the bend?' 'Is he bonkers?' They were afraid that he'd become mentally deranged."

After the film's release, Kubrick himself detailed his thinking. "I wanted something that would express the beauty and the grace that space travel would have, especially when it reached a fairly routine level with no great danger involved," he told the *Toronto Telegram*. "I also wanted something that wouldn't sound too 'spacey' or futuristic. Incidentally, I listened to about twenty-five recordings of 'Blue Danube' before choosing the one for Deutsche Grammophon by Herbert von Karajan, the world's greatest conductor. It's the kind of music that can sound terribly banal, but at its best, it's still a magnificent thing."

· · ·

Well before editing began, everybody associated with the production knew to keep his or her ears open for usable compositions, and in the summer of 1967, Christiane Kubrick and the sculptor and ceramicist Charleen Pederson—Con Pederson's wife—got together regularly at the Kubricks' home at Abbots Mead, where they created alien life-forms in clay and other materials. "They were just slapping together all kinds of weird creatures for Stanley's benefit," Con remembered. "Then one day Christiane went in to see about getting some lunch, and just then this music on the BBC came on. And it was so weird, she thought, 'Boy, Stanley should hear this.'" Charleen ran to fetch Christiane, who was equally struck by the composition. It was a broadcast of "Requiem," by Hungarian composer György Ligeti, performed by the North German Radio Chorus and Symphony Orchestra.

Kubrick immediately sought a copy of the recording, but the BBC said it didn't have rights to lend it, instead offering to play it for him if he came to its studios. Too busy to make that pilgrimage, he then tried to reach Ligeti via Jan Harlan, who determined that the composer was on a trip and his Vienna housekeeper didn't know when he'd be back. Eventually Kubrick managed to hear the piece and was captivated by its power. MGM's production office proceeded to make a deal with the Mechanical Rights Protection Society to use it and several other Ligeti works as "background music" for a given price per minute—"a big mistake," according to Harlan, particularly since this had been done without the composer's knowledge. When Ligeti went to see the film, he was "enraged," wrote music critic Alex Ross. "In a letter recently uncovered by the German scholar Julia Heimerdinger, he called the film 'a piece of Hollywood shit.'"

Apart from the perceived disrespect of not having been paid commensurate with his contribution, or even contacted directly, Ligeti's anger was spurred in part by the fact that one of the four compositions Kubrick used, *Adventures*, had been altered without his permission. "The publisher realized that 'background use' was the wrong clearance and sent a stiff legal letter," wrote Harlan. "The matter was dealt with promptly by MGM in making an appropriate payment." Reports that the composer sued Kubrick for having his music distorted, and accepted an undisclosed sum in an out-of-court settlement, were denied by the composer in 1989—who said, however, that he had to hire a lawyer to negotiate with the studio. "MGM wrote me such nice letters," he said. "They [said] Ligeti should be happy, he is now famous in America."

Charleen Pederson, left, and Christiane
Kubrick with their alien sculptures.

Regardless of these maneuverings, Ligeti's extraordinary compositions proved central to *2001*'s impact—truly one of the film's definitive
features. Clearly influenced initially by his sense of shock and anger,
his position on the film changed with time. Then an obscure figure,
Ligeti would become known as among the most influential avant-
garde composers of the twentieth century, undoubtedly in part because the film did indeed bring him to the attention of millions of new
listeners. Kubrick ended up cutting more than a half hour of his work
into *2001*, including a three-minute prelude played before the curtains
rose during which his eerily meter- and melody-free orchestral composition *Atmospheres* gradually rose in the soundtrack. Like two other
pieces used in the film—*Lux Aeterna* (Eternal Light), a composition for
a sixteen-part mixed choir heard as the Moon Bus glided over stark
lunar terrain, and *Requiem*, a wordless keening by a twenty-part chorus
used over both the man-apes' and then the lunar astronauts' encoun-

ters with the monolith, then yet again when Bowman enters the Star Gate—*Atmospheres* seems to convey an overwhelming sense of both horror and sublimity.

Undoubtedly such generous helpings of Ligeti's masterpieces should have been cleared with the composer himself, however, and it's unclear why Kubrick didn't persist in the effort. Asked by *Die Welt* in the year 2001 if his music had been "correctly placed," Ligeti responded, "My music, in Kubrick's selection, fits these fantasies of speed and space well." Questioned more closely about his feelings concerning the film, he said he "found the way in which my music was used wonderful. It was less wonderful that I was neither asked nor paid."

One of the very rare artists capable of compounding sensations of awe, terror, and reverence in equal measure within his work, Ligeti had in effect met a kind of image-making equal in Kubrick, and though it was hardly a collaboration in the traditional sense, the conjoining of their respective talents yielded something more than the considerable sum of its parts—whether he liked it or not.

Luckily, in the end he did.

■ ■ ■

Among the first pieces of music that David de Wilde transferred to tape for Kubrick when he started work on *2001* in late 1965 was Richard Strauss's thunderous *Thus Spoke Zarathustra*, which like Mahler's Third, pays tribute to Nietzsche's eponymous book. At that time, Jan Harlan was living in Switzerland, and he remembers Kubrick writing and asking for "a big and great piece of majestic music that comes to an end quickly." Harlan immediately rifled through his extensive record collection and, on his next trip to London, "brought a lot of LPs. Bruckner, Wagner, Sibelius, Holst, etc., and the Strauss was part of the package. He loved already the title, and the reason for this I only understood much later when I saw the film."

As with the Mahler composition—only now with greater brevity and sonic punch—the prelude to Strauss's tone poem couldn't have better suited the film's theme. With its indelible passages describing mankind as a kind of rope-dancer balanced between ape and Ubermensch (Superman), Nietzsche's 1883 novel anticipates *2001*'s evolution-spanning trajectory with indelible specificity.

> I teach you the Superman. Man is something that should be overcome. What have you done to overcome him? All creatures hitherto have created something beyond themselves: and do you want to be the ebb of this great tide, and return to the animals rather than overcome man? What is the ape to men? A laughing-stock or a painful embarrassment. And just so shall man be to Superman: a laughing-stock or a painful embarrassment. You have made your way from worm to man, and much in you is still worm. Once you were apes, and even now man is more of an ape than any ape.

Like Ligeti's *Requiem*, Kubrick would use Strauss's opening fanfare three times in the film. It was, in fact, titled "Introduction, or Sunrise" in the composer's notes, after a chapter in the novel—another remarkable coincidence, because by the time Kubrick actually placed the music across the introductory sequence, it had become exactly that: a sunrise over the Moon and Earth.

In fact, evidence exists that *2001*'s spectacular opening shot wasn't originally intended for such prominent placement and probably ended up there because of its powerful resonance with Strauss's prelude. Called "Earth occultation zonk" by the effects crew, it was based on one of Wally Gentleman's sequences from *Universe*, in which the Sun was seen emerging from the *side* of both Earth and Moon. In several attempts to replicate the effect in 1966–67, Bruce Logan produced footage in which the camera panned across a photograph of the Moon (supplied by the Lick Observatory) to the Earth rising beyond, while the Sun—played by a quartz iodide light positioned behind a hole punched in a black aluminum sheet—simultaneously rose over both. But it had never looked quite right, and finally an exasperated Trumbull turned his head at a 90-degree angle and realized the Sun and Earth should rise *above* the lunar horizon—not emerge from its side, as before. The new orientation simply looked right. It solved the problem.★

While playing with Strauss's fanfare in editing, Kubrick realized subsequently that Logan's shot was potentially a perfect match, and sometime in the winter of 1967–68, he asked the animator to shoot it again, this time with opening titles synched to Strauss's brass and kettledrum crescendos: *Metro-Goldwyn-Mayer Presents, A Stanley Kubrick Production, 2001: A Space Odyssey.* Spotting Trumbull in the hallway, he brought him into the editing room for a reaction. "Is this going to work, or is this completely over the top?" he asked with a flicker of anxiety.

As Trumbull recalled it, Kubrick was worried it might seem like overreach. In the edit, Strauss's brass section resounded the moment the director's name appeared on the screen, and he was painfully aware that if it didn't work, he might look ridiculous. With the sound turned up, Trumbull—whose earlier intervention had oriented the shot properly to begin with—now watched its use in the credit sequence intently. "I think it's great," he said at the end, smiling. "I really like it. Go for it."

Remembering the moment years later, he said, "I was really honored to be asked."

★ Much the same issue confronted NASA on the release of the first Earthrise over the Moon picture, taken in December 1968 by the *Apollo 8* crew—the same year as *2001*'s premiere. The first release had the planet's poles in their classic north-south orientation, which meant the lunar horizon ran vertically up and down the picture. It looked wrong, however, and was later rereleased, now with the lunar limb running horizontally across the frame. This is the classic shot we know today.

CHAPTER TEN

SYMMETRY AND ABSTRACTION

AUGUST 1967–MARCH 1968

> *There are certain thematic ideas that are better felt than explained. It's better for the film to get into the subconscious, instead of being pigeonholed by the conscious mind in the form of specific verbal expositions.*

—STANLEY KUBRICK TO THE *TORONTO TELEGRAM*

With Con Pederson subsumed in the flight-control intricacies of the war room and Trumbull preoccupied with his complex visualization engines, *2001*'s animations were languishing during the last quarter of 1967 simply due to the overwork of its top effects people. After consulting with both, Kubrick decided to call in the reserves in the form of another Graphic Films veteran: animator Colin Cantwell. Upon arrival from LA in August, Cantwell, then thirty-five, found Kubrick and his team in "the final stages and trying to bring it together in the best possible way against incredible odds."

He discovered that two-thirds of the film's animations were at varying stages of incompleteness and soon organized twenty-four-hour shifts. As a canny student of film—and in particular, Ingmar Bergman— Cantwell also developed a productive dialogue with Kubrick that he was careful to disguise when at the studio to avoid jealousies or rivalries.* Like Trumbull, he realized quickly that while the director

---

* Many of Kubrick's closest collaborators had a similar feeling that they enjoyed exclusive private creative interactions with the director.

was clearly master of his vessel, true collaboration was possible. "He wasn't wasting any time, he was focused," he remembered of Kubrick during the production's last six months. "He was dedicated to what he was creating, he was listening for it, finding where it was, continually trying to build it to this complete thing. That film is what Stanley is. And the great thing was, the '*is*-ing' of it—the verb of doing it—we all got sucked into that. How can this be done, and the excellence that had to be there. Everything had to have that excellence, or it couldn't happen at all."

Many of the spacecraft visible in *2001* were actually animations of still photographs placed in their cosmic contexts after repeated passes through the animation stand—a technique in which such convincing details as live figures moving in their windows could be added. Even "live" shots of spacecraft models were sometimes combined with cunningly positioned stills: for example, the turning space station with the Pan Am "Orion" shuttle in hot pursuit, a sequence in which the shuttle was actually a receding still. The Moon Bus and other spacecraft were depicted in much the same way, after Brian Johnson shot large-format photographs of them for this purpose.

One of Cantwell's early suggestions influenced the film's look in subtle but meaningful ways. He discovered that as a holdover from storyboarding, a lot of the sequences had been conceived "fundamentally at right angles . . . and one of the things I wanted to do was break things loose from right angles." Although many of the individual components were "superbly usable"—for example, that rotating station—Cantwell saw better ways of depicting the shuttle's arrival there and sought to introduce more diverse camera angles, including two-thirds views.

When he explained his idea to Kubrick, however, the director didn't get it. So Cantwell cut pieces of cardboard into different shapes, making "a little kit . . . Just as a piece of rough art, it's one element and pivots and double pivots . . . you could rotate the stars and rotate the ship's trajectory independently so that you could picture in sequence, frame by frame, what would happen as the ship came up to the docking entry." After using it to demonstrate what he meant, he quickly received the director's green light. Cantwell's visual intuitions were excellent, and Kubrick soon came to appreciate them.

Cantwell realized quickly that while realism was important in *2001*, it was outranked by an ongoing quest for visual purity—for images powerful enough to elide verbal explanation and tell their own stories. "There's a level of abstraction all the way through that Stanley never lost the importance of," he recalled. "He was about images and their impact, and he kept being true to that, and it worked tremendously."

> The film is highly abstract, and that divergence between the words of the book and the experience of the film, I think, was at the core of what Stanley was trying to do. The further he got along in the process, the more he kept on focusing down to that core . . . Stanley was creating a set of experiences that would leap [the audience] past the near present to a further present, and some of the implications of all of this . . . They would feel that they really were experiencing it. There would be no question. It would not look like special effects. It would not feel like special effects.

In Cantwell's telling, during the last months of its production, the identity of *2001: A Space Odyssey* was being forged partly in its differences from the novel, which in critical ways was providing not something to emulate ("If you can describe it, I can film it," Clarke quoted Kubrick saying in 1964) but rather something to part from—and even push against.

It wasn't necessarily an interpretation Clarke would have disagreed with. In a telling echo of Cantwell's language, the author described his role at a press reception prior to the film's Los Angeles premiere in April 1968. "This is really Stanley Kubrick's movie," he said. "I acted as a first-stage booster and offered occasional guidance."

■　■　■

One of the transitions that hadn't been solved on Cantwell's arrival was Bowman's movement from the "real" space of Jupiter orbit to the hyperreality of the Star Gate. Early storyboards had conceived of an actual physical cut in one of Jupiter's moons as the entrance. Another concept had Bowman encountering a giant monolith orbiting the planet, and approaching it in his pod. After seeing his own reflection, he was supposed to reach out a pod arm to touch it—thus echoing the actions of William Sylvester and Dan Richter—only to discover that *it* was the Star Gate entrance. Neither approach had proven easily realizable on film, however, and they hadn't even looked right in drawings.

Another problem was how to convey what had caused the monolith to send its signal to Jupiter in the first place—thus setting the entire second half of the story in motion. Kubrick had already raised this as a topic of concern with Trumbull in their first substantive discussion, at the Tycho Magnetic Anomaly set in Shepperton two years before. In the script, the monolith was a solar-powered device and was supposed to be activated when exposed to sunlight for the first time in four million years. But how to convey this? It had been judged too difficult to accomplish at Shepperton and had been left until postproduction.

In late 1967 Stanley and Christiane were hosting regular weekend movie evenings at their home for the director's inner circle, and Cantwell took to staying afterward for a bite. During a series of bull sessions between Kubrick and Cantwell that fall, talk turned to Ingmar Bergman's use of symmetry, particularly in his 1960 film *The Virgin Spring*, which had particularly struck Cantwell because while most of its shots were asymmetric and highly composed, certain key scenes were almost perfectly balanced. This prompted Kubrick—already a Bergman devotee—to screen the film again, and the two of them discussed it further. According to Cantwell, they agreed that Bergman's symmetries served as symbolic markers. "We chatted, and I suggested that also was an opportunity in the same way Bergman had used it," Cantwell recalled. "[There] could be times where that symmetric departure would then serve as a nonverbal emphasis; a hitching post for people to link their experiences together at important pivot points."

Certainly Kubrick's taste for symmetry and his interest in Bergman predated his discussions with Colin Cantwell. But there's reason to believe the animator's contention that the film's eerie alignments of the

Sun, the Moon, and Jupiter's satellites—always arranged along either the vertical or horizontal axis of *2001*'s central totem, the monolith—were influenced by these late-night discussions. And they, in turn, were catalyzed by that epic opening Earthrise shot, which had predated Cantwell's arrival and already depicted an eclipse of the Sun by Earth as seen from the vantage point of the Moon.

With this sequence and their Bergman discussions in mind, Cantwell proceeded to produce two of *2001*'s most symbolically freighted shots. Filmed on the animation stand, they intentionally mirrored each other. The first was a seemingly low-angle view of the monolith, with one of Pierre Boulat's Namib Desert cloudscapes beyond and the Sun rising above the rectangular object. A crescent Moon rounded out the composition at the top. Kubrick would eventually edit this into the Dawn of Man sequence not once, but twice—first for about five seconds at the end of the man-apes' encounter with the monolith, and then later for a two-second flashback just as Dan Richter conceives to pick up a bone and start banging other skeletal remains with it.

The second composition was virtually identical, this time depicting the lunar monolith and the Sun in identical positions, only now with a crescent Earth above. While many in the audience may have missed it, in principle the latter shot took care of the problem Kubrick had discussed with Trumbull at Shepperton: namely, how to reveal that the monolith's radio beacon to Jupiter took place when the Sun hit it. And it was, in fact, edited into the film at the moment William Sylvester and the other lunar astronauts were seen recoiling from that evidently very powerful signal.

Cantwell's paired shots had a kind of tarot card quality, and were deceptively simple to make. In each case, the monolith was simply a piece of black artist's paper cut at steeply sloping angles and placed on top of large-format eight-by-ten-inch transparencies (in one case, of the Namib sky; in the other, a star field). The crescents Moon and Earth were added respectively in additional animation passes. As with the Earth occultation shot, the sun was a quartz halide light positioned behind a hole cut in black aluminum. Called "Upblock One" and "Upblock Two" in war room parlance, the shots provided a vivid illustration of how in film editing, context is everything. The live-action sequences that preceded each gave the audience a sense of the monolith's scale and importance. The second brief use of the shot in the Dawn of Man signaled the object's power over Moonwatcher's imagination as he gradually intuits his way to the use of a weapon. Although powerful as stand-alone images, Cantwell's symmetrical "pivot points" received their significance from their surroundings—including that opening Earth-occultation shot.

In the case of the Moonwatcher edit, they also provided Kubrick's realization of the promise implicit in his dismissal of Clarke's solution for the scene: the one where the monolith was seen manipulating the man-ape like "a puppet controlled by invisible strings." In June 1966 the director had responded, "The literal description of these tests seems completely wrong to me. It takes away all the magic." His solution now—essentially an attempt to *introduce* magic—was an example of how some sequences of *2001* functioned almost as a rejection of the novel's approach. (It was, however, also a filmic illustration of Clarke's famous third law: "Any sufficiently advanced technology is indistinguishable from magic." Here as elsewhere, Kubrick may at times have had a more intuitive grasp of Clarke's principles than even the author did. For his part, however, Clarke observed in 1969 that he and Kubrick had consciously "wanted to hint at magic, things that we could in principle not understand at this level of our development.")

That still left the transition from Jupiter orbit to the Star Gate as an unsolved problem. On arrival at Jupiter, Bowman was to leave *Discovery* in his pod and approach . . . something remarkable. But what? Here, too, Cantwell's pivot-point symmetries helped provide a solution. Trumbull had already worked out a derivation of his slit scan rig devoted to turning flat artwork representing Jupiter into a realistic, seemingly three-dimensional form. His Jupiter machine involved mounting dual projectors on a soundstage and linking them to an arc-shaped strip of reflective material. The strip, in effect, was a curving version of a slit. The projectors were loaded with two shots of Trumbull's airbrushed painting of Jupiter—one for each hemisphere. When the whole rig rotated—with the projectors, the reflective arc, and the camera all linked by selsyn motors—Trumbull's painting magically morphed into seemingly three-dimensional space as the turning crescent recreated it frame by frame. Today this would be called "texture mapping," and would be conducted digitally.

With a convincing rendition of Jupiter having been accomplished, Trumbull and Cantwell worked together to produce the planet's retinue of moons. Projecting Trumbull's airbrush renderings of them onto white spheres—a less sophisticated version of his Jupiter machine technique—they photographed each at various phases. Cantwell then used the animation camera to experiment with representations of the Jovian system. All of Jupiter's satellites orbit on the same plane, and in late October, Cantwell conceived of a view from the planet's equator. "Once again, we go for the symmetry at the key point," he said. Arranging the moons in a vertical line, he positioned Jupiter itself at the bottom. If alignments of the Sun, Earth, and Moon started *2001*, signaled the power of the monolith, and marked the beginning of its second act, this new celestial arrangement would anchor the opposite end of that axis, opening the film's metaphysical last chapter.

Having left a gap in his chain of Jovian moons, Cantwell proceeded to film a six-foot-long "baby" monolith turning against black velvet, where it seemed to disappear and reappear in deep space. "It was just lit and shot on a stage with a motor running it like a barbecue spit," he remembered. "The black block element was shot to fit a particular location on the screen, and the other elements were just designed as a storyboard." In his final arrangement, the rotating monolith was the horizontal bar of a cross, with a vertical lineup of five moons serving as its upright element, and Jupiter as the ground. A piece of neo-Christian symbolism had found its way into the proceedings, almost by osmosis.★

Because the lineup of moons was reminiscent of a slot machine window, Cantwell's celestial cross was soon dubbed "Fifty Free Games" in war room vernacular. Throughout his colleague's work on the shot,

---

★ Interestingly, the shots that Cantwell had identified as being of interest in Bergman's *The Virgin Spring* included a symmetrical composition reminiscent of Leonardo's *The Last Supper*, with Max von Sydow in the central Christ position.

Trumbull had been making a new kind of slit scan sequence that seemed to emerge out of darkness. Its initial, almost subliminal tendrils of light vaulted toward the viewer, rising to full kaleidoscopic pitch over a period of eight or ten seconds—an effect he heightened by his use of "a little accelerator motor to make the effect keep going faster and faster." The variable speed produced an effect not unlike the neurotransmitted uptick in stimulus characteristic of the psychedelic experience.

On seeing Cantwell's Fifty Free Games shot—which he called "brilliant!"—Trumbull had it loaded into one of two projectors in the screening room, put his new slit scan shot in another, and then he and Cantwell experimented with a kind of dual-projector live mix, arriving ultimately at an understanding that if the animation camera was made to tilt up from Cantwell's Jovian crucifix to the blackness above, Trumbull's new slit scan shot could take over, effectively throwing the Star Gate wide open.

They'd figured out how to take David Bowman "Beyond the Infinite."

. . .

Throughout the fall and winter of 1967, Kubrick pushed himself to the brink of physical and mental endurance as he tried to finalize 2001's music, sound effects, narration, documentary prelude, and credit sequence. Cantwell estimated that during his tenure at Borehamwood, the director "probably worked close to a hundred hours a week, plus or minus, every week. How long he'd been doing that, I don't know, but that's a kind of obsession close to insanity. It's creating something that hasn't been in this world." He compared what they were doing to "vision hacking."

Until quite late in editing, Kubrick evidently still intended to show Roger Caras's documentary prelude featuring interviews with world scientists discussing themes such as extraterrestrial intelligence and interstellar flight. He was also in a constant dialogue with Clarke concerning the film's expository voice-over narration.

Having witnessed the Dawn of Man being shot, Cantwell was surprised and disturbed to discover some of these intentions. During one late-night discussion, the director also revealed his plan to incorporate the woodcut designs of Uruguayan American artist Antonio Frasconi for *2001*'s credit sequence. A master of this printmaking genre, Frasconi was the author of numerous illustrated books. His most recent, *The Cantilever Rainbow*, had featured stylized depictions of solar eclipses and other primitivist depictions of the Sun. Their handmade, retro quality didn't sit right with Cantwell, however. "The lead-in titles were to be over those backgrounds and would have title art superimposed over these almost childlike graphics," he recalled. "That seemed so discordant compared with the Dawn of Man that it bothered me a good bit."

Another thing that struck him as problematic was Caras's interviews. Cantwell seemed to have found diplomatic, low-key ways of transmitting his views to the director. Kubrick's process included a fairly frequent monitoring of his close collaborator's opinions. In effect, their reactions provided navigational markers, helping him understand how entrenched his own positions might be. Cantwell may have intuited that rather than volunteering his opinion—something which didn't necessarily go down well—he should simply wait to be asked. In any case, when his views were solicited, he "recommended to Stanley, the Suns of Frasconi's shouldn't be ahead of the Dawn of Man, and that the wordlessness of the Dawn of Man is exceedingly important," he remembered. "The purity of the start had to not have the scientist's verbal prologue in the wrong half of the brain."

This distinction between verbal and visual, the left and right cerebral hemispheres, colored his recollections of Clarke's visits to Borehamwood as well. "He would come in with a new narration to explain a whole section of the film that he had seen last time," Cantwell said. "He thought that about three or four or five minutes' worth of narration would take care of the uncertainties and puzzlement these scenes would create." Kubrick would then screen new material for his collaborator, and each time, "Stanley would have taken out more of the dialogue, more of the thing was being handled nonverbally . . . So this polarity would be clearly there between them."

When Clarke registered his worries over what he saw as the film's growing opacity, Kubrick effectively steered him back to the novel, saying, "Don't worry about it, Arthur, just put it all in exactly as you'd like to have it. Make the story clear in any way you want, it's your book." At this, Clarke would seem "anxious or subtly frustrated," Cantwell observed. "Arthur was very subtle in what he was expressing, very polite."

As their paths diverged during the last six months of *2001*'s production, Kubrick's aesthetic differences with the novel—and psychological distance from it—appeared to be growing.

. . .

When he started editing in mid-October, however, Kubrick gave every indication of using Clarke's narration, and he'd spent much of 1967 looking for a good voice. As early as February, he'd asked Caras to contact Alistair Cooke, the *Guardian* journalist and BBC radio commentator, to see if he'd consider auditioning. By July, Caras had left for another job, and Kubrick asked his replacement, Benn Reyes, to help him find a voice similar to Canadian actor Douglas Rain—narrator of *Universe*, the film already so influential in *2001*'s look. Rain was then unavailable, but after almost a hundred screenings of *Universe*, Kubrick seemingly couldn't get his voice out of his head. "I would describe the quality as being sincere, intelligent, disarming, the intelligent friend next door, the Winston Hibler/Walt Disney approach," he wrote Reyes. "Voice is neither patronizing, nor is it intimidating, nor is it pompous, overly dramatic or actorish. Despite this, it is interesting."

Throughout his search, Kubrick used Hibler, who'd narrated many Disney films, as a touchstone. By September, he was negotiating with Rain directly, and he wrote New York radio producer Floyd Peterson, who'd been helping him record auditions, to tell him he thought the actor's delivery was "perfect, he's got just the right amount of the Winston Hibler, the intelligent friend next door quality, with great sincerity, and yet, I think, an arresting quality." He asked Peterson not to approach Rain directly, however, as it might cause him to up his price—ending with "Remember this point, it's important."

In early August 1966, Kubrick had brought Oscar-winning American character actor Martin Balsam into the recording studio to play HAL. Although he initially found Balsam's rendition "wonderful," he realized gradually that he'd allowed him to make the computer sound too emotional. Having succeeded in luring Rain to read Clarke's narration, at some point in the fall of 1967, he decided the actor should play HAL instead—and temporarily tabled the narrator question.

Rain flew in to London in late November and recorded HAL's voice in only two days, most likely December 1 and 2. Much to his own dissatisfaction, these recording sessions, totaling only eight and a half hours of work, would become a definitive moment in his career. As an outstanding stage actor, Rain's faceless channeling of *2001*'s computer wasn't necessarily what he had in mind as the role of a lifetime. According to Jerome Agel, the editor and mastermind of the innovative, image-heavy 1970 paperback *The Making of Kubrick's 2001*, the director had hoped Rain would give HAL an "unctuous, patronizing, neuter quality." In this, he wasn't disappointed, though he found ways to help the impression along in postproduction. For his part, Rain found Kubrick to be "a charming man, most courteous to work with. He was a bit secretive about the film. I never saw the finished script; never saw a foot of the shooting." He would play HAL in a vacuum.

The recording sessions gave Kubrick a last chance to refine the computer's character, and he spent the second half of November rewriting key passages. Some of HAL's most memorable lines emerged just before Rain's arrival. One in particular—his response to a question from a BBC journalist—seemingly does double duty as a description of the director's experience over the previous four years: "I am constantly occupied. I am putting myself to the fullest possible use, which is all I think that any conscious entity can ever hope to do." Another set of lines replete with layers of Kubrickian irony seemingly encompasses all of history's disasters. Asked by Bowman to account for his mistaken fault prediction concerning *Discovery*'s antenna guidance unit, HAL replies, "Well, I don't think there is any question about it. It can only be attributable to human error. This sort of thing has cropped up before, and it has always been due to human error."

During the session, Kubrick sat four feet away from the actor, and they went through the script line by line, with the director making small revisions as they proceeded. According to one account, Rain's bare feet were on a pillow throughout, "in order to maintain the required relaxed tone." Occasionally Kubrick intervened, asking the actor to speed up his delivery, act "a little more concerned," or make it "more matter of fact." As with Dullea's responses the year before, quite a few of Rain's Brain Room scene lines would later be lost in editing, including many of HAL's increasingly desperate entreaties. One in particular that didn't make it into the final cut showed multiple penciled revisions by Kubrick, finally becoming Rain saying "I'm sorry about everything. I want you to believe that."

Kubrick could only have been satisfied with how Rain handled the scene where HAL seemingly wanted to confess his knowledge of their mission's true purpose—another part of the session transcript in which various penciled revisions are visible. The actor managed to transmit the computer's sense of divided loyalties as he asked Bowman if he minded being asked a personal question. Hearing "not at all," HAL launched into several observations about events he found "difficult to define" and "difficult to put out of" his mind. "Perhaps I'm only projecting my own concern about it," he said, observing that he'd "never completely freed myself from the suspicion that there are some extremely odd things about this mission." Prompted by Bowman's noncommittal responses, HAL elaborated, citing rumors of "something being dug up on the Moon," and "the melodramatic touch" of the hibernating astronauts having been trained separately from the waking ones and brought on board already in hibernation.

Throughout this sequence, Rain's performance gave the sense that all Bowman really needed to do was give him an excuse, and he'd readily spill the beans about what he knew. Instead, Bowman said, "You're working up your crew psychology report" (a statement, interestingly enough, betraying the inherent assumption that the only crew psychology that could possibly be in question was human). At this HAL skipped a beat, and then retreated, affirming, "Of course I am." (A pause that Kubrick could have inserted in editing.) Immediately thereafter, the computer predicted the failure of *Discovery*'s antenna guidance unit, and the moment had passed. He'd told two overt lies—that the unit would fail and that he was indeed working up such a report—and after that was on a murderous trajectory.

Kubrick had Rain sing, hum, and speak "Daisy Bell (A Bicycle Built for Two)" fifty-one times at various pitches and speeds, sometimes with uneven tempos, sometimes in a mix of hummed and spoken words, and once as a spoken-word piece enunciated in a monotone. In one take, they methodically went through the song five times in five keys: E, G, B, D, and F. Harry Dacre's little ditty would be the yardstick by which HAL's final disintegration would be measured, and the director was determined to give himself every possible interpretive variation.

After the session, Kubrick felt Rain's line readings and singing were excellent but the former still required an extra measure of phlegmatic placidity. Using a new analog recording device called an "Eltro Information Rate Changer," he slowed down the tape speed by between 15 percent and 20 percent without changing the pitch of the actor's voice. In two passes over the device, Kubrick also modified his chosen take of HAL's swan song, this time gradually dropping Rain's pitch almost to zero as he sang "Daisy Bell"—all without changing the tape speed or song length. The second pass did stretch the length, but not as much as the pitch seemingly indicated. The net result was an eerily clear sonic portrait of HAL's dissolution as his consciousness gradually declined to zero.

It's no surprise that Kubrick didn't use his shot of the computer's eye going dark. The sound design did it for him.

Not long after, Cantwell came to see Kubrick in his office and was steered toward the editing rooms in Building 53. Hearing through the flimsy prefab walls that the director was working, he chose not to interrupt and waited outside. With the sound rolling across one head of the Moviola film editor and the picture on the other, Kubrick occasionally rewound back to the middle of the scene and replayed it, gauging its effect. Standing in the drafty hallway, Cantwell heard HAL pleading for his life: "I'm afraid, Dave . . . My mind is going." As he listened, the computer regressed to its first lesson and, with Bowman's encouragement, launched into his song. When Rain arrived at the line

"I'm half crazy, all for the love of you," the scene's cumulative power hit Cantwell, and he discovered to his surprise that tears were streaming down his face.

When he finally entered the room, his face was still wet, and he was unable to speak. Quietly and without comment, Kubrick ran the entire scene for him.

■   ■   ■

Apart from his use of classical music, Kubrick's sound design in *2001* is rarely commented on, though it's no less innovative than the rest of the project. In particular, entire sections of the film have nothing but breathing and a muted feed of air as Bowman and Poole conduct their spacewalks. A similar respiratory soundscape is heard during Bowman's methodical deprogramming of HAL and on Bowman's arrival in the hotel room.

Like the heartbeats sometimes cut into horror films at moments of maximum tension, only with a bit more subtlety, these breaths give the audience a subjective sense of shared humanity. Without necessarily being conscious of it, we monitor Bowman's emotional state by the tempo of his breaths. As virtually the only sound heard during the space walks, this element of *2001*'s sound design also provides a sonic picture of the human organism suspended in the immensity of interplanetary space. Then, just as we get used to his rhythmic, oxygen-fed susurrus, Kubrick introduces another bravura element as Frank Poole struggles to reconnect his severed breathing hose. He does so not by addition but subtraction—definitively breaking the umbilical cord to Earth's atmosphere by cutting the sound to absolute zero. Finally, we see Poole drifting lifeless in the vacuum, a receding figure lost in a star-spangled infinity. By willfully ignoring the implicit sound-design imperative to always keep *something* on the soundtrack, even if it's just ambient "room tone"—by cutting to nothing, rather than something— Kubrick underlines the finality of what's just happened.

Similarly, back on board *Discovery*, sound plays a key role in the weirdly disembodied erasure of the deep-frozen hibernators—perhaps the strangest mass murder ever committed to film. As Bowman flies off to retrieve Poole's drifting body, the medical readouts gradually flatline, intercut with blinking warning messages such as "Computer Malfunction" and "Life Functions Critical" as a panoply of warning beeps and rapid electronic pulses synched to these messages paints a stark sonic picture of incipient emergency. Finally, the sounds stop dead at "Life Functions Terminated"—much as all sound had ceased with Poole's last breath. And yet apart from the messages transmitted by their EKG and heart-rate indicators, nothing untoward appears to have happened to the entombed astronauts. Their transition from life to death seems entirely virtual. As French composer and film theorist Michel Chion has observed, "Death is inflicted like an act outside time. Nothing has changed."

Another kind of execution unfolds during the Brain Room scene, during which Bowman's arrhythmic breathing is the only response HAL hears for most of his final minutes of consciousness, something accentuating the stark difference between living and synthetic forms of life. The intermittent respiration on 2001's soundtrack for the entire last half of the film also helps enforce the first-person effect that was one of Kubrick's overtly stated goals. The breather is meant to be somebody—a character in a film—but also an everybody: all of us in the audience. It's what makes the sudden cut to no sound at Poole's death so visceral. It's the presence of an absence, arrived at with guillotine speed.

That October, David de Wilde took a young assistant editor, John Grover, to meet Kubrick for the first time. With editing now under way, various soundtrack elements needed producing, and Grover had been delegated to supervise that day's recording. He was nervous. Upon arriving at Kubrick's office door in Building 53, he was startled to see the director leaning back behind his desk, a nasal spray inserted in each nostril. This unusual sight failed to faze de Wilde, however. "I was used to Stanley, so it didn't surprise me, and he obviously had a cold and was clearing his head, for the recording."

On seeing his visitors, Kubrick coolly extracted the sprays, picked up the helmet parked on his desk, and the three of them proceeded to Theater 1, which functioned as a large recording studio. There de

Wilde introduced him to J. B. Smith, the studio's chief dubbing mixer. With Grover's help, they ran a microphone from the soundboard in the theater and positioned it next to Kubrick's face, after which he lowered the helmet onto his shoulders. Then time passed, and nothing much happened. Decongested, ready to respirate, the director was breathing as usual—but it wasn't being recorded. De Wilde watched him with amusement. He'd never seen Kubrick's bearded visage in one of *2001*'s helmets before: the creator, now part of his creation. The soundproofed theater was deathly silent. After an attempt to ask what was going on via the microphone, "Stanley lifted the helmet up and asked me what was the holdup," de Wilde recalled. "My answer was that J. B. was deaf. He was not amused at the thought of a deaf sound mixer."

Finally, Smith was ready to roll, Kubrick lowered the helmet, and the small team recorded about a half hour of him breathing. Later, *2001*'s sound editor Winston Ryder looped the tape for use in editing.

In his novel *A Portrait of the Artist as a Young Man*, James Joyce had his hero, Stephen Dedalus, observe, "The artist, like the God of creation, remains within or behind or beyond or above his handiwork, invisible, refined out of existence, indifferent, paring his fingernails." As an example of his own handiwork, Stanley Kubrick's film *2001: A Space Odyssey* bears evidence of his own life within it, a small segment of his human soundtrack on Planet Earth from July 26, 1928, to March 7, 1999.

It was also, undeniably, a form of acting.

■   ■   ■

One of the more macabre documents to survive from *2001*'s years of production was a cable, written on May 31, 1966, from Roger Caras to General Biological Supply House in Chicago. "Please advise return cable availability preserved human embryos indicating stage of development price delays etc. Also availability best quality models of same," it reads. A cable from Ivor Powell to Caras had preceded it, requesting original prints of Lennart Nilsson's sensational 1965 embryo photographs in *Life*.

A terse communiqué soon ricocheted back from Chicago: "Cannot offer quotations or supply hum embryos or model sorry." With no genuine terminated fetuses forthcoming, by the fall of 1967, Kubrick had decided to produce the apparitional "Star Child" of the film's final sequence in Borehamwood, and he recruited a talented young sculptor, Liz Moore, for the task. Moore, who'd helped Stuart Freeborn with his man-ape costumes, had already made something of a name for herself as a student by sculpting clay busts of the Beatles. That summer she produced a clay rendition of a human embryo with features eerily similar to Keir Dullea. As per Kubrick's requirements, it had an abnormally large head, meant to signify humanity's next evolutionary stage. "Originally it was going to be much more complex, mechanically, with arms and fingers that moved," remembered Brian Johnson. "But then Stanley had the idea of surrounding it in a cocoon of light, and in the end decided that all he really wanted to happen was for its eyes to move."

After it had been molded and reproduced as a hollow resin-coated flesh-toned figurine about two and a half feet tall, Johnson set to work on the sculpture, inserting glass eyes into the head through a removable skullcap. Connecting the eyes to small rods, he motorized them by attaching them to selsyn motor-controlled bearings—2001's ubiquitous motion-control tool. "If you look at automatons that were made in the eighteenth and nineteenth centuries, it's the same sort of principle," he observed. Johnson's motors governed the fetus's eerie lateral gaze.

In early November they positioned the Star Child in the studio, surrounded it with black velvet, and did a series of camera passes. The footage was far too crisp, however, and required serious diffusion, so they deployed another of the film's visual effects secret weapons. It had been "provided by Geoff Unsworth, and he called it 'prewar gauze,'" Trumbull recalled. "He had this very limited secret cache of prewar gauze, which was, like, 1938 silk stockings that created the most beautiful glow, which was used on the Star Child and a lot of other shots in the movie." Rumor had it the material originally came from Marlene Dietrich's pantyhose drawer, but in any case, because a given percentage of the light passed straight through the material from the sculpture to the lens, a beautifully atmospheric diffusion could be produced without reducing the underlying sharpness of the subject. Ultimately all the slit scan and Jupiter footage was reshot through Unsworth's precious prewar gauze, producing the unearthly glow much in evidence during the film's final half hour.

Moore's sculpture was filmed again, now through about fifteen layers of gauze, Trumbull recalled, "with about forty thousand watts of backlight—something like four big arc lights to rim-light it, and [we] got this tremendous, overexposed glowing effect . . . Stanley filmed a number of different moves on the Star Child—shots of it entering frame and sliding through frame and so on. Then I airbrushed the envelope that surrounded it onto a piece of glossy black paper, which was photographed on the animation stand and matched in movement to the model, also with a lot of gauze and overexposure."

Of these camera passes, only two were used, plus a pair of stills, with the latter fused via animation stand wizardry to different views of Masters's Louis XIV hotel room deathbed—now evidently the site of a resurrection—then wreathed in airbrushed nimbuses of light. One of the two live-action takes, the last shot seen in the film, revealed the embryonic child gradually turning, its eyes taking in the Earth and then panning farther still, breaking the theater's fourth wall as its gaze swept across the audience. Accompanied by the triumphant fanfare of *Thus Spoke Zarathustra,* it was one of the twentieth century's indelibly powerful cinematic moments.

Cantwell's celestial alignment, Dullea's last supper, and, finally, the death and resurrection of *Discovery*'s surviving crew member all seemingly conspired to put *2001*'s audiences into a theological frame of mind. But even without the surrounding context, Moore's sculpture with its autonomous eyes evidently had an extraordinary, even supernatural presence. The powerful studio lights illuminating the Star Child gave it an otherworldly radiance against the black velvet that surrounded it. The flesh-toned skullcap permitting access to its inner mechanism had been joined seamlessly to the rest of its head with a waxy fixative. Long exposure times per frame were required to achieve the partially overexposed effect Kubrick wanted. Some of those camera passes took eight hours or more to accomplish.

Only one unit camera operator was required to monitor this lengthy process, and he didn't give it his undivided attention. During one long filming day that winter, the heat of the powerful arc lights melted the sculpture's fixative, which gradually spilled down inside its hollow skull, finally emerging at the corners of its eyes as perfectly formed drops. And on one of his periodic checks, the cameraman was amazed to see the Star Child weeping real tears.

Shocked to the core by the supernatural sight, "the operator ran away screaming his lungs out," recalled Daisy Lange, Harry's wife. "Now that I think about it, maybe he was a devout Catholic."

. ■ .

Throughout the fall and winter of 1967–68, Kubrick's overcommitted collaborators conducted a last push to produce credible aliens worthy of their extraordinary film. If deemed successful, they would be seen at some stage during Bowman's vault through Star Gate, or perhaps during the hotel room scene. This was never quite clear, and Kubrick may not have known himself. Just as the prehistoric protohumans at the Dawn of Man couldn't look like men stuffed into ape suits, *2001*'s extraterrestrials couldn't look like that dreaded cliché, the Little Green Men. In one of the production's more bizarre episodes, in September 1967 Kubrick asked Stuart Freeborn to transform Dan Richter into an eerily pointillist creature soon dubbed "Polka Dot Man."

Discussions between Stanley and Christiane concerning what extra-terrestrials might look like were ongoing throughout the production, and led eventually to her late-breaking effort with Charleen Pederson to create them in sculptural form. "I said, 'Have you ever seen a good alien in a film?'" she remembered. "'No.' 'Can you make one up?' 'No.' I did hundreds of drawings of aliens. They'd look like . . . draw-ings of aliens. Ugh!"

Their debate was typically exhaustive. "You can go way out and go so far as to make it into bacteria . . . or a veil of gas," said Christiane. At this, she remembered her husband replying, "Yeah, all right, but that's not interesting. We don't want to watch a veil of gas." And continuing: "That's one interesting thing about aliens. We can't satisfy our imagination, and there's nothing that would [make you] think, 'Oh yeah, it's probably like that.' No, nothing comes to mind that would astonish you beyond everything. Okay, maybe it's on a planet where the gravity is so much that anything can only exist in a gazillionth of a millimeter of gas. Or it's a bird that flies through the air like a stone and smashes everything up—right, okay—or gases or chemicals or bacteria or . . . You know, you go through certain kinds of cells, certain kinds of brain waves that form things."

He worked the problem rhetorically first, in other words. "He said, 'I can't come up with anything that wouldn't bore me to death,'" Christiane recalled. "And the minute you name it, you're already bored, the minute you draw it, the minute you . . . I mean, you can do it comical, that's funny; you can do it beautiful, nice; you know, you can do wonderful, magical things, but it isn't satisfying. What would truly astonish you if you saw an alien? What would it have to be? We don't know, they always fail big-time. I've never seen an alien in a film, or in a book, or in a drawing, or in a description, that made me think, 'Oh yeah, that must be like it.'"

Asked if she was remembering her own thoughts or what Stanley had said, Christiane replied, "We both did. And we talked to Arthur about it a lot. That this is all very frustrating, because our imagination just isn't interesting. The minute we can imagine it, by that token, it loses the charm or interest."

Still, Kubrick wasn't a theorist, he was one of the century's great practitioners, and he wanted to try. He'd kept Dan Richter on the payroll not in friendship—though that may have been part of it—but because he recognized the man's abilities. If anyone could become an extraterrestrial, it was this diminutive, sui generis character who had not only successfully incarnated *Australopithecus africanus* but also had brought an entire tribe of them to gibbering, gesticulating life.

That achievement had required Stuart Freeborn's rather considerable help, too, of course, and in late August Kubrick tromped up the stairs to the makeup man's sunny workshop. Aliens were already a lively topic of discussion at Borehamwood, and the subject needed no introduction. "I've got an idea about this," the director said. "We can do it, not necessarily optically, but . . ." He proceeded to outline a process involving a form of optical illusionism. "He said, 'What I figure is this,'" Freeborn remembered. "He'd seen a dotted pattern somewhere, in front of a dotted-pattern background, and the result was something that was virtually invisible yet somewhat visible just because it was on a different plane than the background. It was an intriguing idea, and Stanley asked me to begin working on something along those lines."

Kubrick huddled with Richter as well. "I want to shoot some footage of aliens," he said.

"How could I help with that, Stanley?"

"Do you know what high-contrast film is?"

"Not really."

"Well, it's black-and-white film with no grey scale. Everything is either black or white. What I want to do is paint you white and then cover you with polka dots. Then I want to put you in front of a white background with black polka dots and shoot you in high-con. If you're still, the screen should be filled with stationary points and you should be invisible. What I want to see is if, when you move, the dots become an image of an alien. Then I can reverse the color, and I will have created an alien out of dots of light."

Reflecting on the man's tenacity, Richter wrote, "I have come to realize that I cannot judge him by the measure I apply to other men and women. What would be compulsion in others is single-mindedness in Stanley." By the winter of 1967–68, four years had passed since Kubrick's first meeting with Clarke, and the clean-shaven kid with the successful Cold War satire had morphed into a bearded figure with fatigue lurking about the eyes. (A beard that in Clarke's estimation made him look "like a slightly cynical rabbi"—a line that Kubrick, who'd experienced overt anti-Semitism in Hollywood and was uncomfortable drawing attention to his ethnicity, crossed out when censoring the author's unpublished *Life* magazine profile.) A contrarian to the end, the director may still have been reacting to Carl Sagan's suggestion that *2001*'s aliens should be hinted at but not shown. More likely, Sagan's views had been forgotten as he wrestled with the challenge of presenting . . . *something*. A creature just weird enough, and different enough, and powerful enough that he, or she, or it didn't risk taking the air out of the whole enterprise.

Accordingly, Freeborn made a bald cap to fit Richter "nice and tight," painted it white, found the biggest paper punch in England, and used it to "stamp out perfect rounds of black paper." He fitted Dan with a pair of form-fitting shorts and made him up in white all over. In some kind of weird premonitory riff on a putative Area 51 medical procedure, his assistant methodically handed him little round dots of black paper with tweezers, and Freeborn dabbed each with glue and pasted it on Richter.

> And we went all over his body, and it gave an incredible appearance—stunning. We covered him completely—right over his feet, all down his legs, everywhere. Then we stood him against a background with the same-size black dots all over it. The effect was stunning. Standing still, he would disappear into the backing; but when he moved, you could just make out a shape. It was an amazingly weird effect—quite extraordinary— but I don't think it really fit in the movie. I could never see how Stanley was going to use it, and, of course, he never did. But it was quite phenomenal.

Reflecting on this last phase of his work on *2001*, Freeborn continued, "I think it was too good, in a way, but it wasn't really getting to what he wanted. Because an alien is—what?"

That was indeed the question, and Trumbull also became embroiled in these attempts, which expanded within his workshop to encompass strange levitating slit scan cityscapes. With the sleet of his third winter in England pounding on the studio roof, Trumbull constructed circles, squares, and hexagons of tiny lightbulbs, which could be switched on and off in sequence while the camera went past, and used them to create ostensibly extraterrestrial architectonic forms. "And so a flat bunch of lights would become a huge streak, like a building, a floating building with no top or bottom," he recalled. "It'd be big in the middle and

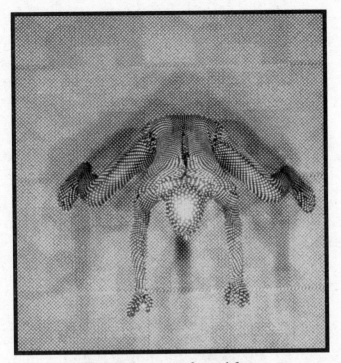

Dan Richter, transformed from
Moonwatcher to Polka Dot Man.

taper down as the lights switched . . . It was really quite beautiful. I
shot some really interesting tests of these floating cities of light."

Trumbull's efforts to create aliens were also composed of translucent, floating light patterns. "I made this really interesting little slit
scan machine that was a kaleidoscope projector of two mirrors and a
little rotating piece of artwork under the mirrors," he recalled. When
turned, the artwork "actually created a humanoid shape, like a head,
shoulders, arms, and a body and two legs of pure light, just by the
patterns . . . And I got pretty far along on that, and I actually got some
beautiful-looking stuff, but we were literally two weeks from having
to shut down, and Kubrick said, 'You've just got to stop. There's no
way. Even if you succeeded with this, I can't cut it into the movie
anymore.'"

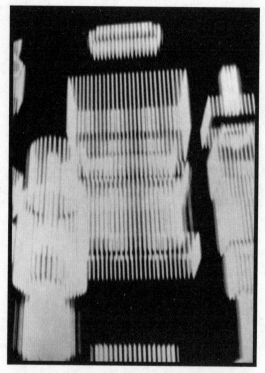

Among Doug Trumbull's last efforts for
2001, an extraterrestrial cityscape.

In the end, the question of producing an unbelievably believable extraterrestrial, or maybe a believably unbelievable one, had a cat-chasing-its-own-tail quality. "If it's so surreal and so crazy that it's unique, then it doesn't work as an astonishment or a scare," Christiane observed. "We have to relate it to something. And if we relate it to something, then it's no longer original."

She remembered Stanley concluding the whole effort with resigned humility. "It would be great if I could think of something that took everybody apart; something that absolutely would cause people to gasp," he told her sadly. "I wish I was talented and I could think of something, but I can't. The monolith leaves that open void that we feel when we try to imagine that which is unimaginable."

. . .

As of late November 1967, Kubrick had chosen music for all of the film's key sequences, but officially, at least, they were "temp tracks"—meaning they were holding space for original new compositions. The director himself referred to them in that way. Few big Hollywood films, let alone hugely expensive Cinerama road-show productions, had been released without an original score. New music was considered crucial to hyping these productions as major cultural events, and thus an important component of any promotional strategy.

By now, *2001* was a good $6 million over budget and years overdue. With MGM having piled an inordinate stack of chips on Kubrick's vision, Roger Caras and Louis Blau had spent much of 1966 and 1967 fending off quietly furious accusations from studio executives that the director's actions—and specifically, his strict security lockdown—were severely hampering their ability to protect their investment. Certainly MGM viewed an original score as part of that protection. So Kubrick wasn't in a good position to argue that his temp tracks could easily be converted into permanent ones simply by clearing the rights.

Much of the pressure the studio exerted on him during the last phases of *2001*'s production remains opaque, but nobody who worked with him doubted that it existed and was consequential. Close collaborators such as Doug Trumbull noted with appreciation that Kubrick functioned as a kind of firewall, never transmitting the heat to his subordinates, even as he grew more preoccupied and less lighthearted.

Certainly MGM president Robert O'Brien absorbed significant in-house and shareholder disquiet over Kubrick's expanding costs and ever-deferred delivery dates. In effect, he ran interference. A well-read and charismatic figure, O'Brien was conspicuously active with liberal causes and a friend of the Kennedy family. Born in 1904, he emerged from rural poverty in Montana, graduated from the University of Chicago Law School in 1933, and was appointed a commissioner at the Securities and Exchange Commission by President Franklin D. Roosevelt.

O'Brien's lifelong interest in films then prompted him to switch careers, and in 1944 he became a special assistant to the president of Paramount Pictures, ultimately moving to MGM's parent company Loews in 1957. On ascending to the position of studio president in 1962, he quickly earned a reputation for unflinching loyalty to the directors he championed.

Kubrick's project wasn't the first time that he'd put the studio's future on the line in standing by a film. David Lean's three-hour epic *Doctor Zhivago*, released in 1965, more than doubled its original budget, with production costs ballooning from $7 to $15 million during ten months of arduous production in Spain and Finland. No expense was spared as nearly a thousand men labored for a half year to build sprawling sets across ten acres of studio ground near Madrid. With wintry Moscow rising improbably on the Iberian Peninsula, O'Brien quelled internal dissent and ensured that the studio provided Lean with the tools he required to realize his vision. And the gamble paid off: a huge, crowd-pleasing hit, *Doctor Zhivago* grossed $112 million in its initial release.

While this success certainly helped O'Brien hold the line for Kubrick, other senior studio executives weren't so sanguine, and neither were the shareholders. As MGM dipped repeatedly into the red, Chicago real estate tycoon Philip Levin—who owned more than five hundred thousand shares in the company, or about 10 percent—waged two expensive proxy fights attempting to oust the MGM president while *2001* was in production, one in 1966 and the other the following year. While Levin's moves weren't necessarily tied to a specific film—rather, they concerned O'Brien's overall business strategy—there's no doubt that Kubrick's project had caused a profound sense of disquiet among those invested in the company. "They were very nervous—quite rightly," Christiane remembered. "If I had that much money floating around, and somebody's promising me something and I see nothing, and he has the arrogance to say, 'No, I don't want you to look'. . . . They were really very gracious in taking this shit from him."

Although Kubrick had a well-deserved—and fiercely protected—reputation for sovereign creative decision making, he relied on MGM to keep the lights on, and as a result, he wasn't immune to persuasion.

In particular, he was attentive to views expressed by O'Brien, whose commitment to his production had proven unshakable and whose future at the studio had become increasingly contingent on *2001*'s success. All of these factors certainly colored his decision in late November to call one of Hollywood's leading film composers, Alex North.

Film music producer Robert Townson has stated that Kubrick's call, in which he offered North the job of scoring *2001*, came at MGM's suggestion and followed the director's proposal that his temp tracks simply be cleared and used. Kubrick had already worked with North, who did the score for *Spartacus* in 1960, and the composer's reputation had only grown since due to his well-received scores for *Cleopatra* (1963) and *The Agony and the Ecstasy* (1965). Initially "ecstatic at the chance to work with Kubrick again," North was particularly happy to discover that the film had only twenty-five minutes of dialogue—something he fondly imagined would give him unprecedented freedom.

He was soon disabused of this notion, however, on flying to London in early December and discovering that Kubrick fully intended to keep some of his "temporary" tracks and wanted him to work around them. After watching the film's first hour with its temp tracks laid in, North said he "couldn't accept the idea of composing part of the score interpolated with other composers. I felt I could compose music that had the ingredients and essence of what Kubrick wanted and give it a consistency and homogeneity and contemporary feel." Reluctantly, Kubrick agreed—thereafter seemingly expecting the composer to emulate and even improve on the existing music.

North knew he was in a difficult position. He'd be competing with some of the top masterworks in the Western canon. Nevertheless, by mid-December a deal had been made. MGM would pay him $25,000 plus expenses and accommodations. A ninety-one-piece orchestra was booked, and orchestrator Henry Brant, who'd orchestrated composer Alec Templeton's "Bach Goes to Town" for Benny Goodman in 1939, and who'd worked with North most recently on the *Cleopatra* score, was hired to do the arrangements. Recording sessions were scheduled for January 15 and 16, and North flew back to London with his wife, Anna Höllger-North, on Christmas Eve. "Alex was treated like a king," she remembered in 1998. "We were given an apartment, a cook, and a car, and he and Henry Brant went straight to work, realizing that Kubrick had gotten used to these temp tracks and that something similar had to be manufactured."

Brant remembered North telling him that Kubrick had said during their December meeting that if he'd been able to get permissions for his temp tracks, *2001*'s soundtrack would have been a "fait accompli." (Whether "permissions" in this context signified the clearing of rights or agreement from MGM to use the tracks in the first place remains unclear, though evidence points to the latter.) In any case, North was plagued throughout by the unsettling sense that nothing he could compose would truly compete with the two Strausses and Ligeti. Laboring under a great deal of psychological pressure, he prepared a major set of new compositions in a remarkably short period of time. "I worked day and night to meet the first recording date, but with the stress and strain, I came down with muscle spasms and back trouble," he remembered.

It all came to a head just prior to the recording sessions, when North experienced an incapacitating physical collapse. Looking down at the gathering orchestra at Denham Film Studios on the morning of January 15, 1968, de Wilde and Kubrick were shocked to see the composer being wheeled in on a stretcher. "He came in an ambulance," de Wilde recalled. "Stanley and myself were flabbergasted." With North compos mentis but horizontal, Henry Brant conducted, and the session proceeded.

Throughout the two days of recording, Kubrick came and went, sometimes providing comments to Brant, sometimes to North. "I was present at all the recording sessions, Kubrick was very pleased and very complimentary, and there was no friction," said Höllger-North. Her husband had a similar sense of the proceedings. "He made some very good suggestions, musically," he recalled. "I assumed all was going well, what with his participation and interest in the recording."

De Wilde read Kubrick's reactions somewhat differently. "He looked at me, and I looked at him, and it was obvious it was never going to go anywhere," he said. Brant also understood that all was not well. He recalled the director listening to the opening movements of

North's composition—probably the part written to supplant Strauss's *Thus Spoke Zarathustra*—and commenting, "It's a marvelous piece of music, a beautiful piece, but it doesn't suit my picture." One of the orchestrator's score annotations tells the story even more clearly: "Stanley hates this, but I like it!" Completing that trajectory, Con Pederson remembered Kubrick returning to Borehamwood with a curtly dismissive, "It's shit."

Nevertheless, after two days, more than forty minutes of new music had been recorded, and North withdrew to his apartment to rest and wait for the second and third parts of the film to be completed so that he could continue. More than a week passed without further word from the director, however, and he found he couldn't reach him directly, instead passing messages via an assistant. He assured Kubrick that his physical duress had been momentary, that he was completely willing to proceed, and that he was working under a doctor's care. Finally, he received word back: no further music would be required. The rest of the film would "use breathing effects."

North's scribbled notes, prepared before calling his agent in mid-January, read like a bulletin from a crisis: "Fulfilled my obligation—recorded over 40 minutes music—delaying me—liked stuff, then changed his mind." They record his assessment of Kubrick's intentions: "clearing rights to temp tracks . . . psychological hang-up." The composer also evidently instructed his agent to contact Robert O'Brien and bring pressure to bear.

Years later, Kubrick confirmed that North's agent had, in fact, called O'Brien "to warn him that if I didn't use his client's score, the film would not make its premiere date. But in that instance, as in all others, O'Brien trusted my judgment. He is a wonderful man, and one of the very few film bosses who was able to inspire genuine loyalty and affection from his filmmakers." In comments made to French film critic Michel Ciment, Kubrick gave an unsparing assessment of North's work.

Although he and I went over the picture very carefully, and he listened to these temporary tracks (Strauss, Ligeti, Khachaturian) and agreed that they worked fine and would serve as a guide for the musical objectives of each sequence he, nevertheless, wrote and recorded a score which could not have been more alien to the music we had listened to, and, much more serious than that, a score which, in my opinion, was completely inadequate for the film. With the premiere looming up, I had no time left to even think about another score being written, and had I not been able to use the music I had already selected for the temporary tracks, I don't know what I would have done.

Was this disingenuous? Höllger-North clearly felt it had been Kubrick's plan from the start. "All along, he was trying to clear the rights to the temp track music, so he really under pretext had Alex compose the score, and I always thought that was unfair," she charged. "Kubrick managed to clear the rights, and Alex was never told that. We went to see *2001* in New York and were very surprised when Alex's music— not a note of it was in the film."

A story from January 1968 undermines her assessment, however. On arrival for one of their late-night talks, Cantwell found Kubrick uncharacteristically despondent. As they assembled sandwiches, "something was on his mind," he remembered. "We just let the silence be there, and then he said he had a problem. He said, 'I just had to fire my fourth composer. I'm starting from scratch. Back to square one.' And 'What to do, and who to consider, in the time we've got left.' By then, of course, I was totally identifying with the scale of what he was trying to do. He said, 'I'm even thinking, should I contact the Beatles?'"

At this question—rhetorically phrased but clearly offered for serious consideration—Cantwell "absorbed that for a little while and just gave him my straight feelings that 'No, that wouldn't be worthy.'"

There's no reason to suppose Kubrick's Beatles idea was anything more than a fleeting straw clutched at momentarily. In any case, Cantwell was struck by the gravity of the situation. "The one thing Stanley communicated to me was a deep disappointment," he said. "For Stanley, that was not something to be devastated by; it had to be resolved, it couldn't diminish the film. And it was this mix of sadness, the urgency, the regret, the adrenaline of 'What's the course from here?' All of that [was] wrapped in. He communicated very quietly so much of the intensity of his feelings."

. . .

In his ongoing quest to improve the Star Gate sequence, Kubrick decided in September that he needed additional aerial footage to augment the material Birkin had shot in Scotland the previous year. He contacted cinematographer Bob Gaffney, who flew to London on September 27, arriving in Borehamwood just in time to witness the director filming bones spinning through the back-lot air. So far, Gaffney's contribution to *2001* had been persuading Kubrick to use 65-millimeter Cinerama stock rather than another, less sensational wide-screen film format back in 1964. Now the director took him into his office and pulled an illustrated book of southwestern desert landscapes from the shelf. "I want scenes in Monument Valley that are shot in very low light, flying over the monument and the terrain as low as you can go," he said. Asked what kind of sequence it was for, Kubrick was characteristically cagey.

By October 10, Gaffney was already in place near Page, Arizona, with a rented Panaflex and plenty of film. He'd decided that a Cessna light aircraft would provide a better shooting platform than a helicopter, and he jury-rigged a camera mount under the wing to get the lens away from the propeller. After several days of filming rugged canyon walls and mesas near Page, he shifted his attention due east to Monument Valley—among the most filmed landscapes in the United States, and practically synonymous with the American West.

Birkin and pilot Bernard Mayer had already courted disaster flying their chopper into the downdrafts and unpredictable wind shears of Scotland the previous November. Gaffney and his pilot had located a dirt strip directly beside a Texaco station not far from the valley, with a motel fortuitously positioned directly across the street. Mules and goats wandered untethered around the area. As with the Dawn of Man stills, their instructions were to film only at dawn or near sunset. Late on the first afternoon, Gaffney loaded the camera, noted that the wind sock was "deader than a doornail," and they took off before sunset in a whirl of dust.

As soon as they cleared the top of the nearest mesa, however, the plane got slammed by a hundred-mile-an-hour wind, which threw its nose higher in the air. Corkscrewing upward and verging on a stall, the pilot wrestled with the shuddering stick and screamed, "I knew you were going to get me killed one of these days, you stupid son of a bitch!"

"Push the stick down, or I will!" Gaffney shouted back, reaching over to help him wrestle it forward. The Cessna dove below the desert jet stream to calmer air. After leveling out, they turned back to where they'd taken off and saw with a new wave of adrenaline that the cloud they'd made was still hanging densely in the air, obscuring their view. There was no wind here, but the setting sun was directly behind the strip, turning the dust into an opaque wall. "You fly, I'll be the radar," Gaffney said, leaning forward and squinting through the windscreen as they made their approach. "Two degrees left . . . Right . . . Let the prop down . . . Chop the power!" And with a *boom*, they slammed down onto the runway, blowing a tire in the process.

They'd made it.

Having lived to tell the tale, they figured out strategies to dodge the wind, and several of their aerial views were subsequently transformed via Purple Hearts processing into vivid shades of maroon, umber, green, and blue, and then edited into the Star Gate sequence. In one, Monument Valley's characteristic sandstone spires are clearly evident. "I almost got killed," the cameraman recalled somberly to Kubrick biographer Vincent LoBrutto in 1997.

A few years after *2001*'s release, Birkin was helping Kubrick research another project and mentioned his regret that one of his Scotland shots—of the summit of Ben Nevis, with its abandoned meteorological station—was perhaps a bit too recognizable in the film. The director replied that in retrospect he probably shouldn't have used some of Gaffney's footage for the same reason.

■   ■   ■

Kubrick edited *2001: A Space Odyssey* from October 9, 1967, until March 6, 1968—the day before boarding the *Queen Elizabeth* bound for New York. By most accounts, he did the cutting himself, with Ray Lovejoy, officially the film's editor, assisting, and David de Wilde coming and going with fresh prints, sound rolls, and cans of film.

Colin Cantwell, who witnessed some of the edit, recalled that the film's structure already existed in Kubrick's head and that in a masterly performance, it was executed in linear fashion—a serial process from beginning to end, with no recutting attempted. The director simply went reel by reel, editing each section complete with the sound and music, and then sent it off immediately—an accelerated schedule necessary so that the time-consuming procedure could start of converting the film to 35 millimeter for general release in late 1968, following its 70-millimeter Cinerama run.

As a result of this imperative, Kubrick had no latitude for error and didn't have the luxury of going back to tweak earlier scenes as he proceeded. The entire picture had to be clearly visualized, with anything that needed echoing from the first reel accounted for in advance, and a comprehensive picture of what the last reel should look like well in mind as he edited the opening sequences. "That was as bold as anything else in the film, and as perfectly executed," Cantwell marvelled.

Despite such constraints—which seemingly disallowed any rethinking, let alone test screenings—"we just had a good time," de Wilde remembered. "It was a bit of fun. It was very relaxed by then, and we were all relaxed. They'd got it in the can."

Cantwell's description would seem to indicate that the winnowing away of the film's more expositional elements had been largely previsualized, with *2001* emerging reel by reel as the largely nonverbal work we know. And indeed, according to de Wilde, the documentary prologue that Roger Caras had labored over in 1966 was never edited, let alone cut into the opening of the film. But the record shows that Kubrick, in fact, continued struggling with the question of a narrative voice-over throughout the five-month editing period, abandoning the idea not once but twice. Meanwhile, throughout September, October, and into November, Clarke conscientiously transmitted revised blocks

Kubrick working with assistant
film editor David de Wilde.

of narration to Borehamwood. His texts covered the Dawn of Man, HAL's neurosis, and *Discovery*'s various features and operations, all rendered in semi-sententious documentary voice-over style.

Kubrick's last big push to finalize the film's narration unfolded in early November. Clarke was then on an American lecture tour, and his copy was either cabled directly to London or phoned in to unit publicist Benn Reyes in New York, who quickly forwarded it on to London. Snippets of his writing reveal how different *2001* would have been if they'd been used. For the Dawn of Man sequence: "They were children of the forest—gatherers of nuts and fruits and berries. But the forest was dying, defeated by centuries of drought, and they were dying with it. In this new world of open plains and stunted bushes, the search for food was an endless, losing battle."

For *Discovery*: "Much of the time you spend inside a giant drum, slowly revolving so that centrifugal force gives you the feeling of normal weight. You can walk, exercise to keep fit, and prepare meals without the inconvenience of weightlessness."

For HAL: "Since consciousness had first dawned for the computer, all of its great powers and skills had been directed to achieving perfection. Undistracted by the passions of life, it pursued this with absolute single-mindedness. But now for millions of lonely miles, it had been brooding over the secret it could not share, and the deception it had been made part of, and it was beginning to experience a profound sense of imperfection, of wrongness. For like his makers, Hal had been created innocent."

While Kubrick's decision in late November to use Rain for HAL's voice was prompted largely by his dislike of Martin Balsam's line readings, some of it may have been due to a growing unease over the effect such expositions would have on the film. A couple weeks before his sessions with the Canadian actor, this feeling had evidently crystallized, and from November 20 to the 22, Kubrick sent a flurry of cables attempting to reach Clarke on his tour. Stop work, he wrote, explaining tersely, "Narration needs drastically reduced."

On the twenty-third, Clarke responded with an alarmed cable: "Just returned from two-week 10,000-mile lecture tour where I spent every spare minute working on narration as requested. Job will be completed in a few days, so very disturbed by your messages. Please clarify." Kubrick's answer that day was even more unequivocal: "Sorry about narration. As more film cut together, it became apparent narration was not needed."

Clarke's follow-up, a handwritten note from the Chelsea Hotel on the twenty-fifth, provides an interesting window into the author's thinking. He was "rather upset," he wrote, but he also referred to "some new purple prosery which it would be a pity to waste"—a revealing comment—and continued with an implicit admission: "I'll be interested to see how you can possibly dispense with much of the narrative material, while at the same time I feel it's a good thing if you can!"

By early December, Kubrick had reevaluated, however, and on the seventeenth he spoke to Clarke, who by then was in London, and said he wanted him to view the edited film in mid-January so that he could finish his voice-over texts. But when Clarke mentioned payment for the additional work—he asked for $5,000—Kubrick balked. "I must confess I've been under the impression that you have been paid to do the work on the screenplay and that this narration work is of a similar nature to the additional year and a half of my own work *for which I have received no additional payment*," he wrote. (Italics in the original.)

At this, Clarke contained his exasperation with a mighty effort and fired off a message to Scott Meredith. Deploying the word "chutzpah," he requested his agent not to "lean on him too hard, as I don't want any unpleasantness or holdups on this last lap." Kubrick's argument might have resonated a bit better if he'd given Clarke a financial stake in the film three years before. In any case, he was clearly only going through the motions, and by January 23 he wrote the author again, now back in Ceylon: "It doesn't look like we'll get to the narration until mid-February. Scott has probably told you the $5,000 is okay. All goes well but desperately short of time—premiere Apr. 2 Wash D.C."

Clarke soon responded that he was anxious to help, but cautioned, "It's impossible to rush the sort of literary composition you'll want—it has to be reworked over and over again, so as soon as you have anything finalized, I want to start thinking about it." He added, however, that he could work from his existing drafts.

He never received Kubrick's summons to London.

.   .   .

As Terry Southern discovered, few subjects were more fraught for Kubrick than the question of credit. Dan Richter remembers a discussion with the director near the end of his tenure at Borehamwood. Summoned for an audience in late September, he arrived to discover

Kubrick standing and "pretending to look busy" behind his desk but "obviously nervous about something." After a detour into an unrelated matter—one that Dan sensed immediately wasn't the reason for their meeting—there was an uncomfortable pause. Which Richter broke. "What's the matter, Stanley?" he asked.

"Well, Dan, I have to settle your credit with you, and I've got a problem."

"What's the problem? I'm the choreographer of the Dawn of Man."

"Well, I can only give you one credit. So I want to give you the fourth starring credit. You can have that or the choreography credit, but you can't have both."

Evaluating the statement, Richter asked why.

"Well, Dan, that would give you more than one credit. And I'm the only one with more than one credit. So you'll have to decide."

Despite playing Moonwatcher, it hadn't occurred to Dan that he might receive star billing. He'd always seen himself as the choreographer. He decided to make an awkward situation easier. "Well, I'll take the starring credit," he said. "Everybody knows I did the Dawn of Man choreography."

"That's what I thought you'd say," Kubrick replied. But he still looked anxious. "Can you live with that?" he asked.

"Believe me, Stanley, I like the idea of being a star," said Richter.

They stood and solemnly shook hands. And Richter's name appears directly after Dullea, Lockwood, and Sylvester on *2001*'s credit roll, appropriately enough.

Asked, decades later, what he thought Kubrick might have said if his answer had been that he deserved both credits and had worked extremely hard for them, Richter replied without hesitation, "He wouldn't have given it to me."

. . .

Doug Trumbull remembers *2001*'s end game ambivalently, as a time of fatigue, low morale, and bitter feelings. Rivalries and tensions that had been implicit were becoming explicit, as they sometimes do when people are worn out and something significant is palpably ending. "Everybody was departing the movie like rats from a sinking ship, because everyone was so exhausted, just completely wiped out and tired of it, and their careers had been stalled," Trumbull said. Work on *2001* had taken much longer than anyone had expected, and "they couldn't get jobs on other movies. They were pissed off . . . They didn't really realize what they had just been involved in—which I did."

Nevertheless, his relationship with Kubrick was deteriorating. Trumbull knew how crucial he'd become to the making of *2001*, and he was proud of it. Apart from his slit scans, he'd built and managed the film's animation department after Wally Gentleman's departure, among other contributions. Trumbull's importance to the film can be judged by something Christiane told him when she ran into him at a film festival years after Stanley's death. According to Trumbull, Christiane recalled asking her husband during the last phase of *2001*'s production if he was worried about getting the film done. The director's response had been, "No, I'm not worried because I have Doug."

Despite this, and although he was now senior staff, Trumbull's salary was frozen at the level it had been two years before. Chafing at this injustice—his peers were making much more than him—he agitated for a raise. But Kubrick, acutely aware that he was way over budget and that others might make the same argument, flatly refused. "That was one of the biggest challenges between me and Kubrick," Trumbull remembered. "He just always said, 'No.' Utterly inflexible."

Then as his March trip to New York and Los Angeles approached, the director decided he wanted to patent Trumbull's slit scan machine. Its designer objected. "I was extremely averse to this idea because . . ." Trumbull hesitated. "I was just a kid. I didn't know at the time that the employer actually does have the right to patent the work of his employees." Kubrick told Trumbull that he'd hired people to write the patent claim.

"Hey, Stanley, do whatever you want, but I'm not going to help these people," Trumbull replied.

"Well, I'm going to do it anyway."

"Okay, do it. Just do it on your own."

Remembering the scene decades later, Trumbull said, "It was a tough moment. And there were engineers who came into my office and into the slit scan machine and tried to follow the wires and understand all the logic of what I had built, and what I heard is that they couldn't figure it out, either, and they gave up and went away."

Apart from that, Trumbull's bonds with his former mentor, Con Pederson, were also unraveling. Although he'd provided significant creative input throughout, and had a hands-on role in many of the film's visual effects, Pederson's responsibilities on *2001* had been predominantly managerial. The film's postproduction process "was epic in its complexity, and Con was the smartest guy in the room," Trumbull said, stating flatly that *2001* "absolutely would not have happened without Con." But as he watched Trumbull ascend and contribute so significantly to the film's visual impact, Pederson began to intimate to Kubrick that much of what Trumbull had achieved derived from slit scan pioneer John Whitney—and that Trumbull was in effect claiming credit for something he hadn't invented.

"I remember some kind of very uncomfortable altercation with him in an office a day or two before I left, where he was pissed off at me for making that claim," said Trumbull of Pederson. "And I said, 'You're wrong.' He was trying to deride me for not crediting John Whitney for the seminal work, which I had gone way beyond. As though I had just copied him and I hadn't invented or done anything, which was bullshit. I'd sweated blood over that machine. I built the whole thing. Even though I was sparked by his work, John Whitney had nothing to do with what we did on *2001*."

Asked about their confrontation a half century later, Pederson graciously took Trumbull's side. "A lot of years have gone by since *2001*," he observed. "As far as I'm concerned, Doug's work on it cannot be overstated. His slit scan project was original and spectacular. Any quibbling about its resemblance to Whitney's work should be disregarded as trivial. It made the movie what it is, and I am certainly grateful for Doug's genius."

Then there was "some other hanky-panky" that Trumbull caught wind of, made him suspicious, and planted the seeds of a bitterness that lasted for decades. "In the last days I was there, we were shooting the end credits for the movie, and I had set the whole thing in motion because I was running the animation department," he said. "And there was always one missing card. And no one knew what the missing card was, but I didn't concern myself with it . . . Well, that was the card that said, 'Special Effects Designed and Directed by Stanley Kubrick.' That really pissed me off."

At that time, Oscar ground rules specified that a maximum of three people could receive a Best Visual Effects award. But *2001* listed four effects supervisors: Wally Veevers, Trumbull, Pederson, and Tom Howard. So Kubrick's sole design and direction credit arguably allowed the film to be considered in the first place. Still, as Trumbull has maintained for years, he could have seen to it that the Academy of Motion Picture Arts and Sciences made an exception, particularly for a film that had advanced visual effects so significantly. Instead, he had Louis Blau contact the Directors Guild of America and ask for its official agreement that his visual effects credit line was acceptable.

On receiving the DGA's consent, Kubrick positioned himself as sole candidate for the Oscar.

. . .

With only days to go, in late February the remnants of Kubrick's inner circle decided that *2001*'s production had to be commemorated, and they asked the Props Department to fabricate a stylized diorama with a shiny metal space pod parked on a faux stone bronze outcropping—a replica, possibly, of the Spitzkoppe Hills. Facing it from another outcrop was the monolith, and the whole thing was mounted on a marble base with an engraved plaque. This memento was accompanied by a large card addressed to the director with a list of 104 names inside, all "members of his unit to mark their association on the production of '2001: A Space Odyssey.' "

With so many people gone, however, the card wasn't signed, and the presentation was evidently a low-key affair, one that hasn't made its way into accounts of the film's production. In a surviving photograph, Kubrick looks palpably exhausted, even haggard, as he stands over his trophy.

Another enduring fragment of *2001*'s last days at Borehamwood is a list of farewell notes, dated March 7, from the director to his crew. In one, addressed to matte cameraman Jimmy Budd, Kubrick wrote, "As you know, everything closed down in a mad rush and I was even looking for someone to do a Purple Heart at 2 o'clock in the morning of March 7th."

Later that day, Cunard Line baggage handlers hoisted a Moviola film editor and several dozen bulky 65-millimeter film cans into the freight window of the RMS *Queen Elizabeth* at Southampton. A little while after that, the Kubrick family could be seen climbing the high gangway to the ship's top deck, trailed by assistant film editor David de Wilde, an angular figure in a wool sweater.

CHAPTER ELEVEN

RELEASE

I can't do anything about things when they're
going well, but I can when they're not.

—STANLEY KUBRICK

The Moviola wasn't on board the *Queen Elizabeth* to continue editing *2001*. That had already been accomplished—or so it was assumed. It was for editing trailers, though de Wilde estimates that in their week's crossing, he and Kubrick spent only six or seven hours in the cutting room, a cabin dedicated for that purpose directly adjoining his on the A deck. The Kubricks, meanwhile, installed themselves in the grandest quarters in the ship, the PM Suite, from which British prime minister Winston Churchill had allegedly planned important military operations while reclining in the bathtub smoking cigars during World War II. The director rarely left it, apart from three times a day for meals, though he and de Wilde did go for occasional walks on the deck.

They'd established a real rapport. De Wilde believes he was on the boat because "I had done three years of working for Stanley and given a hundred percent . . . I think he felt he owed me something." He also sensed that Kubrick simply wanted to have a male friend on board whom he could talk to freely about whatever was on his mind. Every night, they all ate together in the best restaurant on board, the Verandah Grill, with Stanley in his rumpled blue jacket and Christiane and the girls dressed to the nines and looking "brilliant. They were beautiful," de Wilde remembered.

One night they were invited to the captain's table, in this case presided over by the Cunard commodore, Geoffrey Marr—something like a civilian admiral. Although Kubrick mustered up his best tie for the occasion and de Wilde broke out his mohair suit, it wasn't enough. The ladies had gone ahead, and when Kubrick and de Wilde presented themselves at the grand dining room door, they were coolly met. "Sorry, you can't come in," the liveried doorman said flatly. "You're not dressed for dinner." Cunard code required tuxedoes. They looked at each other. Neither owned such clothing. "He couldn't care less," David remembered. "He didn't give a toss." They stood around for a few minutes as word filtered back to the door that it was the occupant of Churchill's suite—one Stanley Kubrick, director—who'd been denied entry. Suddenly with a flurry of "Terribly sorry" and "Right this way," they were escorted to the grand central table, where Kubrick was seated between Commodore Marr and Captain William Law, his blue jacket now framed by their dress whites.

Queen Elizabeth arrived in New York early on a misty March morning. It was the last trip Stanley Kubrick ever made to the United States, and de Wilde joined him on the deck to witness a sight he'd never expected to see: a giant female form with a spiky crown and upraised torch emerging from the fog. Jeremy Bernstein went to greet them at Pier 92 on the Hudson River side of Midtown Manhattan, only to discover that apart from an MGM representative, he was waiting with four or five alert-looking men wielding briefcases and manila envelopes: process servers. When Kubrick finally emerged from Immigration and Customs, they pounced, pushing papers into the startled director's hands and announcing, "You're served," one after the other. At this, Kubrick shot an ironic look at de Wilde, who stood, shocked, a short distance away. "I was so unhappy and devastated that this man who controlled a whole studio was being . . . he wasn't being frog-marched, but he had to swallow it, he had to accept it," David remembered. "I was embarrassed for the poor man, and it hurt me."

One of the processes served that day almost certainly was a notice of legal action by Fred Ordway on behalf of an organization he'd co-founded, the International Space Museum and Gallery. After leaving the production in 1966, Ordway had started the project with his friend the space historian Carsbie Adams and a group of investors. Intended as a Washington, DC, showcase for the production's sets, models, and costumes, it was initially encouraged by Kubrick, who seems to have believed it was a nonprofit associated with the Smithsonian Institution. Following his positive feedback, Ordway and Adams raised funds sufficient to rent a large building in Washington and had already spent more than $200,000 by the time Kubrick became aware it was actually a for-profit company. He subsequently interrogated Lange to find out what he knew and refused any further cooperation. Ordway and Adams's lawsuit, filed on March 12, claimed damages of $1 million from Polaris Productions and MGM for their outlays plus loss of revenue.

After the processes had all been served and their servers had retreated, Bernstein accompanied the Kubrick family and de Wilde to the Plaza Hotel in a small convoy of MGM Cadillacs. Once they'd settled in, the director came to join Bernstein in the lobby, and they tried to get into the Oak Room—but were refused entry because Kubrick wasn't wearing a tie.

His first day back wasn't going particularly well.

·   ·   ·

By March 16, the Kubricks had made their way to Los Angeles via Chicago in a luxurious transcontinental train trip, with the Chicago-LA leg spent in a glass-walled dome car apartment complete with master bedroom, easy chairs, and showers. De Wilde and Caras had flown ahead to meet them, and by then, film editor Ray Lovejoy had also arrived from London. After a meeting at MGM's Thalberg Building in Culver City, Lovejoy and de Wilde began the process of overseeing sound work on the cut reels, which included converting the mono soundtrack they'd made in England into stereo, as well as finishing the trailers started on board the ship.

The first time *2001: A Space Odyssey* was screened at full length, with all visual effects in place and no interruptions, was on Saturday, March 23, in a giant theater in Culver City in front of a tiny audience. In addition to Kubrick, who hadn't seen his own film in its entirety before, only de Wilde, Lovejoy, and the studio's top brass were present:

president and CEO Robert O'Brien; VP and president of MGM International Maurice Silverstein; editorial administrator Merle Chamberlain; and MGM supervising editor Maggie Booth. Then seventy, Booth was one of Hollywood's grand dames. She'd started as a negative cutter for seminal producer-director D. W. Griffith in 1915 before going on to work for Louis B. Mayer well prior to the merger that had produced MGM in 1924. A legendary figure, she was diminutive but commanding, and de Wilde was keenly aware that she was one of the most experienced film editors in Hollywood and had reigned over all the studio's final cuts for three decades.

De Wilde and Lovejoy sat in the back of the theater, as far as possible from the VIPs up front. Both were anxious. "For three years, they had been coming over and we'd been running this show reel, and now we were showing them the whole thing, what it was all about," said de Wilde. "And, of course, the actual film itself is a mystery. A mystery." Asked if he thought Kubrick was nervous as well, the assistant editor said he gave no sign of it.

When the lights faded and the film started, however, the director kept coming back to tweak the sound. The film hadn't yet been fully mixed, and Kubrick wanted de Wilde to increase the volume for *Thus Spoke Zarathustra*, among other places. The running time was then 161 minutes, and from their darkened perch, de Wilde and Lovejoy watched *2001* unfold, simultaneously dazzled by the achievement and uneasy. A radically experimental work of art, it had no narrative voice-overs and few cues given to guide audience understanding. It also betrayed signs of first-draft editing, with certain scenes unnecessarily long and others needlessly repetitive of sequences that had gone before. Apart from the opening credit sequence and the Dawn of Man, the film was without title cards.

When the lights went up again almost three hours later, "they all stood up, and I thought, 'Shit.' I thought it was a disaster," de Wilde remembered. He'd deemed the film brilliant—but on examining the body language of O'Brien, Silverstein, Chamberlain, and Booth at the other end of the theater, he recognized that they might have reached a different judgment. "The way they stood up and shook hands with Stanley and everything, and 'Congratulations,' 'Thank you, Stanley'—it was just the body language . . . I thought, 'Jesus this didn't go down well. *Jesus.*'"

De Wilde was in charge of the print, however, and he couldn't hang around to verify his supposition—he had to go claim it in the booth. Leaving the theater, he realized to his surprise that he was shaking. He was profoundly invested in the film. "Three years of my bloody life, and Stanley's as well," he recalled. "I loved the guy." Climbing the stairs, he replayed Maggie Booth's restrained reaction in particular. In congratulating Kubrick, she'd looked like she was only going through the motions. It didn't bode well.

On reaching the projection area, he saw to his surprise, in addition to the projectionist, a young man of about eighteen—Maurice Silverstein's nephew. He'd evidently watched the whole thing from the booth windows. He had an ecstatic look on his face.

"Did you work on that?" he asked excitedly.

"Yeah, we just finished it," confirmed de Wilde.

"Who are you?" Not a demand but, rather, in wonder.

"I'm Stanley's assistant editor." De Wilde was gradually calming down.

"Well, I just watched it."

"What did you think?"

"It was the most amazing thing I've ever seen."

. . .

A good two years after Clarke had first insisted it was ready, Kubrick finally green-lighted the novel. Following an auction among publishing houses that had been well aware of the property for that period, on March 20 Scott Meredith succeeded in extracting an offer from New American Library for $130,000—$30,000 less than what Dell had offered two years before. Still, after Kubrick's percentage, Clarke would earn the equivalent of over a half million dollars today. Enough to cancel his debts with money to spare.

No directorial revisions had been offered after Kubrick's initial tranche in June 1966, and the suggestions he had made hadn't been incorporated. Nor was he listed as coauthor—though they'd agreed that the title page would contain the line "based on a screenplay by Stanley Kubrick and Arthur C. Clarke." In his incomplete list of proposed changes, Kubrick had questioned whether one could use the word "veld" for a drought-stricken area, if bees could be found in such conditions, and if leopards could be said to growl. In the published book, bees buzzed above a veld populated by growling leopards. "When Stanley approved the book for publication, not a word had been changed," Clarke observed to Jerome Agel, editor of *The Making of Kubrick's 2001*, in 1970. "There seems to be a right way to do things, a wrong way, and Stanley's way."

A revealing few letters between Clarke and two old friends through 1967 and 1968 provide an interesting portrait of the writer on the verge of *2001*'s world premiere. One exchange was with his fellow science fiction author Sam Youd, who was also represented by Meredith's literary agency under his pen name, John Christopher. The other was with rocket scientist Val Cleaver, the chief engineer of the Rolls–Royce Rocket Division.

In June 1967 Clarke had filled Youd in on the failure of his Dell deal, detailing with theatrical relish how far into debt he'd fallen as a result (he cited a figure of $50,000) and concluding with "Apart from that, Mrs. Lincoln, all is going well, and I've no doubt *2001* will be published eventually and I'm not too upset—the movie is going to be fantastic."

Youd answered with a reassuring "I'm sure the deal will reemerge much as before when things get near the film's release. I trust you get a good slice of the proceeds from *that*?"

To which Clarke had replied, "No, I have *no* part of the movie, alas."

The response was both incredulous and indignant:

If I have got it straight in my mind, the position is this: you write the book which sets the whole thing up, and Kubrick collects

50% of the cash and credit on that one, plus the right to bugger the whole thing up as he eventually did.* But you get nothing for the film rights. I'm surprised that Scott OK'd a deal of that kind, and a little surprised that you yourself didn't do a spot of boggling. I hope you did not rely on the good faith of anyone—unwise in any kind of business, suicidal where film people are concerned.

Rationalizing Meredith's failure to leverage the writer's standing during the negotiation with Kubrick back in 1964, Clarke responded on October 21, "The book-film situation is much more complex than I can go into here—I'll still do pretty well out of it *eventually*. But I could have done with that $100,000 in 1966."†

As for Val Cleaver, he'd been Clarke's friend since the 1930s, and had served as the second when the author went for a good-natured verbal duel with C. S. Lewis in Oxford in 1954—a faceoff designed to defend the merits of spaceflight, which Lewis had publicly attacked. (Lewis's second was J. R. R. Tolkien, producing a remarkable—but unfortunately, otherwise undocumented—clash between the leading British fantasy and science fiction writers.) On March 24, shortly before *2001*'s world premiere, Cleaver wrote Clarke a simple, heartfelt message.

My best wishes for every success on April 2nd. I do hope it all lives up to your expectations. I'll even include Stanley in that—I'm feeling in a generous mood tonight! But I have some idea how much this means to you—as a vivid representation of the world of imagination in which you have mostly lived during the past 40 years, emerging from it only occasionally to converse with mid-20th century mortals such as I. Stanley thinks he's made this film for himself, or for MGM, or for the public, but

---

* Actually 40 percent, at least of the cash.
† In fairness to Meredith, it could well have been Clarke, who disliked personal disagreements, who'd caved to Kubrick's intransigence three years before.

of course he's wrong. He's really made it for you. So, above all, I wish you every happiness and satisfaction on The Night.

. . .

Although he made concerted efforts to disguise it, when Clarke finally saw *2001: A Space Odyssey* at a press preview screening on March 31 at the Uptown Theater—an imposing movie palace with a Mesopotamian façade in the Cleveland Park neighborhood of Washington, DC—it failed to live up to his expectations, produced little happiness, and gave no satisfaction. Though he knew in advance that Kubrick hadn't used any of his voice-over narration, he was shocked and disappointed by the film's lack of concession to audience understanding. It was only hours after Lyndon Johnson had announced he wouldn't seek another term in the White House, and Clarke overheard one MGM executive say to the other, "Well, today we lost two presidents"— referring, of course, also to studio chief O'Brien.

The official world premiere took place two days later at the Uptown, and by then, the entire top MGM hierarchy was in attendance, as were Keir Dullea and Gary Lockwood. Kubrick had returned to New York from Los Angeles by train only days before, and because the Manhattan opening was the following day, he'd elected not to attend. It was a rainy evening, and Lockwood, who'd smoked a joint for the occasion, remembered tables set out in the lobby with beautifully produced illustrated programs and a sense of excitement as the lights went down. Before too long, however, people started voting with their feet, and at the intermission, they "were just streaming out. It was a disaster. No one liked it."

Victor Davis of the *Daily Express* had flown in from London for the event and filed a piece on April 3 that led with: "I have never seen a major film preview like it: there was not a single handclap—not even from the studio men. The audience just rose, stunned and thoughtful, and shuffled out to the pavement."

He pointed out that MGM had "gambled their financial health" on the film, writing, "Man's deepest purpose is questioned, and then

Kubrick ejects us on the street baffled." Standing on that damp side-walk was none other than Clarke, however, who soon found himself surrounded by "wet huddles" of confused journalists and was quoted as saying, "You'll have to see it six times to arrive at any sort of under-standing."

Although Davis's piece was actually mixed—he not only called the film "visually staggering" but predicted that the "under-35s" among its audience would be "repeatedly drawn" to it, a prescient comment—other UK press reports were categorically negative, with the lead para-graph of Donald Zec's *Daily Mirror* article the same day reading, in its entirety, "It took four years to complete, £4,000,000 to produce, and precisely two hours and forty-two minutes to confirm that it is uncom-monly painful not to laugh when you feel like it." The Washington papers weren't much better, with a gossipy piece in the *Evening Star* stating, "Despite its marvelous and elaborate sets and the perfection of its technique, it has a plot aimed at 7-year-olds. Cracked a man in the movie business, 'I have never seen such a piece of junk in my life.' Said another observer: 'It took three hours to find out that God is a monolith.'"

Exiting the premiere, Lockwood skirted the press scrum around Clarke and returned to the Shoreham Hotel, where he found himself riding up in the elevator with Robert O'Brien and "one of the syco-phants, who looked at me and said, 'Where did you guys go awry?'" Lockwood, who knew the significance of what he'd just seen, glared at the man. "We didn't 'go awry,'" he said. "Wait, just wait." (By con-trast, O'Brien was "very nice about it all.") He met up with Dullea and they proceeded to the postpremiere party in the ballroom, where they discovered that the Lester Lanin Orchestra was playing to only a handful of people, and made themselves scarce.

Anxious not to antagonize Kubrick, and acutely conscious that his book's future depended on the film's reception, Clarke kept his true views to a small circle. On April 24 he received a letter from screen-writer Howard Koch, who'd written a script based on *Childhood's End* a decade previously that had been in development at MGM when Kubrick came along—after which it was suppressed, evidently with some help from Louis Blau. "George Pal called me and told me you were deeply disappointed with *2001*," Koch wrote. "I saw it and share your disappointment . . . So much money and brains expended on such a cold, unhuman, bloodless work!"

Clarke's annotations to the letter tell their own story. A wavy blue pen line through "deeply disappointed," Koch's phone number in Los Angeles—and a note indicating that he'd called him on May 2, no doubt to ask that his views be kept in confidence.

Decades later, however, Clarke quietly confirmed that he'd been shocked, baffled, and dissatisfied by the film—at least initially.

. . .

Christiane Kubrick remembers *2001*'s premiere as the one she attended on the night of April 3 at Loews Capitol Theater on Broadway—an invitation-only event with a press mob outside and an audience comprised predominantly of entitled representatives of the media and cultural elite, senior- to midlevel MGM staff, and celebrities of various stripes, including Paul Newman, Joanne Woodward, Gloria Vanderbilt, and Henry Fonda. The New York premiere was, in effect, the Kubrick premiere, and against all odds, she'd persuaded Stanley to don a tuxedo for the occasion. He even consented to do an on-camera interview in the lobby—one of the few such clips available. Little Vivian was there, wearing the same red dress as in her space station videophone scene, and Clarke had returned to New York for the occasion. He and Kubrick sat together in the front row of the theater, which seated more than 1,500 people and was completely jammed.

Looking around before the lights went down, Christiane noted the predominance of older people in the audience, and thought to herself, "Lots of *alte kackers* here"—literally "old shitters" in German and Yiddish, but more accurately translated as "old farts." Magician and radio host James Randi attended the screening with a posse of science fiction writers that included Isaac Asimov. Randi sat two-thirds of the way back in the theater with a clear view of Clarke and Kubrick up front, silhouetted by the curving, two-story-high Cinerama screen. He has vivid memories of the event, and recalls audience disquiet erupting

The Kubrick family at the Loews Capitol Theatre on Broadway for the New York premiere of *2001*.

during an extended sequence toward the end of the first half in which Bowman jogged around and around the centrifuge—a scene he suspected was "Kubrick trying to prove how boring it was to be in space. He certainly proved his point."

As Dullea ran in protracted circles, Randi started hearing boos and hisses. People "were saying things like, 'Let's move it along,' 'Here we go,' 'Next scene,' that kind of thing. Not calling out loudly, but you could hear them saying it. And they even started to giggle, because it was a silly thing, with this long, long scene in there." When the lights went up for the intermission, he witnessed Kubrick and Clarke walking silently up the theater aisle together. Grim faced, Kubrick was lost in thought. Tears were clearly visible on Clarke's face. "He was very upset," Randi recalled. "Very, *very* upset."

Throughout the first half, a fretful Kubrick had prowled the theater, commuting regularly up to the projection booth to check focus and monitor sound. Caras remembered him pacing "up and down the side aisles and across the back, looking for the squirm factor." The director would comment later, "I've never seen an audience so restless." He'd stationed someone at the entrance to count the walkouts, which gradually turned from a trickle to a steady rain, and then a deluge at the intermission. When Kubrick returned to his seat, he muttered terse comments to Christiane. "He was suddenly thinking, 'I held certain things too long, either [the] running around the wheel, or . . . ,'" she recalled. "And he felt great hostility from the staff of MGM, and all the executives totally walked out. Bored. It was really frightening."

At the intermission, Christiane felt an oppressive schadenfreude radiating from the house. Escaping to the washroom, she overheard one woman saying to another, "I didn't know there wasn't any air on the Moon." At this, she began to understand that this could simply be the wrong audience. Things might not be as bad as they seemed. "*They* are not going to be the ones that like this film," she told herself fiercely. "They are not going to be who pays for *tickets* for this film. I *know* people love to read Arthur Clarke. I *know* they like stories like that. It's brilliantly done!"

Clarke, who'd already seen the film twice, left at the intermission, retreating to the Chelsea in humiliation and disappointment. Later, he recalled overhearing another comment emanating from the seated phalanx of MGM executives: "Well, that's the end of Stanley Kubrick."

By the end, 241 walkouts had been recorded—more than one-sixth of the audience. Afterward, the Kubricks sent their girls off to the mansion they'd rented on Long Island and made their way back to the Plaza, where they'd organized a postpremiere reception in a large suite. Jeremy Bernstein had attended the screening with his *New Yorker* editor, who commented, as they rose to leave, "Whatever it was, it was a big one." Arriving at the party, which he described as "very gloomy," he saw Terry Southern sitting in a corner "looking rather miserable." A sense of doom hung over the room. "Apart from Louie Blau, who kept on going around saying it was a masterpiece, I think the rest of us, including me, were disappointed," he recalled. "We had expected something else." Asked about Kubrick's demeanor, Bernstein said, "He looked a little bewildered, I think."

Christiane remembers the scene with the indelible clarity that traumatic events can retain for decades. "It was just a room full of drinks and men and tension," she said. "I've never seen such a full room, you could barely move, and it went on and on. And Stanley was so unhappy." People said "poisonous things," and friendships ended that night, with "schadenfreude and a nasty smile. The thing we all fear most: other people being triumphant that you've failed."

She recalls Southern as a hugely reassuring presence, however. "Terry was trembling for Stanley because some of these people were very negative," she said. "I spoke to Terry most of the time because I was lost and afraid of the people." She remembers saying, "I'm so sad, this is so horrible for Stanley," and Southern replying, "This is a great film, don't worry, it'll be fine. Look at these assholes. Do you normally talk to them? Talk to me."

"I've always loved him for that," she said.

Throughout, Kubrick chain-smoked, and by three in the morning, everyone had gone, leaving a nasty vacuum, a stink of cigarettes, and a colossal sense of failure. "Stanley was tearing himself to *shreds*," Christiane said. "Saying, 'Oh my God, they really hated it.' He was heartbroken." She tried to rally him, saying, "They did *not* hate it. Did you see that audience? They're people you never *talk* to. Of course *they* didn't like it. Wait until the audience comes—it will be different."

Impossible to mollify, he paced the room, asking, "What am I going to do?" over and over. Normally her husband was very secure, observed Christiane, but the evening had devastated him, and now he was losing his voice. She decided to take a different tone.

"This is rubbish," she told him. "You're just obeying something you would never obey normally, and now you're folding up. This is crap!" She poured him a scotch. "Have a drink."

By now, it was four in the morning. "And we lay down, and Stanley couldn't sleep and couldn't speak and couldn't do anything, he was just shattered," she said. "He was close to crying. I mean, he didn't cry, but he said, 'Oh God, this is just terrible.' He felt terrible—terrible, *terrible*—and we had rented this house, so very early in the morning, like four, he said, 'Listen, let's drive there, and at least have something to do.'"

> So we went to that house on Long Island—which was splendid—and I remember I had a handbag and an evening dress, and I just fell on my stomach on the bed and fell asleep completely. Only to wake up to the radio, where the guy was reading the news and saying, "They're standing around the block for Stanley Kubrick's film." And they did. There was the first performance of the day, twelve o'clock or something like that, and there were huge queues, and [on the radio] they said, "This is a fantastic film." And from then on, it rained praise.

She woke him up just in time to hear the end of the report—a silver lining concluding one of the darkest nights in Stanley Kubrick's life.

. . .

Apart from Penelope Gilliatt in the *New Yorker*, who described *2001* as "some sort of great film, and an unforgettable endeavor," the city's leading critics lined up to slam the movie, mostly in reviews published that morning. Writing in the *Times*, Renata Adler called it "somewhere between hypnotic and immensely boring," accusing the film of "a kind of reveling in its own IQ." (She did, however, approvingly cite the "carnivorous apes that look real.") In the *New Republic*, Stanley Kauffmann called *2001* "a major disappointment," stating that it was "so dull, it even dulls our interest in the technical ingenuity for the sake of which Kubrick has allowed it to become dull." In the *Village Voice*, then at the height of its influence, Andrew Sarris dubbed *2001* "a thoroughly uninteresting failure and the most damning demonstration yet of Stanley Kubrick's inability to tell a story coherently and with a consistent point of view."

Although he caught wind of her disdain well in advance, Pauline Kael, already one of the most influential film critics in the country, waited almost a year before absolutely strafing Kubrick. In an article for *Harper's* that didn't even give *2001* the respect of considering it alone, Kael called the film "monumentally unimaginative," denounced it as "trash masquerading as art," and characterized it as "Kubrick's inspirational banality about how we will become as gods through machinery." She even accused him of theft, on the grounds of similarities between the Star Gate sequence and the work of experimental filmmaker Jordan Belson—something like accusing Faulkner of stealing from Joyce because both used stream of consciousness techniques.

We don't know what MGM's supervising editor, Maggie Booth, said to Kubrick or her colleagues after that first Culver City screening less than two weeks before. But she'd left her imprint on all of the studio's productions for more than three decades, and we can be sure her views were taken into consideration. By April 4, the consensus at MGM was that the film was an epic disaster, with some going so far as to say it would sink the studio. Kubrick's contract gave O'Brien the right to request changes that if disputed by the director would be subject to an audience reaction test. But after O'Brien's unwavering support over the past four years, Kubrick wasn't about to initiate such a confrontation. In any case, by April 4, Kubrick had already conducted several excruciating assessments of the audience's "squirm factor," not to mention counting the walkouts. That reaction test had already been made, and the results were clear.

With the film now playing daily in Washington and New York and the negative reviews pouring in—albeit ameliorated by a handful of positive ones—a decision was taken to make cuts. While Kubrick would later state they were made "at no one's request," we can be sure the studio exerted significant pressure in this regard. Participants at the April 4 trim meeting in MGM's headquarters on Sixth Avenue included Kubrick and top studio officials, including O'Brien; MGM VP and director of advertising and promotion Dan Terrell; VP and general sales manager Morris Lefko; and the president of MGM International, Maurice Silverstein—whose teenaged nephew had provided the first "under-thirty-five" reaction. The next day, Kubrick and Ray Lovejoy met in a basement editing facility at the MGM building and sat up all night working. Editing continued across that weekend, ending on the ninth.

Because 70-millimeter prints with optical soundtracks integrally bound to them had already been distributed to eight theaters, Kubrick's edits were artificially constrained to places where impact on the film's sound was minimal—a highly unorthodox situation, and far from ideal. As a result, Aram Khachaturian's moody "Gayane Ballet Suite (Adagio)"—which plays across the opening shots of *Discovery* on its way to Jupiter—had to be trimmed, though the effect on the composition was subtle. On the ninth, a list of cuts was sent to the Loews and Uptown theaters, the Warner in LA, and also to theaters in Boston, Detroit, Houston, Denver, and Chicago, all of which started screening *2001* on April 10 or 11. After years of excruciating, finicky exactitude concerning all aspects of the production, crucial cuts to *2001: A Space Odyssey* were being entrusted to eight unknown projectionists working on tabletop splice rigs—at least until new prints could be struck.

According to an unnamed studio executive quoted in an unpublished draft of Agel's *The Making of Kubrick's 2001*, top MGM brass "were all present until the decision of the trim was finalized. None of them stayed for the mechanics, however." Kubrick insisted this line be removed from the book. Still quoting the executive, the text continued:

> The basement in the MGM building where we have our editing facilities was generally crowded during trim sessions. Lots of people looking for Stanley—big people, little people, fat people, thin people. Stanley's youngest daughter, Vivian, was present during much of the editing. She's a chocolate donut nut—always asking for a chocolate donut. Stanley hated the Sixth Avenue Delicatessen across the street. He said it couldn't make a good outgoing sandwich.

This passed muster but was followed by another line that Kubrick crossed out: "He did not seem happy to make the cuts—trims—but I might be wrong. I certainly got the feeling that what was trimmed was what would show in the least in projection." Kubrick defended his changes to the book with scribbled comments. About the trim meetings: "Not true. No MGM exec ever suggested the film be cut." Concerning being unhappy to make the cuts: "Bullshit, again this is anonymously attributed. It's a lie."

In fairness to the unnamed executive—who himself indicated he might be wrong—all the evidence suggests that by April 4, Kubrick had fully recognized the necessity of trimming his film. This includes what he'd said to Christiane at the premiere the night before. Apart from exhaustion, Kubrick's demeanor may have stemmed from concern over how to accomplish this, given the unusual stricture that any trims be undetectable in a sound mix now permanently wedded to the picture.

As for the director's categorical statement about MGM not suggesting trims, it's certainly wrong and reflects his determination that no hint of the studio's true influence be permitted beyond a small circle. As even Kubrick loyalist Roger Caras put it in 1989, "Stanley will deny anything, no matter what, he will deny anything he thinks will reflect less than sensationally on the mythic Kubrick."

The cuts, made on MGM publicity man Michael Shapiro's Moviola, amounted to about nineteen minutes—or about 12 percent of the film's original running time. They included the removal of Dullea's jogging sequence—the one which had triggered boos and catcalls on the third—a redundant scene anyway, since the first interior views of *Discovery* featured Lockwood doing exactly the same thing. Also tossed was a lengthy, almost shot-by-shot repetition of Bowman's preparations to exit *Discovery*, this time by Poole—an excision that reportedly reduced some of the shock of his subsequent murder. Other edited sections included a request by HAL to replay Mission Control's message concerning his own malfunction; trims to the lunar monolith scene; a sequence in which HAL switches off the radio link to Poole just before killing him; and some edits to the Dawn of Man sequence.

Two new title cards were added as well. One was positioned just prior to the first view of *Discovery* and read "Jupiter Mission 18 Months Later." The other, "Jupiter and Beyond the Infinite," signaled the film's last two acts, namely the Star Gate and hotel room sequences. Finally,

Kubrick inserted the brief reprise of Cantwell's sunrise-over-the-monolith shot, just prior to Moonwatcher's bone-weapon epiphany.

The changes were controversial in some quarters. A letter from NYU graduate film student Jon Davison to the *New York Times* on April 28 slammed MGM for the edits and "meaningless title cards." Davison wrote that "Stanley Kubrick's magnificent work has been butchered; the sad result of the critical abuse heaped upon it," concluding, "The bastardization, complete with sloppy splices and uneven pacing, is now being viewed by even more confused audiences than met the original. But the most confused of all is MGM, whose lack of artistic faith in its own film led it to cut what it couldn't comprehend, thus destroying what it hoped to save."★

There are good reasons to suppose that the initial hostility to the film, particularly among New York critics, was at least to some degree due to that first edit with its avoidable redundancies. Contrary to Davison's supposition, however, Kubrick's trims were both self-directed and well founded. The bad splices he referred to must mean that the print he saw was one of those trimmed by projectionists in response to the cut list sent out from MGM on April 9. They would soon be replaced by seamless new prints.

While *2001*'s negative reviews weren't limited to New York, they were predominantly by critics who'd seen the first edit. Hollywood's leading industry paper, *Variety*, gave the film a decidedly mixed and ultimately negative review on April 3, singling out "some wholesale and rather hasty cutting decisions on the part of Kubrick" and stating flatly, "*2001* is not a cinematic landmark." (The cutting decisions referred to were in the premiere print, not the shorter version.) Remarkably, the writer believed the makeup in the Dawn of Man sequence was "amateurish compared to that in *Planet of the Apes*." A more sympathetic *Variety* piece published two weeks later pointed out that "Kubrick didn't see a final cut of *2001* until eight days before the press preview," and meditated on the damage the longer first edit may have done: "Con-

---

★ Davison would go on to produce such high-profile science fiction films as *RoboCop* (1987) and *Starship Troopers* (1997), both for director Paul Verhoeven.

sidering that most viewers, regardless of their initial reaction, agree that *2001* is substantially better in the tighter version, it seems a shame that Kubrick's first, not final cut, was the one subjected to national reviews."

As if to prove the point, a sidebar story quoted *Boston Globe* critic Marjorie Adams, who'd seen the second cut and called *2001* "the world's most extraordinary film. Nothing like it has ever been shown in Boston before, or, for that matter, anywhere . . . This film is as exciting as the discovery of a new dimension in life."

Despite the problematic launch, Kubrick and Clarke's cosmic epic wasn't just holding its own. It was starting to prevail.

■   ■   ■

Contrary to the myth that *2001* faltered at the box office and was on the verge of being withdrawn when younger audiences came riding to its rescue, box office data reveal excellent ticket sales from Day One. Within a week of its premiere, the April 10 *Variety* was already recording advance ticket sales 30 percent better than they were for MGM's 1965 hit *Doctor Zhivago*.

But as the reaction of Silverstein's nephew had already hinted, how *2001* was received largely depended on which side of the late-sixties generational divide the audience fell. Like "Mr. Jones" in Bob Dylan's 1965 song "Ballad of a Thin Man," something was happening, and older audiences didn't necessarily know what it was. Most film executives didn't, either, and the *Variety* piece on April 10 led with what today might seem a ridiculously self-evident observation: "Because today's filmgoers are predominantly under 25, it would seem vital for the industry to learn something about this market and their tastes." Citing Marshall McLuhan, the article cut to the heart of the Clarke-Kubrick left-brain, right-brain split that Colin Cantwell had identified months before, pointing out that to the "visual-oriented" youth of 1968, "visual and aural sensations have replaced words." It also quoted one of Kubrick's earliest public comments on *2001*: "I wanted to make a nonverbal statement, one that would affect people on the visceral, emotional, and psychological levels. People over forty aren't used to breaking out of the straitjacket of words and literal concepts, but the response so far from younger people has been terrific."

So while younger people were, in fact, largely responsible for the film's success, this happened almost immediately. By mid-May, *Variety* was reporting that in its first five weeks *2001* had already grossed more than $1 million from only eight theaters, with Loews Capitol in New York instituting five o'clock screenings on weekend afternoons to keep up with the demand. Meanwhile, as popular acclaim continued mounting, a number of critics had second thoughts. *Newsday*'s Joseph Gelmis, for example, was intrigued enough to attend a repeat screening, this time of the second cut. His first review, on April 4, had concluded, "The film jumps erratically. The episodes aren't structured logically until the very last moments of the film. It is a mistake. Instead of suspense, there is surprise and confusion, and, for many, resentment."

In his second piece—a remarkable mea culpa published two weeks later—Gelmis proclaimed the film a "masterwork," comparing initial critical reactions, including his own, to the reception that had greeted Herman Melville in 1851.

> About 100 years ago, *Moby-Dick* was eloquently damned and devastatingly dismissed by one of Britain's most influential and erudite literary critics. He argued persuasively that the book was a preposterous grab bag. He ridiculed its self-indulgent lyricism and poetic mysticism. He said it was an unconditional failure because it didn't follow the accepted canons of how a 19th-century novel should be written. He was impeccably correct. Yet today there are perhaps a half-dozen scholars who can recall the critic's name, while every college freshman knows the name of the maligned novelist.

He went on to observe, "A professional critic is sometimes trapped by his own need for convenient categories, canons, and conventions . . . He is the upholder of the familiar, the promoter of the status quo." Clarke was so tickled that he wrote Gelmis a note from the Chelsea on May 6.

I was fascinated, and most impressed, by your article in the 20 April *Newsday*, as the parallels with the initial receptions of *2001* and *Moby-Dick* had already occurred to me. In fact, I had remarked that the rapid *volte-face* of the critics (3 weeks, instead of some 80 years) is an indication of today's rate of progress. It's a point for McLuhan too . . . However, what makes your comment doubly interesting to me is the fact that during the last couple of years of work on the project, I was quite consciously aware of another *Moby-Dick* parallel—i.e., the use of "hard" technology to set a background for metaphysical and philosophical speculations.

Although he wasn't a critic and hadn't seen the trimmed version, one viewer who remained unconvinced was none other than Fred Ordway, who'd attended the Washington premiere and counted fifty-two people leaving. Already agitated by Kubrick's turnaround concerning his International Space Museum project, Ordway may not have been in an ideal state of mind to view the film. His single-spaced, eight-page letter to Kubrick on April 9 vented exasperation at *2001*'s opacity and contained an extraordinarily detailed and comprehensive negative assessment. Coming from one of Kubrick's closer collaborators, it was a remarkable document.

■　■　■

With ticket sales booming and most of the later reviews positive, the Kubricks largely withdrew to their large rented mansion on Long Island's North Shore, where crowds of "big people, little people, fat people, thin people" descended on them, mostly from the media. "The house had a landing with a light facing Long Island Sound," recalled Mike Kaplan, a young marketing executive brought in to handle *2001*'s promotion once the median age of its audience became clear. "Such was the interest in anything Kubrick that rumors quickly spread it was the actual setting of Jay Gatsby's mansion in *The Great Gatsby*."

A talented visual artist, Christiane tried her hand at depicting the bay and house in oils while simultaneously providing tea and coffee for an endless stream of unwanted guests. She recalled female journalists jostling one another in an effort to get close to Stanley. "It's the only time in my life that I experienced a kind of catty atmosphere where my husband was concerned," she said. "They wanted to spend time with him, and I was inconvenient." Disapproval at her national origins rose to the surface in disgusting ways. "There were some really nasty remarks. You know, my English wasn't as good then, and I had a German accent. And the war was much more recent. And they just didn't want me around. I remember one time I had a nice suede jacket on. And one of them looked at me—I don't remember her name—and she said, 'What kind of skin is that you're wearing?'"

Recalling the incident with a shudder, Christiane said, "I praised myself for one thing: that these kinds of horrible experiences that I had—several times—I didn't tell him. I thought I was strong in the sense that my little German hang-up didn't need to be his fate."

Although the trimmed film—and the excellent box office—was helping to facilitate *2001*'s acceptance, in late April Kubrick decided that a further explanation might be in order, and he chose the *New York Times*'s movie editor Abe Weiler, then writing a Sunday filmmaking column, as the conduit. Weiler quoted the director at length.

What happens at the end must tap the subconscious for its power. To do this, one must bypass words and move into the world of dreams and mythology. This is why the literal clarity one has become used to is not there. Here is what we used for planning. In Jupiter orbit, Keir Dullea is swept into a Star Gate. Hurled through fragmented regions of time and space, he enters into another dimension where the laws of nature as we know them no longer apply. In the unseen presence of godlike entities—beings of pure energy who have evolved beyond matter—he finds himself in what might be described as a human zoo, created from his own dreams and memories . . . His entire life passes in what appears to him a matter of moments. He dies and is reborn—transfigured. An enhanced being, a Star Child. The ascent from ape to angel is complete.

It took the Kubricks time to recover from their ordeal on the night of the third. At one point, Stanley confided to Christiane, "I was so shocked that I still think any minute now something horrible is going to happen." Gradually, though, the tension of four years' sustained effort started to ebb. While he raged at the nasty reviews, he also received a flood of awestruck and supportive letters, and not only from young people. As time passed, the positive assessments outweighed the negative ones. Asked about the initial reactions that September, Kubrick responded, "New York was the only really hostile city. Perhaps there is a certain element of the lumpen literati that is so dogmatically atheist and materialist and Earthbound that it finds the grandeur of space and the myriad mysteries of cosmic intelligence anathema."

In addition to its landing and light, the Gatsbyesque house with its enormous porch on Long Island Sound also had a shooting range in the basement, and that summer Roger Caras and family would regularly drive over from East Hampton to visit. He and Kubrick shared a love of guns, and Roger would bring a small arsenal to match Stanley's own. According to Christiane, both were excellent shots. And that's as good a scene as any to leave the director of *2001: A Space Odyssey* and his New York consigliere: the two of them blazing away at a target wreathed in a blue haze of gun smoke, side by side in the subterranean recesses of a magnate's mansion beside the darkening sea, with the sun sinking over distant Hollywood and Christiane wincing in disapproval at the noise and stink from the top of the stairs.

CHAPTER TWELVE

AFTERMATH

> *Two possibilities exist: either we're alone in the*
> *universe, or we're not. Both are equally terrifying.*
>
> —ARTHUR C. CLARKE

Many decades later, when he was relegated to a wheelchair and rarely left his adopted country, now called Sri Lanka, Arthur Clarke saw a photograph of Stanley Kubrick on his TV and was deeply moved. The author was seated in his study on the second floor of his spacious house in Colombo, watching a taped documentary about his old collaborator. The photograph appeared quite suddenly. In it, Kubrick was sitting on the floor in front of his own television in Childwickbury Manor—the sprawling eighteenth-century estate house with acres of surrounding greenery he'd purchased in 1977— and holding a microphone up to the screen.

At first glance, this mirroring and sense of remove might seem to evoke the distanced, screen-based world the two of them had conjured into being between 1964 and 1968 in *2001: A Space Odyssey*—one in which human interactions occur across vast, echoing, alienating gulfs of space and time. But for Clarke, the sudden appearance of that televised photograph inverted any sense of remove, bringing a decisive friendship—and certainly the most important professional relationship in his life—crashing back into the room without warning.

It's not surprising that the photo produced an upwelling of emotion in the writer. Apart from providing another example of the director's familiar intensity—Kubrick's lifelong determination to absorb and record everything important, without missing an implication or word—it revealed that his microphone was, in fact, recording none other than his old ally and intellectual sparring partner Arthur Clarke, who, in turn, was being featured in a BBC documentary.

"In those days, you couldn't record television very easily, and so he sat there throughout the film," Christiane recalled of her husband, who was most likely recording a 1979 broadcast of *Time out of Mind*, a program focusing on science fiction. "Stanley really admired Arthur and valued his opinions on everything and had great respect for him."

A scene, then, of an aging science fiction writer watching a screen, seeing an aging film director also watching a screen and recording that writer's every word. Taking it all in, Clarke blinked owlishly behind his glasses several times and burst into tears.

▪   ▪   ▪

Finally published four months after the film's release, Clarke's novel was dedicated "To Stanley" and became an instant bestseller. Although he went out of his way to stress that his version of their mutually conceived narrative didn't necessarily reflect Kubrick's views, the book quickly became a kind of code breaker's handbook, pored over by baffled viewers who attempted to decipher *2001*'s cryptic meanings.

One of Clarke's early comments following the film's Washington premiere was, "If anyone understands it on the first viewing, we've failed in our intention." Given what we know of his own initial reaction, it was his way of making lemonade. Kubrick quickly expressed his disapproval. "I don't agree with that statement of Arthur's, and I believe he said it facetiously," he said that September. There followed an exchange in the media, with Clarke standing by his remark and pointing out that it didn't mean you couldn't enjoy *2001*, only that it rewarded repeated viewings. In any case, following the novel's July 1968 release, his new sound bite for confused viewers was "read the book, see the film, and repeat the dose as often as necessary."

The box office remained extraordinary, and *2001* became the highest-grossing film of the year—the only Kubrick picture ever to achieve such a standing. That June, Clarke wrote his friend Ray Bradbury, who'd written a negative review of *2001*, urging him to see the second cut. Commenting on the ticket sales, Clarke observed, "Stanley is now laughing all the way to the bank." The following year, he amended this in an interview to "Stanley and I are laughing all the way to the bank." By then, the novel was manifestly helping his bottom line as well, evidently to considerable merriment. The lemonade had become lucrative.

A decade later, Clarke was disarmingly candid about what his expectations had been following the film's disastrous initial screenings. "The success of *2001* was a great surprise to me," he told a BBC journalist, probably in the very program that his collaborator would later record with a handheld microphone, "and I suspect to Stanley Kubrick as well."

> Of course, we hoped it would be successful, but we never imagined it would become a cult movie and have such tremendous sustaining power. It may have had something to do with the timing. It came out just before the first Apollo flight around the Moon—Apollo 8; the Christmas flight in 1968—and then, of course, the landing on the Moon in July of 1969. But I don't think the people who were interested in spaceflight as such were the ones who made up the audience of *2001*, which included a large number of hippie types, and perhaps even people who were antagonistic to technology.

He wasn't wrong. *2001* swept the entire sixties counterculture into theaters worldwide, inspiring raves from some of its leading figures. Asked about the film in 1968, Beatle John Lennon quipped, "*2001*? I see it every week." The following year, David Bowie released "Space Oddity," the single that introduced his astronaut alter ego "Major Tom" to the world—a nod, clearly, to *2001*. But, of course, everyone even remotely interested in spaceflight and technology went to see it as well.

When he'd had time to recover from his initial reaction, Clarke spoke candidly about the fate of his narrative voice-overs. "Stanley very wisely realized that using narration, although it might have made

Clarke on a rare visit with the Kubricks in the late 1970s.

it simpler and clearer, would have been intolerable," he told Joseph Gelmis, the *Newsday* critic who'd publicly reevaluated his own initial review. "It would have destroyed much of the mystery."

Within a few weeks of the premiere, Clarke even found himself enjoying the heated debates engendered by *2001*'s willful ambiguity, sometimes positioning himself near theater doors just to overhear them. "It's creating more controversy than any movie I can think of," he told a Berkeley radio station in May 1968. "I used to have great fun standing outside the theater and listening to the crowds, the people coming out and arguing all the way down Broadway . . . And this is fine. We want people to think, and not necessarily to think the way that we do."

Asked if he felt the pervasive spread of technology was beginning to dehumanize us, Clarke replied, "No, I think it's superhumanizing us."

∎   ∎   ∎

A half century later and almost two decades after its eponymous year, *2001*'s influence remains so pervasive that it's hard to overestimate. The film's fusion of scientifically informed speculation, industrially supported design, technofuturism, and kaleidoscopic cinematic abstraction brought art and science together in ways never seen previously. *2001*'s ongoing sway over design alone can be seen across filmmaking, advertising, and technology. Its impact on contemporary discourse includes the ubiquitous name checking of HAL in increasingly relevant and even alarming discussions of artificial intelligence. Richard Strauss's *Thus Spoke Zarathustra* is forever associated with the film, so much so that it's hard to consider it in isolation from Kubrick's epochal opening sunrise over the Earth and Moon. *Zarathustra* has been used many times to reference *2001*, including in Hal Ashby's 1979 film *Being There*, in which Peter Sellers, playing the middle-aged simpleton Chauncey Gardiner, leaves his house for the first time while Brazilian composer Eumir Deodato's Grammy-winning funk arrangement of the composition booms on the soundtrack. In late 2017, Todd Haynes revived Deodato's version to similar effect in his film *Wonderstruck*.

Other recent references include overt tributes in various episodes of *Mad Men*, including one titled "The Monolith"; repeated riffs in Matt Groening's animated series *The Simpsons* (in one episode, Bart threw a felt-tip marker in the air, which becomes a satellite); Tim Burton's 2005 film *Charlie and the Chocolate Factory*, in which shots of Dan Richter and his capering man-apes were actually inserted into a scene (the monolith morphs into a chocolate bar, a bit too obviously); and an homage to *2001*'s final sequence in Bill Pohlad's moving 2014 docudrama about the Beach Boys' Brian Wilson, *Love & Mercy*. These and other examples attest to the film's continuing power to affect contemporary culture—certainly one measure of a masterpiece.

Apart from such overt references, *2001*'s epic validation of the science fiction genre stands at the wellspring of all the big-budget, effects-laden movies to follow. Just as Kubrick and Clarke had intuited in their opening conversations in New York, the film marked the end of the Western as the predominant Hollywood genre and its replacement by stories set in a more expansive new frontier. Among the first cinematic responses, ironically, was Russian master Andrei Tarkovsky's 1972 film *Solaris*. Although he publicly disparaged *2001* in 1968, the wheel-shaped space station, disembodied alien intelligence, and metaphysical concerns of *Solaris* betray the clear influence of its predecessor—even if Tarkovsky intended his film as a kind of riposte. Worth watching and reasonably well received in the West, *Solaris* had its flaws, and the director himself later let it be known that in his view it was a failure—so he ultimately had problems with both *2001* and his own rejoinder.

In American filmmaking, *2001*'s visual power and excellent box office gave an emphatic green light to Hollywood, which set about funding successor projects. From *Close Encounters of the Third Kind* to the *Star Wars* and *Alien* franchises—and ultimately such comparatively recent films as James Cameron's *Avatar* and Christopher Nolan's *Interstellar*—a series of landmark movies was influenced and enabled by Kubrick's and Clarke's achievement. Asked if a lightbulb ever went off that inspired him to become a filmmaker, Cameron, the director of the two highest-grossing movies ever made, was unequivocal.

> The first one was when I saw *2001: A Space Odyssey* for the first time. And the lightbulb there was, "You know, a movie can be more than just telling a story. It can be a piece of art." It can be something that has a profound impact on your imagination, on your appreciation of how music works with the images. It sort of just blew the doors off the whole thing for me at the age of fourteen, and I started thinking about film in a completely different way and got fascinated by it.

For his part, Steven Spielberg has called *2001* "the big bang" that inspired his generation of directors. And in the late 1970s, George Lucas assumed a tone of humility when considering the film. *2001*, he said, is the "ultimate science fiction movie," continuing, "It is going to be very hard for someone to come along and make a better movie, as far as I'm concerned. On a technical level, [*Star Wars*] can be compared, but personally I think that *2001* is far superior."

More recently, he grew almost tongue-tied when asked about *2001*.

"I'm not sure . . . I would have had the guts to do what Stanley did," he said in the documentary *Standing on the Shoulders of Kubrick: The Legacy of 2001*. "I don't know whether I—I'm trying to get up the, the fortitude to do something like what he did."

· · ·

As of 2018, Clarke's novel *2001: A Space Odyssey* has gone through well over fifty printings, selling in excess of four million copies. Before his association with Kubrick the author was considered one of science fiction's "big three," along with Robert Heinlein and Isaac Asimov. But *2001* vaulted Clarke into another category altogether. He was now truly world famous, and wealthy as well. While he may not have had a direct financial stake in the film, he had a significant indirect one, and his manifest loyalty to Kubrick, coupled by his patient tenacity in wrestling with *2001*'s innumerable plot revisions, paid off hugely. His other work benefitted as well, with three new printings of *Childhood's End* in 1969 alone. As Kubrick had predicted, everything came out all right in the end.

One of the deals that Clarke's protégé Mike Wilson thought he'd made during an ill-fated attempt to take on the role of film producer during *2001*'s final year of production concerned *Childhood's End*. In 1967 he informed Clarke that he'd managed to sell the property to director John Frankenheimer, but like so much else with Wilson in the late 1960s, the deal soon fell through—and eventually, so did his relationship with an exasperated Clarke. The definitive break finally occurred in 1972, when Wilson filed a lawsuit trying to recover what he claimed as his property and cash from his erstwhile partner. The suit was settled out of court, and Hector Ekanayake took over as the primary beneficiary of Clarke's extraordinary largesse. The James Bond parody that Clarke had been forced to finish was the last film Wilson ever made.

· · ·

When the time came to finally vacate his studios and depart Borehamwood in February 1968, Doug Trumbull was told he couldn't bring any of his working materials with him, and he flew back to Los Angeles with a bitter taste in his mouth. "I was to go back with my clothes," he recalled incredulously. The following year, *2001* was largely shut out of the Oscars, having been nominated for four, including Best Picture. As previously mentioned, its sole award, for Best Visual Effects, went to Kubrick alone. It was the only Oscar the director ever received.

Trumbull has long maintained it was inappropriate for Kubrick to do what he did. "Those effects were not designed and directed by Stanley Kubrick," he said categorically. Questioned further, he gave a little.

> One directs the movie. You don't direct the visual effects. You know, that's a double-crediting thing that's really not appropriate. Stanley was extraordinarily involved in directing the visual effects. There's no question about it. He was on the set, telling the cameraman where to put the camera on a miniature shot. That's directing special effects. Okay, so I don't deny him that. But he didn't design them, and he didn't do them. There was a whole crew of people that did them. It's like saying he did the costumes, or he did the whatever. So it was just really inappropriate, and I let it go by because I didn't want to play out some contention in the press or the trades or anything like that.

There's little doubt, however, that the first footage shot for *2001: A Space Odyssey*—the Manhattan Project cosmological material filmed in noxious tanks of paint thinner, ink, and paint in early 1965—was designed by Kubrick, who also manned the camera. Produced before Trumbull's direct involvement, it comprised about a third of the Star Gate sequence and was clearly a point of pride for the director.

In any case, throughout his post-*2001* career, journalists and others would sometimes refer to Trumbull as having "done" *2001*'s visual effects—a description that failed to mention the contributions of Wally Veevers, Con Pederson, Tom Howard, and Kubrick himself. This always irritated the director, who tended to blame Trumbull. At first, he sent warning notes, which his erstwhile protégé always responded to politely, stating that it was typical journalistic oversimplification, that he'd never be so crass as to try taking sole credit, and that Kubrick himself had been misquoted and mischaracterized often enough in the media that he should understand.

It all came to a head in 1984, when Hewlett-Packard published an ad featuring the lines "The year was 1968. But for the audience, the year was 2001. And they were not in a movie theater, they were in deep space—propelled by the stunning Visual Effects of Doug Trumbull." At this, Kubrick and MGM threatened HP with legal action, and the company quickly withdrew its ad. This wasn't enough for the director, however, who subsequently paid for a full-page message in the *Hollywood Reporter*. Quoting the offending copy in full, it said the ad had been withdrawn and that "Mr. Trumbull was not in charge of the Special Effects of '2001: A Space Odyssey.'" Published under the names Kubrick and MGM, the message proceeded to list *2001*'s lead visual effects credits—with Kubrick first, Veevers second, Trumbull third, and Howard last—stating that the order reflected "the comparative contributions of the people principally responsible for the Special Effects work."

This public slapdown, so similar to Kubrick's disavowal of Terry Southern after *Dr. Strangelove*, was, of course, a humiliation for Trumbull, and led to a break between them that lasted for years. Finally, after about a decade, Trumbull—who considers the director "an absolute genius, and he was my friend, and he was my mentor and my teacher and my buddy, and I did a lot of really good stuff for him, and I was grateful to him"—decided to cold-call Childwickbury Manor. On reaching Kubrick, he said, "Stanley, I'm calling you because I wanted to tell you to your face that working with you was the most important thing that ever happened in my life. And I want to thank you."

Kubrick responded with "Wow, thanks."

Doug continued, "I just want you to know I'm here. I'm so grateful because everything I do every day harks back to that opportunity."

It was the last time they spoke.

· · ·

Kubrick never again attempted anything of the ambition, complexity, and scale of his eighth dramatic feature, *2001: A Space Odyssey*. He did produce and direct five more manifestly innovative films, however: *A Clockwork Orange, Barry Lyndon, The Shining, Full Metal Jacket*, and *Eyes Wide Shut*. All were exceptional, with *Barry Lyndon* in particular breaking new ground in its pioneering use of low-light cinematography. Deploying fast Zeiss lenses originally developed for the Apollo program, Kubrick and director of photography John Alcott, who'd graduated from the role of Geoffrey Unsworth's camera assistant during *2001*'s last year of production, took advantage of their extremely wide apertures to shoot interior scenes illuminated almost entirely by candlelight. The result was the first accurate representation of what eighteenth-century interiors looked like before the advent of electricity, giving the film the remarkable aspect of a period oil painting come to life. Alcott would win an Oscar for his work on the film.

Kubrick's most controversial work by far was 1971's dystopian, ultraviolent *A Clockwork Orange*, based on the Anthony Burgess novel—a film the director characterized as "a running lecture on free will." Shot on a budget of only $2.2 million, it was a commercial success, but its graphic portrayals of murder and rape earned it an initial X rating in the United States. After episodes of supposed copycat violence and multiple threats to the Kubrick family, the director had the film withdrawn from distribution in the United Kingdom, where it remained effectively banned until 1999. It was another example of his unmatched clout with the studios.

The initial published responses to *2001: A Space Odyssey* remain among the most widely cited examples of critical misfire, and the film is now regarded as among the most important motion pictures ever made. Yet its rise in critical estimation was surprisingly slow. Every decade *Sight & Sound*, the magazine of the British Film Institute, polls an international group of critics, programmers, academics, and cinephiles, asking them to rank the ten most important films of all time. Almost a quarter century passed before *2001* showed up in its widely respected decadal list, considered the best such survey in the world. In 1982 the film was a close runner-up but didn't quite make the cut. *2001* finally entered the listing in 1992 at number ten. By 2002, it was ranked the sixth most important film in history, a position it retained in the last such survey, in 2012. (Hitchcock's *Vertigo* is currently number one, and Orson Welles's *Citizen Kane* second.)

That's the contemporary critical establishment. At the same ten-year intervals, *Sight & Sound* also asks Kubrick's peers—the world's top directors—to render the same judgment. In this case, it took until 2012—a whopping forty-four years—for the film to make an appearance on their decadal Director's Top Ten Poll. When it did, however, it landed with a bang, in second place. (Yasujirō Ozu's *Tokyo Story* is the current number one, and *Citizen Kane* number three, after a long reign at the top.)

The American public seemingly arrived at a similar conclusion less than ten years after *2001*'s release, however. In 1977 National Public Radio's top news show, *All Things Considered*, conducted a poll among its 1.5 million listeners. Seeking to tabulate American cinematic tastes, the program asked which should be considered the ten greatest American films of all time. The results placed *Citizen Kane* first, *2001: A Space Odyssey*, second, and *Gone with the Wind*, third.

Stanley Kubrick's last public statement was in 1998. It came in the form of a video address transmitted to the Directors Guild of America on the occasion of his receiving its D. W. Griffith Lifetime Achievement Award. (Long considered among the most important figures in the history of feature filmmaking, Griffith, who lived from 1875 to 1948, was also highly controversial due to his use of racist caricatures in films such as *The Birth of a Nation*, which depicted the Ku Klux Klan in a heroic light.) Observing that Griffith's career was "both an inspiration and cautionary tale," Kubrick said the director "was always ready to take tremendous risks in his films and in his business affairs." He meditated on how the man who'd transformed movies from a "nickelodeon novelty to an art form" had spent the last seventeen years of his life shunned by the very industry he'd helped create. For Griffith, Kubrick said, the wings of fortune "had proven to be made of nothing more substantial than wax and feathers."

I've compared Griffith's career to the Icarus myth, but at the same time, I've never been certain whether the moral of the Icarus story should only be, as is generally accepted, "Don't try to fly too high," or whether it might also be thought of as, "Forget the wax and feathers and do a better job on the wings."

It was a statement he was uniquely qualified to make.

· · ·

Stanley Kubrick died of a massive heart attack on March 7, 1999, less than a week after screening a fine cut of *Eyes Wide Shut* for his family and the film's stars, Nicole Kidman and Tom Cruise. Close associates contend he undoubtedly would have made further edits after preview screenings, just as he'd done with *2001* and other pictures.

On hearing the news, Doug Trumbull—who, like many who'd worked with Kubrick, was profoundly shocked, and couldn't easily conceive of the director as mortal—realized there would inevitably be a memorial service. He called Jan Harlan to see if he might be invited, and, on receiving the okay, packed his clothes, hopped a plane, and made it to Childwickbury Manor just in time. By all accounts, the service, which took place under a huge canopied tent on a cold and rainy afternoon, was beautiful and moving. It unfolded among an extraordinary display of flowers, most of them rooted in soil directly below a cover that had been stretched over the lawn. The family had received permission to bury him on the estate, and the service was conducted beside an open pit at the base of a large, symmetrical evergreen—Kubrick's favorite tree. The coffin was carried in by a group of men that included Tom Cruise, after which several eulogies were delivered. These were punctuated by musical interludes performed in part by the Kubrick grandchildren.

Jan Harlan's eulogy came first, and as film critic Alexander Walker recalled, it set the tone perfectly. Surveying the 150 or so people gathered, Harlan said, "I had no idea a week ago that I would be addressing the world's greatest assembly of Stanley Kubrick experts."

"So we all chuckled," Walker said, "and we all relaxed."

Other speakers included Cruise, Kidman, and Steven Spielberg. After almost two hours, the mourners were invited to take a rose from on top of the casket, drop it into the open grave, take a pinch of earth from one of the bowls arranged for that purpose and do the same, and say their goodbyes. Then the coffin was lowered into the grave.

Afterward, everyone was invited into the house for refreshments, and gathered around tables in the warmth of the Kubricks' large kitchen. With night falling, at one point Spielberg leaned over to Walker and commented, "You know, this is extraordinary. In Beverly Hills, there would have been cops and bodyguards and velvet ropes and VIP enclosures. And here we are, eating supper in an English kitchen." It was, Walker observed, "really one of the most intimate and affecting farewells to any friend that I could possibly hope for."

Throughout the event, Doug Trumbull felt lucky to be there and found he was weeping openly at several points. Soon after retreating with everyone else to the kitchen and its "beautiful buffet, drinks, conversation," he realized that Stanley was still outside. Slipping quietly back out the door, he took a chair, sat down by the open grave, and spoke to his old mentor. "Stanley, all this crap that happened was stupid, and that's not what it's about at all," he said, tears in his eyes. "We've had our disagreements, and that's been challenging, but I don't care. I don't care. None of that is important to me. I'm here because I love you, and I think that what you did was so important to cinema, and to my art form and to my life, and I'm honored to be here. Thank you for changing my life."

After that, several workmen came and started filling in the hole with soil.

.   .   .

Arthur Clarke outlived Stanley Kubrick by almost a decade. Although he wrote nine novels following *2001*—including three that took the *Space Odyssey* story forward, to mixed critical reception—he remained best known for the narrative they had produced together during their four intense years of collaboration between 1964 and 1968.

In 1984 a sequel to *2001* based on his novel *2010: Odyssey Two* was released by MGM. Directed by Peter Hyams and titled *2010: The Year We Make Contact*, it starred Roy Scheider and featured Keir Dullea and Douglas Rain reprising their roles as Dave Bowman and HAL. Passable as entertainment, it was ultimately forgettable.

In 1988 Clarke was diagnosed with the progressive neural disorder called post-polio syndrome, and he spent much of his last two decades in a wheelchair. It didn't visibly affect the morale of a man whose curiosity and wonderment about the universe was matched only by his ability to convey it in clear, concise language. In 1994 Kubrick sent him a note apologizing for not being able to attend the taping of *This Is Your Life*, a BBC biographical program focusing on the author. "You are deservedly the best-known science fiction writer in the world," he wrote. "You have done more than anyone to give us a vision of mankind reaching out from cradle earth to our future in the stars, where alien intelligences may treat us like a godlike father, or possibly like the 'Godfather.'"

In 1998 Buckingham Palace announced Queen Elizabeth II's intention to make Clarke a Knight Bachelor—the oldest such honor, dating back to King Henry III in the thirteenth century. The investiture was delayed, however, at Clarke's request so he could clear his name after a British tabloid, the *Sunday Mirror*, took the opportunity to accuse him of pedophilia. The paper's charge was backed by dubious quotes allegedly from Clarke himself, though he vehemently denied making them. When the Sri Lankan police requested tapes of the supposed interview, however, they weren't forthcoming, and a subsequent police investigation found the accusation baseless. The *Mirror* subsequently published an apology, after which the author decided not to sue for defamation.

Arthur C. Clarke was duly knighted on May 26, 2000. As with his disability, the episode failed to quell his inherently hopeful, forward-looking character, nor did it interrupt his lifelong quest to understand and describe the human situation within a vast and cryptic cosmos—all of it in prose suffused by the specifically Clarkean optimism that was one reason for his worldwide fame.

A tendency toward self-aggrandizement wasn't foreign to him—in fact, he mocked himself for it, sending out an email circular to his friends every winter that he called an "Egogram." When considering his role in *2001: A Space Odyssey*, however, he was curiously humble. "I would say *2001* reflects about ninety percent on the imagination of Kubrick, about five percent on the genius of the special effects people, and perhaps five percent on my contribution," he had said in 1970. It was a remarkably self-deprecating comment, one unsupported by the evidence. He may have had Terry Southern in mind.

Asked in the year 2001 if he was disappointed that the grand vision of human expansion into the solar system that he and Stanley Kubrick had realized decades before hadn't come true, Clarke mulled the question over for a minute, then replied that he really wasn't, pointing to the robotic explorations that had opened the solar system to human eyes after millennia of speculation. Much more had been accomplished than he'd ever expected to see in his lifetime, he observed. A couple years later, he expanded on this, writing, "We are privileged to live in the greatest age of exploration the world has ever known."

In his last "Egogram," sent in January 2008, Clarke ruminated on the events of his ninetieth year. Although a typically cheery missive, he clearly sensed that something was coming. "I've had a diverse career as a writer, underwater explorer, space promoter, and science popularizer," he wrote. "Of all these, I want to be remembered most as a writer—one who entertained readers, and, hopefully, stretched their imagination as well.

"I find that another English writer—who, coincidentally, also spent most of his life in the East—has expressed it very well. So let me end with these words of Rudyard Kipling:

> *If I have given you delight*
> *by aught that I have done.—*
> *Let me lie quiet in that night*
> *which shall be yours anon;*
> *And for the little, little span*
> *the dead are borne in mind,*
> *seek not to question other than,*
> *the books I leave behind."*

Arthur C. Clarke died in Colombo on March 19, 2008, of respiratory complications leading to heart failure. He left explicit instructions that no religious rituals of any kind should be performed at his funeral, but a few hours after his death, a gamma ray burst of unprecedented scale reached Earth from a distant galaxy. More than two million times brighter than the most luminous supernova ever recorded, its energy had taken seven and a half billion years to arrive at the solar system—about half the age of the observable universe. Having traveled through space and time since long before our planet formed, for about thirty seconds this vast cosmic explosion became the most distant object ever seen from Earth with the naked eye.

It was the kind of salute even a lifelong atheist might have appreciated.

ACKNOWLEDGMENTS

In Cocteau's film Orpheus, *the poet asks what he should do.
"Astonish me," he is told. Very little of modern art does that—
certainly not in the sense that a great work of art can make you
wonder how its creation was accomplished by a mere mortal.*

—STANLEY KUBRICK

As one might expect given its manifest astonishments, many writers have shared my fascination with *2001: A Space Odyssey.* Less predictable is the sense of fraternity that prevails among those who've examined how the film was realized, and I couldn't be more grateful for their support as I strove to understand how *2001* came about.

No person was more indispensable in this regard than Dave Larson, who signed on officially as my consultant and sounding board. Having spent the better part of two decades researching *2001* in granular detail, Dave is probably the most knowledgeable person alive today concerning the film. That's not my observation but Doug Trumbull's, categorically stated even as he advised me to get in touch with the man. (I've since had many opportunities to confirm Doug's opinion.) A gent and a scholar, Dave graciously agreed to provide me with access to his voluminous archive of all things *2001*—a formidable nested set of matryoshka folders containing innumerable photographs and contact-sheet scans, not to mention rare correspondence and other documents unavailable elsewhere. Dave also uncomplainingly transcribed dozens of taped interviews that he'd conducted between 2001 and 2006 with people who'd worked on the film, most of whom are no longer alive.

After fielding an endless barrage of emailed queries from me, he then read *Space Odyssey* in manuscript, checking facts and offering valuable suggestions. Happily his own book about *2001* is in the works, and promises to be a memorable tome indeed.

No less important to this project was Don Shay, the founder and publisher of the industry's leading visual effects magazine, *Cinefex*. Even before its first edition in 1980, Don had embarked on an effort to interview veterans of *2001* with an eye to writing a book focusing on its visual effects. Instead he founded the magazine—but meanwhile continued taping discussions with the film's principals. The results ultimately materialized in April 2001 within a *Cinefex* issue dominated by a long-form oral-history masterpiece, "2001: A Time Capsule." On reading it with blinking astonishment in early 2017, I sought out Don, who not only gave me access to his full-length interviews, but sent them to me as precious original transcripts in hard copy. Being pre–personal computer, they'd never been digitized.

As expected, they contained a gold mine of material that Don and his coauthor, Jody Duncan, hadn't quite managed to squeeze into their *2001* issue. And I would be eternally grateful for his collegiality in this regard alone, but there's more. That March, Don—a highly experienced editor with more than forty years of experience translating the language of film into that of prose—offered to read my draft chapters as they emerged. His subsequent suggestions were always direct, substantive, and to the point. Don's contribution to this book was significant, and I thank him for it.

Also standing in solidarity throughout was Arthur C. Clarke's authorized biographer, Neil McAleer. The very definition of a mensch, Neil not only emailed me transcripts of his many interviews with Arthur's friends and associates conducted between 1988 and 1990, he also sent me a priceless package of original tapes, since not everything had been transcribed. Neil's labors as he assembled his magisterial four-hundred-page biography in the early 1990s inevitably encompassed matters he felt he couldn't use in a book officially sanctioned by its subject, and it says much about his integrity that he unhesitatingly trusted me with these materials, which enabled me to understand much that had previously been opaque. He followed up by checking in regularly to find out how I was faring, always answering my many email queries promptly and with brimming good humor. Look for his revised, reissued biography, *Arthur C. Clarke: Odyssey of a Visionary.*

As if this wasn't already an embarrassment of riches, in 1999 American mime Dan Richter—who didn't play so much as he incarnated *2001*'s lead man-ape, Moonwatcher—conducted upward of thirty discussions with key people involved in the production. They helped inform his 2002 book, *Moonwatcher's Memoir*—an excellent and entertaining read—and when I spent a couple of revelatory days interviewing him in Provincetown in August 2016, I asked if he'd kept the transcripts. Well, he had—and like Don and Neil, immediately transferred them to me. I soon discovered that veterans of *2001* had opened up to this manifest insider in ways they probably wouldn't have to an outside researcher, and Dan's interviews, handed to me with unconditional generosity, were a substantial boon.

Other irreplaceable supporters of this project include Christiane Kubrick, Stanley's irrepressibly witty wife; her thoughtful and forthcoming brother Jan Harlan; and Stanley's long-serving personal assistant, the inimitable Tony Frewin. Each judged me worthy of their confidence, and for that I'm very grateful. This book is no hagiography, but luckily all three appreciate the intrinsic truth that projects of this kind are worth nothing if not honest, and they trusted me to be fair. I believe I have, and thank them for their friendship and faith.

As for Doug Trumbull, whom I first met more than a decade ago while both of us were working with Terrence Malick to help him realize cosmological sequences for his film *The Tree of Life*, he's been an unwavering supporter and advocate of my work in general and of this project in particular, and I'm honored by his fellowship. My invaluable agent, Sarah Lazin, helped me channel an initially inchoate concept toward something I'd like to think is far more focused and precise, then took my final proposal to the best publishers in the field with excellent results. Her erstwhile assistant, Julia Conrad, provided candid and incisive reader's notes to my initial pitch, and this book is certainly better for her input. Prior to all this, my friend and mentor of many years, Eric Himmel, encouraged me to pursue this project. As did another good friend, the writer and cultural showman extraordinaire Ren Weschler. I thank them all.

Scrupulously fair and open-minded throughout, my editor at Simon & Schuster, Bob Bender, tolerated a certain brinksmanship on my part regarding deadlines, then patiently saved me from myself many times as we negotiated multiple editing phases together. His judicious suggestions and unerring instinct for pacing and tone benefited this book in countless ways. His assistant, Johanna Li, was a model of imperturbable professionalism throughout. Art Director Alison Forner likewise fielded my various design suggestions with forbearance and flexibility. And Manager of Copyediting Jonathan Evans patiently accepted late-breaking revisions and saw to it that the writing was given every chance to speak clearly and concisely. I'm grateful for their expertise and tact.

Sri Lanka–based writer Richard Boyle kindly gave me access to his personal archive of materials connected to his friend Mike Wilson, and I'm grateful also to Mike's daughter Damani for her kindness and insights. I also want to thank Ashley Ratnavibhushana, the distinguished Sri Lankan film critic and historian, for his invaluable help in connecting me to key players in that nation's film industry during my fourth trip there in February 2016. Additionally, I am indebted to longtime Clarke assistants Nalaka Gunawardene and Rohan de Silva, as well as noted Colombo astrophysicist Kavan Ratnatunga, Arthur's friend, and finally Angie Edwards, his niece, for their insights.

The Stanley Kubrick Archive at the University of the Arts London was an essential resource throughout the writing of this book. Senior Archivist Richard Daniels and Assistant Archivist Georgina Orgill steered me to relevant materials, went well beyond the call of duty in forwarding important information my way, and did everything possible to ensure that my experience at the archive was a positive one. I'm grateful to them and their colleagues for their kindness and professionalism.

Likewise Arthur C. Clarke's Papers at the National Air and Space Museum's Steven F. Udvar-Hazy Center provided a crucial window into Clarke's meditations and methods as he worked for four years to help realize *2001*'s narrative in both the film and novel. NASM

Space History Curator Martin Collins deserves particular praise for his achievement in bringing these documents from tropical Colombo to climate-controlled storage near Dulles, where they will be preserved for researchers.

Ex–Kubrick assistant—and later award-winning film director in his own right—Andrew Birkin, aka the Kokerboom Bandit, provided particularly incisive insights into *2001*'s production, checked in regularly, and transferred important images from his personal archive. Piers Bizony, author of two pioneering and authoritative books on the film, has been an invaluable supporter throughout, providing information, encouragement, ideas, and image scans. A leading writer in the aerospace field, Piers arrived at the subject long before I did, illuminating it with probing intelligence. The author of the best biography of Stanley Kubrick, Vincent LoBrutto, likewise gave me the benefit of his insights and helped me understand his sourcing. Toronto-based writer Gerry Flahive kindly forwarded useful gleanings from his research into Canadian actor Douglas Rain, the voice of HAL. And Christopher Frayling, author of an excellent monograph on production designer Harry Lange, responded to my queries with collegial patience.

Writer Andy Chaikin, my longtime space-geek coconspirator and the preeminent Apollo historian, provided encouragement along the way. Andy helped me get the manuscript to Tom Hanks, with gratifying results, and I'm indebted to both. Likewise my old friend Jill Golden and the talented film editor David Tedeschi kindly worked to bring the book to Martin Scorsese's attention; I thank all three. And Carter Emmart, the wizard of the Hayden Planetarium dome, is a keen student of Kubrick and Clarke's masterpiece, which we saw together in December 2012, each for the umpteenth time. Carter's extraordinary free-form image-slinging at the Hayden, which amounts to a kind of guitar improvisation with the universe, represents the first time since my initial exposure to *2001* that I truly felt transported *out there*—among the galaxies. And yet his dome projections also provide a reminder that our planet remains the most beautiful destination in known space.

My friend and exhibitions agent, Jernej Gregorič, and his talented wife, Natalie, unhesitatingly put me up in their London townhouse as I researched this book, despite the distractions associated with raising two beautiful children. Ljubljana film theorist Nace Zavrl steered me to film historian Billy Brooks, who conducted additional research for me on a tight deadline at the Kubrick archive; I thank them all. And my London gallerist, Matthew Flowers, helped me connect with Christiane Kubrick, among other favors that I gratefully acknowledge.

Top journalist Diane McWhorter, my filmgoing buddy from a year spent at MIT, provided collegial research tips and advice as I set to work in Dulles. Her upcoming book on Wernher von Braun will undoubtedly rewrite our understanding of that controversial figure. David Mikics weighed in late in the game with some useful suggestions; look for his upcoming book on Kubrick. And my longtime crony Stuart Swanson tossed me the keys to a clean, well-lighted office at his regional corporate HQ, requesting only a few planetary prints as compensation. Accordingly I wrote this book comfortably ensconced at Amicus Pharma in Ljubljana, and hereby thank general manager Zeljko Čačić—whose regular wine runs to Belgrade helped take the edge off long writing days—as well as the singularly good-humored Amicus team for their friendship and tolerance.

When it comes to good-humored tolerance, however, nobody exceeds my top supporter in all things for well over two decades, my extraordinary wife, Melita Gabrič, who gives new meaning to that English-derived yet somehow entirely indigenous Slovenian phrase, "da best." She is.

Finally, in my corner for a lifetime has been my father, Ray Benson, who even as his body began yielding to the gravitational pull of ninety-three orbits around the Sun, retained the keen, outgoing intelligence for which he was renowned. When I asked about his health, he always switched the subject immediately to my writing progress. His stoic courage and perennial faith in me took none of the sting out of losing him on November 12, 2017. These pages are dedicated to his memory.

MICHAEL BENSON
LJUBLJANA, SLOVENIA
JANUARY 2018

NOTES

Research for this book drew from a wide spectrum of primary sources, the most important being Arthur C. Clarke's papers at the Smithsonian National Air and Space Museum's Steven F. Udvar-Hazy Center in Chantilly, Virginia, and the seemingly bottomless holdings at the Stanley Kubrick Archive at the University of the Arts London. Other primary wellsprings include the voluminous private archive of *2001* researcher David Larson, which includes invaluable personal papers and other materials from such key figures as Frederick Ordway, Doug Trumbull, and Stuart Freeborn, among many others, and invaluable tranches of original, unedited interviews kindly provided by Dave, by Don Shay of *Cinefex* magazine, by Dan Richter (Moonwatcher), and by Clarke biographer Neil McAleer. In addition, the personal archive of Sri Lanka–based UK journalist Richard Boyle allowed me to better understand Clarke's relationship with his partner during the fifties and sixties, Mike Wilson.

Augmented by Dave Larson's and Richard Boyle's holdings, the capacious Clarke and Kubrick archives enabled me to collate together a 313-page single-spaced document grandly titled "Grand Synthesis Timeline," jammed with quotes from various letters, cables, telexes, and telegrams among the principals. In many ways, it served as a backbone for the book and certainly sourced most of the conclusions presented here.

As one might expect, *2001: A Space Odyssey* has received much attention over the years, and a large number of secondary sources were also regularly consulted and drawn from, none more so than Clarke's own *The Lost Worlds of 2001* (New York: Signet, 1972); Jerome Agel's invaluable, graphically vibrant *The Making of Kubrick's 2001* (New York: Signet, 1970); Dan Richter's excellent *Moonwatcher's Memoir: A Diary of 2001: A Space Odyssey* (New York: Carroll & Graf, 2001); Vincent LoBrutto's exceptionally comprehensive *Stanley Kubrick: A Biography* (New York: Donald I. Fine Books, 1997); Neil McAleer's no less thorough and thoughtful *Visionary: The Odyssey of Sir Arthur C. Clarke* (Baltimore: The Clarke Project, 2010), previously published as *Arthur C. Clarke: The Authorized Biography* (Chicago: Contemporary Books, 1992); Clarke's own career-spanning 555-page anthology of essays, *Greetings, Carbon-Based Bipeds!* (London: Voyager, an imprint of HarperCollins, 1999); and finally, Piers Bizony's beautifully illustrated, diligently researched pair of tomes on the subject, *2001: Filming the Future* (London: Aurum Press, 1994) and *The Making of Stanley Kubrick's 2001: A Space Odyssey* (Cologne: Taschen, 2014). The notes that follow source mostly direct quotes. Various conclusions presented *outside* of quotation marks in these pages thus aren't necessarily linked to individually cited sources. In this case, absence of citation shouldn't be taken as evidence of absence, however. Any facts or observations not enclosed by quotation marks invariably derived from my own processing of the sources contained herein, including my own interviews. Suffice it to say that as I wrote, I regularly engaged in a kind of self-interrogation concerning various conclusions and statements, the aim being to ensure that they derived from exposure to primary sources (including three extended visits with Arthur C. Clarke himself in the early aughts) and, fortuitously, also an ongoing interaction—sometimes vigorously ongoing, with many emails whizzing back and forth daily across weeks of writing—with many of the people who actually worked with Kubrick and Clarke on *2001: A Space Odyssey*.

One might think that nobody alive today is in a better position to support or refute my interpretations of the events that unfolded between 1964 and 1968 in New York and MGM's UK studios in Borehamwood, north of London, than those who actually worked on the film. While this is largely true, an exception exists: Dave Larson, a consultant on this project, who apart from serving as an invaluable reality check throughout the writing process kindly agreed to go through the manuscript at the end, looking for mistakes of fact or interpretation. Having said that, of course any errors in these pages are mine and mine alone.

CHAPTER 1: PROLOGUE: THE ODYSSEY

1 *The very meaninglessness*: Eric Norden, "*Playboy* Interview: Stanley Kubrick," *Playboy*, September 1968, 195.

3 "separation–initiation–return" . . . "*might be named the nuclear unit of the monomyth*": Joseph Campbell, *The Hero with a Thousand Faces* (Princeton, NJ: Princeton University Press, 1949), 30.

3 *a term he borrowed from Joyce*: James Joyce, *Finnegans Wake* (New York: Viking Press, 1939), 581.

3 "*from ape to angel*": A. H. Weiler, "Kazan, Kubrick and Keaton," *New York Times*, April 28, 1968. The "ape-angel" dialectic undoubtedly derives from Robert Ardrey, who used similar wording in *African Genesis* (New York: Dell, 1961), 354.

6 "*NASA East*": Frederick I. Ordway III and Robert Godwin, *2001: The Heritage and Legacy of the Space Odyssey* (Burlington, Ontario: Apogee Prime, 2015), 35.

6 *Clarke's 1945 paper*: Arthur C. Clarke, "Extra-Terrestrial Relays: Can Rocket Stations Give World-wide Radio Coverage?," *Wireless World*, October 1945.

7 "*Earth is the cradle of the mind*": Konstantin Tsiolkovsky, *Vestnik vozdukhoplavaniia* (*Journal of Aeronautics*), 1911–12.

7 "*phony on many points*" . . . "*the emotional foundation of a film*": Andrei Tarkovsky, 1970 interview by Naum Abramov, reprinted in *Andrei Tarkovsky: Interviews* (Jackson: University Press of Mississippi, 2006), 36.

7 "*run through the chopper, heartlessly*": Ray Bradbury quoted his own review in a letter to Arthur C. Clarke, June 3, 1968.

8 "*paying attention with their eyes*": Jerome Agel, ed., *The Making of Kubrick's 2001* (New York: Signet, 1970), 7.

10 *the surest sign that intelligent life exists out there*: Arthur C. Clarke, interview by author, December 19, 2001.
10 *"This is the last big space film"*: Clarke, interview by Joseph C. Gelmis, *Camera Three*, CBS-TV, January 3, 1969, available online at Creative Arts Television, www.dailymotion.com/video/x3ilajz.
11 *"Who do you think wrote it?"*: Clarke, interview by author, December 19, 2001.

CHAPTER 2: THE FUTURIST

14 *For every expert, there is an equal and opposite expert*: Arthur C. Clarke, *Profiles of the Future: An Inquiry into the Limits of the Possible* (London: Victor Gollancz, 1999), 143.
15 *quite rumbustious and irresponsible and most unpalatable . . . Herschel still had not attained the caliber*: Harry Pereira to Arthur C. Clarke, April 4, 1964.
16 *and turned as his assistant Pauline entered*: Clarke's secretary in Colombo in the midsixties was Pauline de Silva; Rohan de Silva to author, email, October 20, 2017.
17 *Her lawyers had sued him in a New York court*: Clarke to Len J. Carter, March 26, 1964.
17 *"It's about boat building"*: Clarke to Major R. Raven-Hart, February 18, 1964.
17 *effectively holding Clarke to ransom*: Clarke to Scott Meredith, April 8, 1964.
18 *it was no good explaining that he'd spent almost all of it*: Clarke's funding of Mike Wilson's projects and lifestyle is documented in numerous letters, among them Clarke to Wilson, February 27, 1971.
19 *"Those parameters are poorly known:"* Carl Sagan, "Direct Contact Among Galactic Civilizations by Relativistic Interstellar Spaceflight," *Planetary and Space Science* 11, no. 5 (May 1963).
19 *a meeting that Sagan had attended*: Lee Billings, "The Alien-Life Summit," *Slate*, September 27, 2013, www.slate.com/articles/technology/future_tense/2013/09/green_bank_conference_seti_frank_drake_s_equation_for_estimating_the_extraterrestrial.html.
20 *"electromagnetic communication does not permit"*: Sagan, "Direct Contact Among Galactic Civilizations."
20 *It is not out of the question*: ibid.
20 *I was particularly interested in your suggestion*: Clarke to Carl Sagan, November 12, 1963.
21 *"with some stimulation towards my present line of work"*: Sagan to Clarke, December 2, 1963.
21 *"a recluse"*: Roger Caras to Clarke, February 17, 1964.
22 *A gay, bisexual, straight, Anglo, Asian group*: Clarke was gay, though he never came out publicly: Jeremy Bernstein, interview by author, September 17, 2016; Christiane Kubrick, interview by author, September 26, 2016.
22 *"frightfully interested in working with enfant terrible"*: Stephanie Schwam and Martin Scorsese, ed., *The Making of 2001: A Space Odyssey* (New York: Modern Library, 2000), 15.
22 *"I was talking with Stanley Kubrick" . . . Since you insist on escaping*: Caras to Clarke, February 17, 1964.
23 *an abusive, mean-spirited egomaniac*: Thomas Flanagan, "Ray Bradbury's Fearsome Encounter with Film Director John Huston," *Chicago Tribune*, May 31, 1991.
23 *"I remain the most successful writer"*: Clarke to Sam Youd, October 24, 1963.
23 *had done a creditable job*: Howard Koch to Clarke, August 1, 1958; Clarke to Koch, August 13, 1958; Koch to Clarke, February 2, 1961; Clarke to Koch, March 6, 1961; Koch to Clarke, September 21, 1967.
24 *Dr. Strangelove is, in a word, a masterpiece*: Val Cleaver to Clarke, February 12, 1964.
24 *"Kubrick is obviously an astonishing man" . . . "passed by the censors"*: Clarke to Caras, February 22, 1964.
24 *"matrimonial trouble" . . . "To put you into the picture" . . . "completely incapacitated by polio" . . .*

There is also a possibility: Clarke to Robert Rubinger, February 26, 1964, and March 11, 1964.

25 *more than a million people seeing it*: Nalaka Gunawardene, "From Great Basses Reef to Ran Muthu Duwa," *Ceylon Today*, August 12, 2012.

26 *when Wilson left the set early "to relax"* . . . *"I saw he starts writing like this"*: Tissa Liyanasuriya, interview by author, February 24, 2017.

27 *"the possibility of doing the proverbial"*: Stanley Kubrick to Clarke, March 31, 1964.

28 *"I have a big organization here"* . . . *As to the main point of your letter* . . . *"We have our own organization here"*: Clarke to Kubrick, April 8, 1964.

29 *at least Buddhism and Hinduism, of all the Earth's religions*: from various discussions with Clarke, 2001–08.

30 *Mike and his wife, Liz*: Mike Wilson, described as "Clarke's boyfriend" by Satyajit Ray in a letter to Andrew Robinson dated November 2, 1984, was Clarke's companion and protégé from 1950 to 1970. He was clearly bisexual and married Elizabeth Perriera in 1958, as documented in Clarke to Sam Youd, October 21, 1958.

30 *Opening: Screen full of stars . . . The soundtrack singles out . . . "I have thought of a nice opening"*: Clarke to Kubrick, April 9, 1964.

CHAPTER 3: THE DIRECTOR

32 *You seldom get what you pay for*: Anthony Frewin, "The Sayings, Maxims & Aphorisms of Chairman Stanley," undated compilation sent by Frewin to author via email, September 22, 2016.

32 *whose name was a source of fascination for the director*: In the early 1990s, Kubrick's assistant Anthony Frewin asked him about the title, confirming that it originated with the director's anagrammatic play with "Lovejoy," and that Kubrick was "fascinated by these odd English names." Anthony Frewin to author, email, July 24, 2017.

33 *"to any preview you may be considering"*: Ray Lovejoy to "Mr. Fenson," Cinerama, April 13, 1964.

33 *In answer to your question*: "Notes for a Dutch Magazine," circa April 13, 1964 (undated but surrounded by other documents from that date at the Kubrick Archive, University of the Arts London). The magazine's name is unknown.

34 *"Any longer, and they'll think"* . . . *"Saw ship sank same"*: Frewin, "Chairman Stanley."

34 *he wrote eleven letters . . . A week later, on April 6*: All letters are at the Kubrick Archive, UAL.

35 *the second-highest-grossing film that year*: Vincent LoBrutto, *Stanley Kubrick: A Biography* (New York: Donald I. Fine Books, 1997), 244.

35 *a bit of a control freak*: Caras to Stanley Kubrick, November 3, 1966.

35 *Kubrick had enjoyed Forbes's praise of his work . . . "little green men stuff"*: Christiane Kubrick, interview by author, October 22, 2016.

36 *"Oh, Stanley, for God's sake!"* . . . *"you really can't walk around like that"* . . . *"I guess it's an English thing"* . . . *"No, Stanley"* . . . *Later, she remembered the falling-out*: Christiane Kubrick, interview by author, June 5, 2016.

37 *It caused humans to be impervious to the cold*: information on *Shadow on the Sun* gleaned from the BBC's *Radio Times*, no. 1979, October 12, 1961; Jon Ronson, "Citizen Kubrick," *The Guardian*, March 27, 2004.

37 *good books make bad films*: Frewin, "Chairman Stanley."

37 *"a weird, hydra-headed"*: *Dr. Strangelove* script dated March 31, 1963, Kubrick Archive, UAL.

38 *"I was disgusted"*: Christiane Kubrick, interview by author, September 22, 2016.

38 *it emerged that Shaw was also a science fiction fan . . . "the first science fiction film that isn't considered*

trash" . . . *Shaw suggested he read Arthur C. Clarke*: Christiane Kubrick, interview by author, June 5, 2016. Shaw's recommendation of Clarke is further substantiated by Clarke's timeline for "Son of Dr. Strangelove," his unpublished draft article for *Life* magazine, dated September 22, 1966, at the Sir Arthur C. Clarke Papers, Smithsonian Institution National Air and Space Museum, Steven F. Udvar-Hazy Center, Chantilly, VA.

38 *"You gotta read this"* . . . *"We had to take turns staying awake"*: Christiane Kubrick, interview by author, June 5, 2016.

39 *"at heart, Stanley was a peasant"* . . . *"You'd laugh"* . . . *" 'I want to do a film' "*: Caras, interview by Dan Richter, September 7, 1999.

40 *"Fantastic. As a matter of fact"*: Caras, interview by Neil McAleer, September 19 and October 21, 1989.

40 *"Why are you going through all that?"* . . . *" 'Yeah, but I understand he's some kind of a nut' "* . . . *" 'Not really, he lives in Ceylon' "*: Caras, interview by Richter, September 7, 1999.

41 *"Jesus, get in touch with him, will you?"*: Caras, interview by McAleer, September 19 and October 21, 1989.

41 *the flight had been delayed not by hours but by two days*: Clarke to Meredith, April 11, 1964.

41 *"Its impressive technical virtuosity"*: Arthur C. Clarke, "Son of Dr. Strangelove," unpublished draft, January 23, 1967.

41 *where he mingled with*: Neil McAleer, *Visionary: The Odyssey of Sir Arthur C. Clarke* (Baltimore: Clarke Project, 2010), 143.

42 *"lovely office on the thirty-second floor"*: Clarke to Youd, June 19, 1964.

42 *"It was strange, being back in New York"*: Clarke, "Son of Dr. Strangelove."

42 *Kubrick arrived on time*: LoBrutto, *Stanley Kubrick*, 261.

42 *"a rather quiet, average-height New Yorker"*: Clarke, "Son of Dr. Strangelove."

42 *"the somewhat bohemian look"*: Jeremy Bernstein, "How About a Little Game," *New Yorker*, November 12, 1966.

43 *"pure intelligence"* . . . *"Kubrick grasps new ideas"* . . . *"Please tell Wernher"*: Clarke, "Son of Dr. Strangelove."

43 *"I never did, because (a) I didn't believe it"*: Arthur C. Clarke, *Astounding Days: A Science Fictional Autobiography* (New York: Bantam Books, 1989), 183.

43 *"in camera," as Caras put it*: Caras, interview by McAleer.

44 *"Stanley had a great ability to concentrate"*: Christiane Kubrick, interview by author, June 5, 2016.

44 *"Somewhat to my chagrin"* . . . *"This is a bit ungenerous"*: Clarke, "Son of Dr. Strangelove."

45 *"Even from the beginning"* . . . *"When I met Stanley for the first time"*: Arthur C. Clarke, *The Lost Worlds of 2001* (New York: Signet, 1972), 29.

45 *eight hours of talking*: Clarke, "Son of Dr. Strangelove."

45 *"perhaps the most intelligent person"*: *Stanley Kubrick: A Life in Pictures*, a documentary film directed by Jan Harlan (Burbank, CA: Warner Bros. Home Video, 2001).

46 *an incinerator chimney on the building's roof*: setting based on Piers Bizony, *The Making of Stanley Kubrick's 2001: A Space Odyssey* (Cologne, Germany: Taschen, 2014), 17.

46 *Kubrick's study was jammed*: description based on Bernstein, "How About a Little Game."

46 *"The preoccupation with hi-fidelity equipment"*: Kubrick marginalia is from annotated draft version of "How About a Little Game," Jeremy Bernstein's November 12, 1966, *New Yorker* article, sent by JB to the author on September 8, 2016.

46 *"Every time I get through a session with Stanley"*: Agel, ed., *Making of Kubrick's 2001*, 136.

47 *"I don't control easily"*: Arthur C. Clarke, interview by McAleer, November 25, 1989.

47 *"Arthur was really like talking to the ultimate"*: Christiane Kubrick, interview by author, June 5, 2016.

48 *"I was working at Time-Life during the day"*: Jeremy Bernstein, "Out of the Ego Chamber," *New Yorker*, August 9, 1969.

49 *"exclaimed in anguish, 'What are you' "*: Clarke, "Son of Dr. Strangelove."

49 *"busily cribbing"* . . . *"Still spending every spare minute with Stanley K."*: Clarke to Mike Wilson, May 14, 1964.

50 *"A very good plot is a minor miracle"* . . . *How the first cavemen sat around a fire* . . . *"Because of the film form"*: transcript of Heller–Kubrick dialogue, May 14, 1964, Kubrick Archive, UAL.

50 *"restaurants and Automats"*: Arthur C. Clarke, *Report on Planet Three and Other Speculations* (London: Corgi Books, 1973), 247.

51 *Having discussed the matter at some length*: LoBrutto, *Stanley Kubrick*, 289.

51 *Clarke agreed that the struggle*: Arthur C. Clarke, "Rocket to the Renaissance," *Greetings, Carbon-Based Bipeds!, Collected Essays, 1934–1998*, ed. Ian T. Macauley (New York: St. Martin's Press, 1999), 211.

51 *But in his view, interplanetary travel*: ibid., 37.

51 *"What we had in mind was a kind of semidocumentary"*: Clarke, "Son of Dr. Strangelove."

51 *"We seldom stop to think that we're still creatures"*: Clarke, "Rocket to the Renaissance," 216.

52 *"The old idea that man invented tools"*: ibid., 218.

52 *"from the fog of words"*: Clarke, "Son of Dr. Strangelove."

53 *"a whole microcosm of living creatures"*: Arthur C. Clarke, "Before Eden," in *The Collected Stories* (London: Victor Gollancz, 2001).

54 *Before we start, I'd like to point out something*: Arthur C. Clarke, "Out of the Cradle, Endlessly Orbiting," *Dude*, March 1959.

54 *"Earth is the cradle of the mind"*: Tsiolkovsky, *Vestnik vozdukhoplavaniia.*

54 *Now its signals have ceased*: "The Sentinel," in Agel, *Making of Kubrick's 2001*, 15–23.

55 *"savage cousins waiting at the dawn"* . . . *"a long line of light slanting upward"* . . . *"on which, more than a thousand centuries"*: Arthur C. Clarke, "Encounter in the Dawn," first published in *Amazing Stories*, June/July 1953.

56 *"It's impossible"* . . . *"This is altogether too much of a coincidence"*: Clarke, "Son of Dr. Strangelove."

56 *The whole episode had lasted*: Stanley Kubrick to US Air Force Colonel Jacks, June 17, 1964.

56 *"That's the most spectacular"* . . . *"So Stanley had seen his first artificial satellite"* . . . *"Why this wasn't listed in the* Times*"*: Clarke, "Son of Dr. Strangelove."

58 *Clarke didn't like to get too involved*: substantiated by Clarke to Koch, February 26, 1961, and comments made to the author by Clarke in 2002.

58 *While the contract they eventually signed included*: contract details from copy at the Clarke Papers, Smithsonian, May 28, 1964. Signatories are Clarke and Louis Blau as president of Polaris Productions.

59 *MGM had inflexible ground rules*: Koch to Clarke, February 26, 1961.

59 *"The deal is a complicated one"* . . . *"Stan, who is a ball of fire"* . . . *"If she finds out about this deal"* . . . *You realize just as well as I do.* . . . *"Will try to get some money"*: Clarke to Wilson, May 22, 1964.

CHAPTER 4: PREPRODUCTION: NEW YORK

61 *You can never have enough information*: Frewin, "Chairman Stanley."

61 *"Stan's a fascinating character"*: Clarke to Youd, June 19, 1964.

61 *"When your children are young"*: Christiane Kubrick, interview by author, June 5, 2016.

62 *The schedule they'd drawn up* . . . *"hilariously optimistic"*: Clarke, *Lost Worlds*, 32.

62 *On the way there on the late morning* . . . *"put Stan onto me"*: Clarke, timeline for "Son of Dr. Strangelove."

62 *about a thousand words a day*: Clarke, *Lost Worlds*, 33.

62 *"We will not sit down and write a screenplay"*: Clarke, interview by Gelmis, *Camera Three*.

63 *"Writing a novel is like swimming"*: Clarke, *Lost Worlds*, 31.

63 *I decided trying to do an original story*: from draft notes by Stanley Kubrick for *The New York Herald Tribune* interview, November 23, 1965, Kubrick Archive, UAL.

63 *"It is extremely difficult to represent"*: Clarke to J. B. S. Haldane, May 23, 1964.

64 *"might be machines who regard organic life"* . . . *"hilarious idea we won't use"*: Clarke, *Lost Worlds*, 32–33.

64 *"As she must do often in eternity"*: Clarke, "Encounter in the Dawn."

65 *"I argued that the number of individually"* . . . *"I suggested that any explicit"* . . . *"The film's title, by the way"* . . . *"this fairly important plot line"*: Carl Sagan, *The Cosmic Connection, An Extraterrestrial Perspective*, ed. Jerome Agel (New York: Doubleday, 1973), 183.

66 *"Get rid of him"*: Clarke, interview by author, December 19, 2001; Arthur C. Clarke, review of *Carl Sagan's Universe*, ed. Yervant Terzian and Elizabeth Bilson, *The Times Higher Education Supplement*, December 12, 1997.

66 *"Kubrick is absolutely brilliant"*: Clarke to Haldane, May 23, 1964.

66 *In this, he was fueled by a diet of liver paté*: Peter Arthurs, interview by McAleer, May 27, 1989.

66 *"nb Stan's enthusiasm for material"*: Clarke, timeline for "Son of Dr. Strangelove."

66 *"Finished the opening chapter"*: Clarke, *Lost Worlds*, 32–33.

67 *"And they would remember"*: ibid., 13.

67 *"roughly the size and shape of a man"* . . . *"generating the words himself"*: undated "Man and Robot" rejected novel drafts, Clarke Papers, Smithsonian.

67 *"We've got a bestseller here"* . . . *"He's fascinated"* . . . *"coming apart at the seams"*: Clarke, *Lost Worlds*, 33.

68 *"Chose hero's name—D.B"*: Clarke, timeline for "Son of Dr. Strangelove."

68 *"It was literally months"*: Clarke, interview by Gelmis, *Camera Three*.

69 *"The Odyssean parallel was clear"*: Clarke, "The Myth of 2001," in *Report on Planet Three*, 253.

69 *"auto-highway"* . . . *"the great Washington–New York"*: undated "Goodbye to Earth" rejected novel drafts, Clarke Papers, Smithsonian.

69 *"What we want is"*: Clarke, *Lost Worlds*, 33.

70 *"[D]uring this period"* . . . *"This sounds pompous"* . . . *There were times* . . . *"Again, implication you do"*: Clarke, "Son of Dr. Strangelove."

70 *"We spent the better part"*: Hollis Alpert, "Offbeat Director in Outer Space," *New York Times*, January 16, 1966.

71 *"We're in fantastic shape"* . . . *"I was a robot, being rebuilt"* . . . *"Joe Levine doesn't do this"* . . . *"Lots of actors standing around"*: Clarke, *Lost Worlds*, 34.

71 *a superb Oscar-nominated black-and-white short*: *Universe* (1960) can be seen online on YouTube or in higher quality at Internet Archive, https://archive.org/details/TheUniverseNational FilmBoardOfCanada.

72 *"The Predatory Transition from Ape to Man,"* Raymond Dart, *International Anthropological and Linguistic Review* 1, no. 4 (1953).

72 *"Came across a striking"*: Clarke, *Lost Worlds*, 34.

72 *And they changed their working title*: Clarke, timeline for "Son of Dr. Strangelove."

72 *We were born of risen apes*: Ardrey, *African Genesis*, 354.

73 *"A hero ventures forth"* . . . *"Clashing Rocks"* . . . *"power to wrest"*: Campbell, *Hero with a Thousand Faces*, 30; see also Clarke to Purdy (no first name available), September 9, 1968.

73 *"crystal slab"* . . . *"inquisitive tendrils creeping"*: undated rejected *Dawn of Man* novel drafts, Clarke Papers, Smithsonian.

74 *"sense of wonder"* . . . *"into a sky as wild"* . . . *"the charred corpse"* . . . *"nova—as must all suns"* . . . *"glorious apparition"* . . . *"no hotter than a glowing coal"*: ibid.

75 *"Now the turning wheels of light"*: Clarke, *Lost Worlds*, 238.

75 *"a most peculiar ocean"*: ibid., 200.

75 *"offered mental security"*: ibid., 225.

75 *"something that could not possibly"*: ibid., 217.

75 *"Stanley Kubrick and I"*: ibid., 199.

75 *"Stan's idea 'camp' robots"*: Clarke, timeline for "Son of Dr. Strangelove."

76 *"Stanley has invented"*: Clarke, *Lost Worlds*, 34.

76 *"Stan, I want you to know"* . . . *"Yeah, I know"* . . . *"like a schoolteacher"* . . . *"He was very pleased"*: Christiane Kubrick, interviews by author, June 5, 2016, and October 22, 2016.

77 *"Stanley tries to refute"*: Clarke, *Lost Worlds*, 33.

77 *"the passports of the lazy"* . . . *"Keep asking the question"*: Frewin, "Chairman Stanley."

77 *"He hoped he was good"*: Christiane Kubrick, interview by Charlie Rose, *Charlie Rose*, PBS, June 15, 2001.

78 *who had once been the country's flyweight boxing champion*: Ekanayake was the youngest-ever flyweight boxing champion of Ceylon in 1956. Rohan de Silva to author, email, October 20, 2017.

78 *ending up in the hospital*: Liyanasuriya, interview by author.

79 *Hector attacked the tree*: Christiane Kubrick, interview by author, October 22, 2016.

79 *"in search of inspiration"*: Clarke, *Lost Worlds*, 35.

79 *"We've extended the range"*: LoBrutto, *Stanley Kubrick*, 268.

79 *"We were, indeed"*: Clarke, *Lost Worlds*, 35.

80 *"Delivered complete copy"*: Clarke, timeline for "Son of Dr. Strangelove."

80 *"The first version of the novel"* . . . *"Confusing. No one will know"*: Clarke, "Son of Dr. Strangelove."

81 *a small film visual effects outfit*: "Talking With Con Pederson," interview by William Moritz, *Animation World Magazine*, Issue 4.3 (June 1999).

81 *The overcranked camera*: descriptions of process are in part from John Alcott, unpublished interview transcript by Don Shay for *Cinefex*, June 1984.

81 *"many of the truly cataclysmic"*: Wally Gentleman, interview transcript by Don Shay, May 3, 1979.

81 *"unspeakably disgusting"* . . . *"The difference between a lot of us"*: Christiane Kubrick, interview by author, January 15, 2016.

82 *He had published numerous articles*: Christopher Frayling, *The 2001 File: Harry Lange and the Design of the Landmark Science Fiction Film* (London: Reel Art Press, 2015), 22.

83 *"It happens we've just published"* . . . *"How extraordinary"*: ibid.; Harry Lange, interview by David Larson, September 25, 2002.

83 *"Arthur went off and"* . . . *"mentally exhilarating"* . . . *"He had immersed himself"*: Frederick Ordway, interview by Larson, August 14, 2003.

84 *"Harry, your work makes money"* . . . *"Well, I can get better illustrators"* . . . *"Yes, yes, I've seen enough"*: Lange, interview by Larson.

85 *"They had a two- or three-day deadline"*: LoBrutto, *Stanley Kubrick*, 268–69.

85 *accompanied by beautiful but highly experimental*: Although Clarke doesn't mention this, Wally Gentleman stated that Kubrick's Manhattan Project footage was shown to MGM along with textual material; Gentleman, interview by Shay; also, Con Pederson echoed this in Aladino Debert, "2001: A Space Odyssey—A Discussion with Con Pederson," *Visual Effects Headquarters*, last modified April/May 1998, www.vfxhq.com/spotlight98/9804c.html.

86   *a backbiting chorus that would only grow*: ibid.

86   *by its end over 150 had been released*: Bradley Schauer, *Escape Velocity: American Science Fiction Film, 1950–1982* (Middletown, CT: Wesleyan University Press, 2017), 11–12; see also Patrick Lucanio, *Them or Us: Archetypal Interpretations of Fifties Alien Invasion Films* (Bloomington: Indiana University Press, 1987).

87   *For Kubrick and Clarke*: Among several sources for this see for example Frayling, *2001 File: Harry Lange*, 29.

87   *"When I first saw Kubrick and the apartment"*: Jeremy Bernstein, "Memories of Stanley Kubrick," *Scribd*, https://www.scribd.com/document/220043494/Memories-of-Stanley-Kubrick-By-J-bernstein.

87   *A surviving cover of the* Journey Beyond the Stars: Kubrick Archive, UAL; cover is also described in Pederson, interview by Moritz.

88   *"a motion picture now tentatively entitled"*: the January 14 date is from McAleer, *Visionary*, 142.

88   *"He made his deals very carefully"*: Christiane Kubrick, interview by author, October 22, 2016.

88   *The contract, which noted*: All information on the contract is from MGM contract draft with Kubrick annotations, dated May 22, 1965.

89   *Thus to call a spade a spatulous device*: spade riff unearthed from Joseph Devlin, *How to Speak and Write Correctly* (New York: Christian Herald Bible House), 1910, 4.

91   *"Without wanting to seem unappreciative"*: Stanley Kubrick to Robert Shaw, February 17, 1965, www.telegraph.co.uk/culture/film/3555933/The-letters-of-Stanley-Kubrick.html.

91   *Went up to the office with*: Clarke, *Lost Worlds*, 37.

92   *"He was an odd person"*: Ordway, interview by Larson.

92   *"We had a very good time"* . . . *"It was that banal"*: Lester Novros, interview transcript by Don Shay for *Cinefex*, August 18, 1984.

92   *Was it interesting?*: Pederson, interview by Debert, Visual Effects Headquarters, April 1998, www.vfxhq.com/spotlight98/9804c.html.

93   *"He was very coy about it"*: Novros, interview by Shay.

94   *he'd worked closely with*: Laurie N. Ede, *British Film Design: A History* (London: I. B. Tauris, 2010), 136.

94   *"And, well, I am"* . . . *"just sitting in an office"*: Tony Masters, interview transcript by Don Shay, April 8, 1977.

94   *When you do a science fiction film* . . . *"anchorman"*: Don Shay and Jody Duncan, "2001: A Time Capsule," *Cinefex*, April 2001.

96   *"We decided to use a lot of white material"* . . . *"Well, what color would it be?"*: Masters, interview by Shay.

96   *"would walk over you with"*: Gentleman, interview by Shay.

97   *"I was engaged as the director"*: Shay and Duncan, "2001: A Time Capsule."

98   *"We were particularly intrigued"*: Frederick Ordway to Dr. Walter N. Pahnke, June 7, 1965.

99   *"the most powerful cosmic homecoming"*: Huston Smith, *The Huston Smith Reader*, ed. Jeffrey Paine (Oakland: University of California Press, 2012), 73.

99   *"awash in a sea of color"* . . . *"I had to die to become"*: "Tune In, Turn On, Get Well?," Jeanne Malmgren, *St. Petersburg Times* (FL), November 27, 1994.

100   *"completely overwhelmed"*: Clarke, *Lost Worlds*, 37.

100   *An "important personality"* . . . *Arriving at an Earth-orbiting* . . . *When delivered "very quickly"*: *2001* production notes, June 1, 1965.

101   *"2001 newspaper to be read"*: Tony Masters to Caras, June 29, 1965.

101   *"A rig should be made for the newspad"*: production meeting 6 notes, December 11, 1965.

102 *an acorn that would grow into an oak*: Quoted in *Heavy Metal* 7, no. 10, January 1984; see also Arthur C. Clarke, *2001: A Space Odyssey* (New York: New American Library, 1968; repr., with new introduction, 1999), x.

102 *"Later, I had the quaint experience"*: Clarke, "Son of Dr. Strangelove." Note that Clarke meekly modified this to "a nominal fee" in the published version; see Clarke, *Report on Planet Three*, 248.

102 *"I said, 'You bastard!'"*: Caras, interview by Richter.

102 *"That's typical Stanley"*: Caras, interview by McAleer.

103 *"Then one night"* . . . *"and I said, 'Do you want to move'"*: Caras, interview by Richter.

104 *forty-eight steamer trunks . . . books precious to the director*: Anthony Frewin, interview by author, September 21, 2016.

104 *"People who deal with Stanley"* . . . *He could be endlessly thoughtful*: Caras, interview by Richter.

105 *Within a month or so*: Frewin, interview by author, September 21, 2016.

CHAPTER 5: BOREHAMWOOD

106 *Perhaps our role on this planet*: Clarke, *Report on Planet Three*, 145.

106 *"I don't want to get involved"*: Frewin, interview by author, November 11, 2016.

107 *It had shattered Tony's picture* . . . *"The only way I'm ever going to"*: Frewin, interview by Richter, June 29, 2000.

107 *"I wouldn't mind working here"*: Frewin, interview by author, November 11, 2016.

108 *"Shouldn't he be off with his duster somewhere?"* . . . *"Listen, I'm sorry"* . . . *"I'm Stanley"*: Frewin, interview by Richter.

108 *"Gee, you have a wonderful"* . . . *"Well, one of the problems"* . . . *"Are there any other artists"* . . . *"Yeah, I've read the novels"* . . . *Concerning extraterrestrials* . . . *"I need a runner"*: Frewin, interview by author, November 11, 2016.

111 *the complex's considerable overhead costs . . . already considered something of a financial liability*: LoBrutto, *Stanley Kubrick*, 271.

112 *"I'd like you to make a large piece"* . . . *"But we'd like to do it"* . . . *"Let's make it that shape"* . . . *"And then it takes a month"* . . . *"Oh, right, let's go"* . . . *"Oh God"* . . . *"Well, it is nearly"*: Shay and Duncan, "2001: A Time Capsule"; Masters, interview transcript by Shay.

113 *over two tons of it*: weight and present location from http://londonist.com/london/secret/see -the-original-monolith-from-2001-a-space-odyssey.

113 *"File it"* . . . *"It looks like a piece of glass"* . . . *"So, let's just make a black one"*: Shay and Duncan, "2001: A Time Capsule"; Masters, interview by Shay.

114 *"How the underwriters managed"*: Clarke, "Son of Dr. Strangelove."

114 *"unbelievably graceful and beautiful humanoid"*: *2001* script draft, June 7, 1965.

115 *"fighting hard to stop Stan"*: Clarke, *Lost Worlds*, 37.

115 *"sleeping beauties"*: 2001 script draft, December 14, 1965. This phrase was to have been used by astronaut Frank Poole in conversation with Dave Bowman, but was not in the final film. Kubrick brother-in-law Jan Harlan has used the term "Popsicle people" to describe the same.

115 *"I've put my finger on a flaw"* . . . *"You can construct"* . . . *"This element of suspense"*: Clarke to Kubrick with annotations from latter, August 24, 1965.

116 *It's amazing how long it takes . . . "I prefer present nonspecific"*: Ibid.

118 *"put the authenticity into it"*: Shay and Duncan, "2001: A Time Capsule."

118 *To the extent a division of authorship existed*: see Ordway quote in Frayling, *2001 File: Harry Lange*, 56.

118 *"The strange thing was"*: Shay and Duncan, "2001: A Time Capsule."

119 *"We weren't working to any"* . . . *We'd get together with Stanley*: ibid.; Masters, interview by Shay.

120 *"To capture that, we had a camera"* . . . *"I think I see it!"*: Shay and Duncan, "2001: A Time Capsule."

120 *"Sweat was now running off"*: Masters, interview by Shay.

122 *"making a joke of it"*: Frayling, *2001 File: Harry Lange*, 32.

122 *a scale model of a V-2*: from original unedited manuscript of Agel, *Making of Kubrick's 2001*, a story Kubrick conspicuously fails to deny in his revisions dated February 1, 1970; Arthur C. Clarke also alludes to the incident in his notes for a *2001* essay or talk dated November 1967.

124 *"they say that a computer of the complexity"*: Caras to Kubrick, July 27, 1965.

124 *"[T]he IBM Athena drawings are useless"*: Kubrick to Caras, July 29, 1965.

125 *"fairly large, flat, and rectangular"*: John Pierce, Bell Labs report, June 29, 1965.

126 *"quasi-independent"* . . . *"Let us look down the road"* . . . *Suppose in the interest*: Ordway to Eugene Riordan, August 24, 1965.

127 *"a considerably more experimental"* . . . *"an experimental research and development"*: Ordway to Riordan, October 26, 1965.

127 *The terms behind the acronym*: Ordway suggested that Minsky authored the acronym itself, in Ordway, interview by Larson, August 14, 2003.

127 *"Heuristics are, of course"*: Marvin Minsky, interview by David Stork, in David G. Stork, ed., *HAL's Legacy: 2001's Computer as Dream and Reality* (Cambridge, MA: MIT Press, 1997), 27–28.

128 *"If we wish, we can make the 'accident' "*: Clarke to Kubrick, October 12, 1965.

128 *"One evening (or whenever)"* . . . *"The computer says 'No' "* . . . *"even when they play over their heads"* . . . *"suddenly computer asks about paradox"* . . . *"Computer tries to talk Bowman out"*: "The Computer" notes by Stanley Kubrick, Kubrick Archive, UAL.

130 *"If you went down to the office"* . . . *"I'm one of the illustrators"* . . . *"Absolutely great"*: Doug Trumbull, interview by Larson, May 7, 2001; Trumbull, interview by author, September 1–2, 2016.

130 *"I was twenty-three at the time"*: Shay and Duncan, "2001: A Time Capsule."

132 *"Wally was a very scientific"*: Trumbull, interview by author.

133 *"Doug would sort of wander"*: Gentleman, interview by Shay.

133 *"Wally said, 'We can figure' "*: Trumbull, interview by author.

133 *"a million random sources"*: Gentleman, interview by Shay.

133 *"strange little acronyms"* . . . *"We'd sit there"*: Shay and Duncan, "2001: A Time Capsule."

134 *"autonomous, 24-hour-a-day"* . . . *"In reality, it was"*: Trumbull, interview by Larson.

135 *The Eady cap on non–UK citizens*: LoBrutto, *Stanley Kubrick*, 202.

135 *"There was no studio management"*: Trumbull, interview by author.

136 *"pointing out the beauty"*: Frayling, *2001 File: Harry Lange*, 73.

137 *"We are supposed to wind up"*: Gentleman to Raymond Fielding, November 29, 1965.

137 *"My entire career"* . . . *I enjoyed Kubrick the man*: Shay and Duncan, "2001: A Time Capsule."

137 *"If Stanley thought someone was slightly"*: Brian Johnson, interview by author, February 3, 2017.

138 *following an "altercation" with them*: Ordway, interview by Larson.

138 *In addition to carousing and getting into fights*: Gary Lockwood, interview by author, September 11, 2016.

139 *"I think it's a combination of chess"*: Lockwood, interview by Justin Bozung, *Shock Cinema*, no. 42, 2012.

139 *"It is most inappropriate for Bob O'Brien"*: Kubrick to Caras, September 26, 1965.

140 *Special Operations branch*: David Lister, "Queen's Tailor Hardy Amies Was a Wartime Hit-man," *Independent* (UK) online, August 23, 2000.

140 *Although not as celebrated as some*: Some valuable observations about Amies were adopted from Lauren Cochrane, "2001: A Space Odyssey—The Fashion Power of Designer Hardy Amies," *Fashion blog, The Guardian* (US), November 28, 2014.

140 *"We were dealing with a period"*: Hardy Amies, *Still Here: An Autobiography* (London: Weiden-field and Nicholson, 1984), n.p.

141 *"I was skeptical about the major changes"*: Bob Cartwright, interview by Larson, March 4, 2001.

141 *"emerged stunned"* . . . *"Actors at the far end"*: Clarke to Wilson, September 10, 1965.

142 *The perspective switch was too much*: description of set derives from transcript of Masters inter-view by Shay.

142 *"In about six weeks"*: Clarke to Wilson, September 10, 1965.

143 *"Earth is the cradle of the mind"*: Tsiolkovsky, *Vestnik vozdukhoplavaniia.*

143 *"To all outward appearances"*: Arthur C. Clarke, *Childhood's End* (New York: Ballantine Books, 1953; New York: Del Ray Books, 1990), 171. Citations refer to the Del Ray edition.

143 *"Stanley phoned with* another*"*: Clarke, *Lost Worlds*, 38.

144 *"age of . . . Machine Entities"* . . . *"Lords of the galaxy"* . . . *"In a moment of time"*: *2001* script draft, December 9, 1965.

144 *"suddenly clicked: Bowman"* . . . *"Stanley called again later"*: Clarke, *Lost Worlds*, 38.

145 *It looked exactly like an unborn baby*: I owe this insight to Robert Poole, specifically "2001: A Space Odyssey and the 'Dawn of Man,'" in *Stanley Kubrick: New Perspectives*, ed. Tatjana Ljujic, Peter Kramer, and Richard Daniels (London: Black Dog, 2014), 182–83, though given Konstantin Tsiolkovsky, *Childhood's End*, and Lennart Nilsson, unlike Poole, I don't think the illustration is the sole progenitor of the Star Child.

145 *"had been surgically removed"*: "The Drama of Life Before Birth," *Life*, April 30, 1965.

145 *caught the director's attention*: Ivor Powell to Caras, May 26, 1966, requesting that issue of *Life*.

145 *"Back to brood over the novel"*: Clarke, *Lost Worlds*, 38.

147 *"Stanley, we've got these changing circumstances"* . . . *"Trumbull, get the fuck out"* . . . *"Okay, leav-ing now"*: scene compounds two accounts, Trumbull, interview by Larson, and Trumbull, interview by author.

147 *"Stan has decided to kill off"*: Clarke, *Lost Worlds*, 38.

148 *"My upbringing in films"* . . . *"Pay attention, Stanley"* . . . *"After that, he was very cross"* . . . *"He's such a nice guy"*: Cartwright, interview by Larson.

149 *110 cubic yards of a particularly fine sand*: undated Hawk Films press release circa mid-January 1966.

149 *Various kinds of wood . . . about fourteen had been made . . . "He'd stand like this"* . . . *"Foomp, it'd be covered with dust"* . . . *"It was unbelievable"*: Shay and Duncan, "2001: A Time Capsule"; Masters, interview by Shay.

150 *"We've got to do it"*: from Ordway, in Frayling, *2001 File: Harry Lange*, 75.

CHAPTER 6: PRODUCTION

151 *Never let your ego get in the way*: Frewin, "Chairman Stanley."

153 *"Jesus Christ, they were good"*: Ivor Powell, interview by author, September 20, 2016.

154 *"Well, there it is"*: *2001* script draft, December 9, 1965.

155 *Despite an excellent*: ibid.; *Daily Continuity Report*, December 17, 1965.

156 *On January 1 Kubrick noticed that one*: ibid., January 1, 1966.

156 *Another day, a bat . . . A much-touted "magnetic induction"*: undated Hawk Films press releases circa mid-January 1966.

156 *"At first glance, black"* . . . *"a piercingly powerful"*: quotes are from November pages in *2001* script draft; otherwise dated October 13, 1965.

157 *"Rumors about some kind of trouble"*: *2001* script draft, December 9, 1965.

157 *It was the morning of January 6*: all dialogue from *Daily Continuity Report*, January 6, 1966.

159 *After greeting Dullea*: Dullea arrival based on descriptions given in Powell, interview by author.

160 *gone on a jaunt to Rome*: undated Hawk Films press releases circa mid-January 1966.

161 *From the very beginning*: all Moon Bus scene dialogue from *Daily Continuity Report*, January 13, 1966.

163 *"Real is good, interesting is better"*: Frewin, "Chairman Stanley."

165 *Sylvester's monologue was about*: All Clavius conference room scene dialogue from *Daily Continuity Report*, January 13, 1966, and from the November pages in *2001* script draft; otherwise dated October 13, 1965.

166 *"I can't do it anymore"* . . . *"never got it"* . . . *"He wasn't nasty"*: Johnson, interview by author.

167 *Kubrick was "beside himself"* . . . *"Is he on drugs?"* . . . *"We're gonna announce"* . . . *"'Jesus, oh Jesus'"* . . . *"Whatever Stanley needed"*: Caras, interview by Richter, September 7, 1999.

170 *"I want you to redesign it"* . . . *"Right, show me what you've got"*: Johnson, interview by Larson, June 19, 2003.

171 *"Wally, file your stuff"*: Johnson, interview by author.

172 *the reels all fell off*: Trumbull, interview by author.

174 *"I think he saw things differently"* . . . *he would change the setup*: Shay and Duncan, "2001: A Time Capsule."

175 *"If he were not a director"*: Michel Ciment, "Working with Stanley Kubrick—John Alcott," the Kubrick Site, www.visual-memory.co.uk/amk/doc/0082.html.

175 *"Make that a five-six"*: memories of Unsworth-Alcott-Kubrick on-set dynamic from Johnson, interview by author.

176 *"Geoffrey, why so deeply"* . . . *"You know, Roger"* . . . *"Are you aware of this?"*: Caras, interview by McAleer.

177 *"I could have been killed!"*: Minsky, interview by Stork, *HAL's Legacy*, 24.

177 *"Here you are, sir"*: *Daily Continuity Report*, January 19, 1966.

180 *"rotated like a clock"*: centrifuge camera operations meeting notes, January 15, 1966.

180 *We would go up with the wheel*: Shay and Duncan, "2001: A Time Capsule."

182 *"Kelvin! What are you doing?"*: Kelvin Pike, interview by Larson, March 3, 2003.

183 *After a few days of jogging*: Caras to columnist Louis Sobol, February 23, 1966.

183 *"It was a scary place to work"*: Trumbull, interview by Larson.

184 *never seeing a recreational drug*: Frewin, interview by author.

184 *"I felt this was so not-solid"* . . . *"He played, and he never won big"*: Christiane Kubrick, interview by author, June 5, 2016.

185 *At one point, he was owed* . . . *"It was very simple"* . . . *"I'm a bar fighter"* . . . *"He had the most beautiful snooker table"* . . . *"How did you get the way you are?"* . . . *"He was the epitome"*: Gary Lockwood, interview by author, September 11, 2016.

187 *"may be guilty"* . . . *"feasibility study"*: dialogue is from *Daily Continuity Report*, January 26, 1966.

187 *"a little bit redundant"*: Gary Lockwood, interview by Larson, December 8, 2003.

187 *"You're not showing your normal"* . . . *"You know, Stanley, I grew up"* . . . *"It's a wrap"*: Lockwood, interview by author.

188 *his conviction that the scene wasn't right*: Lockwood, interview by Bozung.

188 *"I guess there was a side of me"* . . . *"The Guv wants to see you"* . . . *"Candidly, I like tequila"*: Lockwood, interview by author.

188 *"Yeah, Chopin's fine"*: Pete Thornton, "Gary Lockwood and Keir Dullea—2001: A Space Odyssey Interview," *Front Row Reviews*, December 1, 2014, www.frontrowreviews.co.uk /features/keir-dullea-and-gary-lockwood-2001-a-space-odyssey-interview/31914.

188 *"Well, Mr. President, I'm helping"*: Lockwood, interview by Larson.

189 *"tighten the screws on the computer"* . . . *"cut to the marrow"*: Lockwood, interview by author.

189 *he thought Kubrick was the best*: Lockwood, interview by Bozung.

189 *"Then I want you to go home"*: Lockwood, interview by author.

189 *experience a very human*: Lockwood, interview by Bozung.

190 *It was a subzero February night*: Lockwood, interview by author.

190 *even though they were all standing*: Heather Downham sketches this scenario in a slightly different context in the 2001 documentary *2001: The Making of a Myth*, directed by Paul Joyce, 2001. (Warner Bros. Home Entertainment Group, DVD).

190 *"like trying to write* War and Peace*"*: Kubrick used this comparison throughout his career, but for one example, see his D. W. Griffith Lifetime Achievement Award video message, www .youtube.com/watch?v=tBYJJzpxH9Q.

191 *"I was projecting my awe"* . . . *He felt badly because*: Keir Dullea, interview by Richter, April 1, 2000.

192 *"the greatest film ever made"* . . . *"Can you* believe *this?"*: Lockwood, interview by Richter.

192 *"Stanley was simply much more"* . . . *"He needed people"*: Bizony, *The Making of Stanley Kubrick's 2001*, 371.

193 *In the midst of trying to persuade*: description of Clarke's travails in completing Wilson's film taken from Clarke to Blau, March 19, 1966; Clarke to Wilson, March 27, 1966.

194 *"geophones in the soft"*: Clarke to Kubrick, March 3, 1966.

194 *"He could just read your lips"*: Lockwood, interview by Bozung.

194 *"God, that's a great idea!"* . . . *biggest single contribution*: Lockwood, interview by author.

195 *"Open the pod bay doors, HAL"* . . . *"I'm sorry, Dave"*: film *2001: A Space Odyssey*, Stanley Kubrick, dir. (Metro-Goldwyn-Mayer, 1968).

196 *After the pod arms open*: Kubrick to Clarke, February 10, 1966.

196 *"it's quite okay"*: Clarke to Kubrick, February 13, 1966.

197 *movements rendered jolt-free*: wire flying technique is from Peter Brook, *Peter Brook's Production of* A Midsummer Night's Dream (Woodstock, IL: Dramatic, 1974), 79.

198 *Unsworth had Pike undercrank the camera*: *Daily Continuity Report*, June 20, 1966.

198 *"'Why were you willing'"*: Keir Dullea, interview by author, September 4, 2016.

199 *whose cockney accent, Dullea observed*: ibid.

199 *they nailed the* . . . *"Happy birthday, Frank"*: *Daily Continuity Report*, March 7, 1966.

200 *"I'm sorry, Frank, I think you missed it"*: ibid., March 27, 1966.

200 *Kubrick was subtly transmitting*: chess game analysis helped by the breakdown in Bill Wall, "2001: A Chess Space Odyssey," Chess.com, June 22, 2007, www.chess.com/article /view/2001-a-chess-space-odyssey; also Murray S. Campbell, "'An Enjoyable Game': How HAL Plays Chess," in Stork, *HAL's Legacy*, 75.

200 *"unearthly and brilliant contribution"*: Stanley Kubrick to Ingmar Bergman, February 9, 1960.

201 *"The next time the sound mixer's late"*: Andrew Birkin, interview by author, November 10, 2016.

201 *"If you can't explain"* . . . *What he wanted was the truth*: Bryan Loftus, interview by Richter, November 16, 1999.

201 *"Look at that, Bryan"*: Loftus, interview by author, September 22, 2016.

203 *"Stanley drank tap water"*: Johnson, interview by Larson.

203 *A heavy smoker since the age of twelve*: Christiane Kubrick, interview by author, June 5, 2016.

203 *a self-imposed limit of twenty-nine miles per hour*: Caras, interview by Richter; Clarke, *Lost Worlds*, 47.

203 *being brave was stupid*: Caras, interview by Richter.

203 *"Never get into a fair fight"*: Bernstein, interview by author, October 17, 2016.

203 *"I am now fighting"* . . . *Last week I borrowed* . . . *"Everything I have earned"*: Clarke to Wilson, March 12, 1966.

205 *he didn't want the book coming out*: Frewin, interview by author.

205 *"His relationships, one after the other"* . . . *"I think he found it all very frustrating"* . . . *He couldn't really finish*: Caras, interview by McAleer.

206 *kept a nervous breakdown at bay*: Clarke to Meredith, March 16, 1966.

207 *"a lot of unnecessary dead wood"* . . . *"hints that the fault may be in HAL"* . . . *"sticks by his diagnosis"* . . . *"It is also more logical"*: Clarke to Stanley Kubrick, March 11, 1966.

208 *"I don't know what I'm doing"* . . . *"Why am I critical?"* . . . *"This isn't doable"* . . . *"I'm gonna have a hamburger"* . . . *"He had these 'I'm just an asshole' moments"*: Christiane Kubrick, interview by author, September 22, 2016.

209 *Temperatures in the interior spiked*: temperature estimate is from Bill Weston, interview by Larson, December 19, 2001.

209 *the inside of a toaster*: Pike, interview by Larson, March 3, 2003.

209 *"I want this to be a murder"* . . . *"so it looked fleshy"*: Christiane Kubrick, interview by author, June 5, 2016.

210 *"Gee, that's terrible"*: Birkin, interview by author.

211 *"Yes, I'd like to hear it, HAL"*: *2001: A Space Odyssey*, Kubrick, dir.

212 *Dullea's best performance* . . . *"Dave, look. I'm really sorry"* . . . *"Shut up, HAL"* . . . *"Say, Dave, you know"* . . . *"Yes, that's right, HAL"*: *Daily Continuity Report*, June 29, 1966.

213 *"could hardly be the basis"*: Tony Frewin, ed., *Are We Alone? The Stanley Kubrick Extraterrestrial Intelligence Interviews* (London: Elliott & Thompson, 2005), 11.

213 *"one of Stanley Kubrick's few really bad ideas"*: quoted by Arthur C. Clarke in his preface to Frewin, *Are We Alone?*, 7.

213 *In his first response in February*: Sagan to Caras, February 9, 1966.

214 *"high frequency of quotations"*: Sagan to Caras, March 10, 1966.

214 *editorial control would not be possible*: Caras to Sagan, March 15, 1966.

214 *"an extraterrestrial in your bottle"*: Christiane Kubrick, interview by author, January 15, 2016.

215 *"The situation is roughly this"*: Caras to Craig W. Moodie Jr., July 29, 1965.

215 *"a series of lumps"* . . . *"this floor would glow"*: Caras to Masters, August 6, 1965.

216 *"Why not have a French bedroom?"*: Masters, interview by Shay.

218 *The eight days spent filming the hotel room*: Apart from *Daily Continuity Report*, June 23, 1966, I owe these insights to David Larson, who carefully studied still photographs of these actions, Larson to author, email, August 19, 2017.

218 *"Very good acting!"*: *Daily Continuity Report*, June 24, 1966.

219 *"Stanley, do you mind"* . . . *"Okay, that's fine"*: Dullea, interview by author, September 4, 2016; Thornton, "Gary Lockwood and Keir Dullea."

219 *"Very Good"*: *Daily Continuity Report*, June 24, 1966.

220 *Consider, for example, this koan*: "Zen Koans," AshidaKim.com, www.ashidakim.com/zen koans/zenindex.html.

220 *"tiny contribution"*: Thornton, "Gary Lockwood and Keir Dullea."

220 *"Understand you are considering"*: Stanley Kubrick to David Wolper, December 15, 1966.

CHAPTER 7: PURPLE HEARTS AND HIGH WIRES

221 *Any sufficiently advanced technology*: so widely quoted a single source is almost immaterial, but it was in "Sir Arthur C. Clarke, Visionary, Died on March 18, Aged 90," *The Economist*, March 27, 2008.

222 *"A secondary mystery"* . . . *"I think it makes the introduction"*: Kubrick to Clarke, April 11, 1966.

223 *"Please do not come here"*: Kubrick to Clarke, April 19, 1966

223 *Clarke responded*: Clarke, *Lost Worlds*, 46.

223 *A message from his agent*: Meredith to Clarke, undated but circa April 17–18, 1966.

223 *"A portentous spectacle"* . . . *"be running only in a few Cinerama"*: Clarke, *Lost Worlds*, 46.

224 *"the sheer North Face"*: Michael Herr, *Kubrick* (New York: Grove Press, 2000), 61.

224 *Clarke was off with Roger Caras*: Clarke, *Lost Worlds*, 46.

224 *"that dickhead George the Second"*: Birkin, interview by author.

225 *"Stanley's chaps"*: Powell, interview by author.

226 *One day, for example, he set off . . . he read the whole script . . . "I'm interested in it" . . . Birkin was summoned for a "ghoster" . . . "It's a raid, fellows" . . . "What did he say?" . . . "I just can't believe there isn't a desert" . . . "I know where there's a desert!" . . . "This could be your chance" . . . "Okay, well, go and photograph it" . . . "I wanted it to look like magic" . . . "Oh, hi, Andrew" . . . "Okay, gentlemen"*: Birkin, interview by author.

231 *"We thought, 'Well, that's a good start' " . . . "Oh, jeez, fellows" . . . Kubrick approached, suddenly intrigued*: Masters, interview by Shay; story confirmed by Bill Weston, interview by Larson, December 19, 2002.

232 *"That's all right—either I'll shoot"*: Shay and Duncan, "2001: A Time Capsule."

233 *"these little plastic things"* . . . *"We did it pretty well"* . . . *But the problem was*: ibid.; Stuart Freeborn, interview by Jordan R. Fox, circa 1976.

234 *"a very vivacious assistant"* . . . *But the worst part of all . . . "It really is beginning to look"* . . . *"Stanley, I have no idea"* . . . *"We'll get Geordie"* . . . *"Sure, Stanley"* . . . *So I had this whole outfit on*: Keith Hamshere, interview by Richter, November 11, 1999.

236 *Fine's cable to Meredith*: Fine to Meredith, June 5, 1966.

236 *"congratulations and wow"*: Meredith to Clarke, June 5, 1966.

236 *The figure was substantially higher*: Meredith to Clarke, undated but circa April 17–18, 1966.

237 *Kubrick felt upset and betrayed*: see, for example, Meredith to Clarke, June 13, 1966, and Kubrick to Clarke, July 12, 1966.

237 *"the blunt truth"*: Meredith to Clarke, June 13, 1966.

237 *"During one of my more frantic"*: Clarke, *Lost Worlds*, 47.

237 *"Tears, hysterics, sulks"*: Clarke, "Son of Dr. Strangelove."

238 *"The publishers are so pleased"*: Clarke to Kubrick, June 15, 1966.

238 *"some very acute"* . . . *"[s]ince the book will be coming"* . . . *"a puppet controlled by"* . . . *"The literal description"* . . . *"This scene has always seemed unreal"* . . . *"I think this is a very bad chapter"*: Clarke, *Lost Worlds*, 47–48.

239 *"I have bullied Stanley"*: Clarke to Tom Buck, June 22, 1966.

239 *"He is working about 20 hours a day"*: Clarke to Alfred Lansing, June 22, 1966.

239 *Finally, on July 4*: Clarke, timeline for "Son of Dr. Strangelove."

239 *"I may be leaving in disgust"*: Clarke to Willson Hunter, July 5, 1966.

239 *Clear evidence exists*: Wilson to Clarke, March 11, 1986; see also Clarke, "Son of Dr. Strangelove," and Stanley Kubrick's mail to Caras warning about a potential lawsuit, August 15, 1966.

239 *"an impressive two-page"*: Clarke, *Lost Worlds*, 49. More on Dell's expenses are in Clarke's letter to Sam Youd, June 14, 1967.

240 *an offer that he defer "all or part"*: Kubrick to Clarke, July 12, 1966.

240   *"discuss nothing with Arthur"*: Kubrick to Caras, August 15, 1966.

240   *"fighting back corporate tears"*: Clarke, *Lost Worlds*, 49.

240   *by mid-September, Clarke had, in fact*: Clarke to Blau, September 14, 1966.

240   *Eventually he located . . . "You can't leave!"*: Birkin, interview by author.

241   *"Well, I'm terribly sorry, Stanley"*: Johnson, interview by author, February 3, 2017.

241   *"We still haven't got a design" . . . "something that looked like" . . . Though drawn quickly*: Birkin, interview by author.

241   *"Tony Masters was one of the absolute"*: Trumbull, interview by author.

242   *Working with the new machine*: Loftus, interview by author.

242   *after Wally Gentleman had independently*: Gentleman, interview by Shay.

243   *But Loftus wasn't as unabashedly . . . "What are you suggesting?"*: Birkin, interview by author; see also Birkin, interview by Justin Bozung for TV Store Online, April 1, 2015.

243   *Loftus has a different memory*: Bryan Loftus to author, email, September 14, 2017.

243   *Now he got out maps . . . They settled on an Alouette . . . Birkin had planned a series of low-level runs*: Birkin, interview by author; shoot planning documents provided to author by Birkin, November 3, 2016.

244   *He presented Kubrick with twenty pages*: Birkin, interview by Bozung, TV Store Online Blog, April 1, 2015, http://blog.tvstoreonline.com/2014/09/2001-space-odyssey-interview-series _23.html.

244   *"I just thought of all the places" . . . "Having come up with the idea" . . . "If you've been in a helicopter" . . . "Well, I have three" . . . "You're risking my life" . . . "Don't worry. My lips are sealed" . . . "So having felt somewhat deflated"*: Birkin, interview by author.

246   *Kubrick had quietly commented*: Clarke to Youd, October 21, 1967.

246   *John Esam, who'd followed the beatnik . . . In February 1966 Esam became*: Andrew Phillip Smith, "John Esam Obituary," *The Guardian*, September 11, 2011.

246   *"I know a mime" . . . he'd been the lead performer . . . In contrast with European mime . . . "that looked like a place you'd leave the garbage" . . . "this ruffled guy"*: Dan Richter, interview by author, August 23, 2016.

248   *"Dan, they can't look like men"*: Dan Richter, *Moonwatcher's Memoir: A Diary of 2001: A Space Odyssey* (New York, Carroll & Graf, 2002), 4.

248   *Richter immediately said . . . A classic example of a convention . . . "Your problem is that" . . . "end up hiring some famous" . . . "this unpretentious little office" . . . "He was stocky" . . . "Well, this all sounds great" . . . "I just need twenty minutes" . . . a dim-witted, pushy, paranoid type . . . "a dumb thing to do" . . . "tough, nosy, tentative way" . . . "Oh, that's wonderful!" . . . "delicate, nervous little thing" . . . "Well, I've seen enough" . . . "The character is going to do" . . . Kubrick flooded Richter with questions . . . "I can't do that" . . . Kubrick asked him to write*: Richter, interview by author.

252   *"Passing through the door"*: Richter, *Moonwatcher's Memoir*, 13.

252   *"a beautiful, educated speaking voice"*: Frewin, interview by author, September 21, 2016.

253   *"some freelance soldiering in Africa" . . . Kubrick had turned down the suggestion . . . a directive the stuntman soon evaded . . . Given the complexity of the shots . . . One of the first space-walk sequences . . . Kubrick had insisted that Weston wear . . . he descended headfirst . . . Whipping 180 degrees forward . . . spun downward . . . "If it had caught him straight" . . . "One of the great things about Stanley" . . . One of Weston's shots called . . . "That sort of amused me" . . . "I was doing the alphabet backward" . . . "because he was still terrified" . . . "The first time I went out" . . . "We've got to get him back" . . . "I went looking for Stanley" . . . "Elizabeth Taylor's dressing room" . . . "Stanley, if he got involved"*: Weston, interview by Larson.

258   *The single most complex shot . . . Both Weston and the camera would move . . . High on his platform . . . the stuntman used the aluminum ring*: ibid.

259   *"He's going to get someone killed"*: Weston to Larson after formal interview, December 19, 2002, as told via email from Larson to author, December 14, 2016.

259   *"I was part of a group that actually made history"* . . . *"I just found myself brimming"* . . . *"It's really the story of the blue tail fly"*: Weston, interview by Larson.

260   *"kind of running by the side"*: Pederson, interview by Debert.

260   *"like a spare prick at a wedding"*: Birkin, interview by author.

260   *"Was he before Nero, or after?"*: Birkin, interview by Bozung.

260   *He'd been handed the keys* . . . *"Here you are, Moonwatcher"* . . . *They spoke for about forty-five minutes* . . . *He was off to Ceylon the next day*: Richter, interview by author; Richter, *Moonwatcher's Memoir*, 17–21.

CHAPTER 8: THE DAWN OF MAN

262   *No individual exists forever*: Arthur C. Clarke, "The Obsolescence of Man," in *Greetings, Carbon-Based Bipeds!*, 225.

262   *Half Jewish himself*: background information is from Ryan Gilbey, "Stuart Freeborn Obituary," *The Guardian*, February 8, 2013; Charles Drazin, "Dickens's Jew—from Evil to Delightful," *Jewish Chronicle*, May 3, 2013; Mark Burman, "Stuart: A Face Backwards," BBC Radio 4 online, June 27, 2013, www.bbc.co.uk/programmes/b01k2df5.

263   *"What do you think?"* . . . *"He looks like a character"*: Richter, *Moonwatcher's Memoir*, 15.

263   *"It's just useless"* . . . *"This is really beautiful"* . . . *"I can't do this"*: Richter, interview by author.

264   *Dan could see him covertly fighting*: Richter, *Moonwatcher's Memoir*, 43.

264   *"Here's this American kid"* . . . *"He's very polite"* . . . *"Dan's onto something"*: Richter, interview by author.

265   *a leopard, two hyenas*: from list titled "Animals," dated approximately July 1966, Kubrick Archive, UAL.

266   *"Now just go out and do all the research"*: Richter, *Moonwatcher's Memoir*, 43.

266   *He comes to a stop by a large Victorian*: ibid., 47.

267   *"He'd be, 'Hey, Dan' "*: Richter, interview by author.

267   *"with a calm look"* . . . *"I can't help feeling"* . . . *What great control!*: Richter, *Moonwatcher's Memoir*, 70–71.

269   *In 1964 Kubrick had methodically watched*: Lovejoy to Padashi Yonemoto, June 23, 1964; see also Alexander Walker's memory of Japanese science fiction films being delivered to Kubrick in *It's Only a Movie, Ingrid: Encounters on and off the Screen* (London: Headline, 1988), 286—though Walker evidently misremembered when the incident happened by about a decade.

271   *"How will I know"* . . . *"I don't know, Stanley"* . . . *"Here's what we do"* . . . *"You can't do something like that!"*: Shay and Duncan, "2001: A Time Capsule."

272   *Kubrick asked Birkin to book a ticket* . . . *"for their own use"* . . . *"I go to meet the airplane"*: Birkin, interview by author.

273   *We've been in tents for over a week*: Andrew Birkin to Commander D. L. Birkin, February 14, 1967.

275   *"The pain is so intense"* . . . *"She told me he'd taken her"* . . . *"Everyone's naked running tangled"*: Birkin, interview by author.

276   *"Well, I'm very fond of those trees"* . . . *"What if I get caught?"* . . . *"Make sure you don't mention MGM"* . . . *"I had to do a lot of bribing"*: story and quotes derive from Birkin, interview by Richter, November 16, 1999, though the story also stems from Birkin, interviews by Bozung and by author.

279   *Still intrigued by the* kokerbooms: Birkin, interview by author.

279   *"uncanny physical freedom"*: Richter, *Moonwatcher's Memoir*, 27.

279   *"They have a longer torso"* . . . *"The problem was these big human legs"*: Richter, interview by author.

280   *The walk is interesting* . . . *"It's uncanny"*: Richter, *Moonwatcher's Memoir*, 64.

281   *"an aristocratic lady"* . . . *"Lady Frankau is convinced"*: ibid., 38–39.

281   *estimates he was injecting thirty to forty times* . . . *"It's part of my nature, I guess"* . . . *"nobody else could do it"*: Richter, interview by author.

282   *"It's you. We built the costume"*: ibid.

283   *"As I walk in* . . . *I have to work hard"*: Richter, *Moonwatcher's Memoir*, 90–91.

283   *"Stanley stopped his pay the minute that the crash occurred"* . . . *"I think this is more interesting"* . . . *"pissed off with all these extra shots"*: Birkin, interview by author.

284   *"We now have another photographer here"*: Andrew Birkin to Commander D. L. Birkin, April 27, 1967.

285   *"interesting makeup film"* . . . *The latter didn't encase the performer's head*: Shay and Duncan, "2001: A Time Capsule."

286   *"I made a little acrylic cup"*: Freeborn, interview by Fox.

286   *"It was lightweight, airy, comfortable"* . . . *"longer, shiny, crispier hair"*: ibid.

287   *For the performer to open the jaw*: Shay and Duncan, "2001: A Time Capsule."

288   *"I saw him look at it"* . . . *"There's no reason for it"* . . . *"He was just pushing me"*: Freeborn, interviews by Richter, November 9, 1999, and by Fox.

288   *"He never yelled at Stanley"* . . . *"other people would have just said"*: Richter, interview by author.

289   *Kathleen, who was working just as hard*: Frewin to author, email, October 15, 2017.

289   *"Stuart, in a surge of inventiveness"* . . . *It works!*: Richter, *Moonwatcher's Memoir*, 82.

289   *"Believe me, it was revolting"*: Burman, "Stuart: A Face Backwards."

289   *"What I had was the positive"* . . . *The two things I had to think of were the age and the character*: Shay and Duncan, "2001: A Time Capsule"; Freeborn, interview by Fox.

290   *"Stuart and I are working together"*: Richter, *Moonwatcher's Memoir*, 78.

290   *"I think that both of us understood"*: Richter, interview by author.

291   *"playing monkey amongst the trees"*: Trumbull, interview by author.

292   *"You were quite strict"*: David Charkham, interview by Richter, November 14, 1999.

292   *"awful burden"* . . . *"The only benefit"*: Richter, *Moonwatcher's Memoir*, 42.

292   *"It was a pressure cooker"* . . . *"you couldn't hide"* . . . *"They also recommended"*: Richter, interview by author.

293   *"Dan, we've had a very serious complaint"* . . . *"It's bullshit, Stanley!"* . . . *"Roy is hurt because we let him go"* . . . *"Well, yes, I'm a drug addict"* . . . *"You're legal?"* . . . *"How does it feel?"* . . . *"Suddenly we have an intimacy"*: Richter, *Moonwatcher's Memoir*, 102; Richter, interview by author.

295   *After much trial and error*: screen dimensions are from Alcott, interview by Shay.

295   *"very collegial, very friendly"* . . . *"you're flying at such an altitude!"*: Richter, interview by author.

296   *" 'Well, how are we going to do it?' "* . . . *So I hung mushroom reflective-type globes*: Shay and Duncan, "2001: A Time Capsule"; Alcott, interview by Shay.

297   *Woods calculated that the Dawn of Man sets were lit by* . . . *temperatures in Studio 4* . . . *losing seven pounds during the first week* . . . *"They brought in a big wind machine"*: Richard Woods, interview by Richter, February 10, 2001.

298   *"You start to die"* . . . *"You only really have seconds"*: Richter, interview by author.

299   *"specialized breathing apparatus"*: Freeborn, interview by Fox.

299   *"We had to wake up, run or jump around"*: Charkham, interview by Richter.

299 *"I had to be in a squat"* . . . *"No, no, I need you to touch it"* . . . *"They'll clean it up"* . . . *Afterward, he showed Richter:* Richter, interview by author.

300 *"He was directing Dan"* . . . *"The intensity of that is really something":* Colin Cantwell, interview by Larson, January 16, 2004.

301 *they were filmed separately:* "Notes on Special Effects," Stanley Kubrick, January 9, 1969.

301 *"I was just wondering"* . . . *"Well, I remember the leopard being on the set"* . . . *"He was in a cage, yes"* . . . *"It was sort of rounded on the top":* Stuart and Kathleen Freeborn, interview by Richter, November 9, 1999.

301 *"everybody was very nervous"* . . . *"I remember being very nervous"* . . . *"I trusted Terry"* . . . *"Terry said, 'Don't worry' ":* Richter, interview by author.

302 *"Cut!" Then, "Okay, that's it"* . . . *"Maybe we could work on his movement"* . . . *"Look, I got what I needed":* Richter, interview by author; Richter, *Moonwatcher's Memoir,* 126.

303 *"I think Stanley was probably afraid"* . . . *"The film had enough problems":* Richter, interview by author.

303 *"What the hell's that!"* . . . *"He was bouncing all over the place with delight":* Woods, interview by Richter.

303 *"it was cheating—and Stanley didn't like cheating"* . . . *"I wasn't about to bring that up first"* . . . *"I hope you haven't forgotten"* . . . *"I want to see them suckling"* . . . *"Well, I'll put a little bit of honey on"* . . . *"So here we were, getting deeper and deeper":* Shay and Duncan, "2001: A Time Capsule."

304 *"I went through my reference books"* . . . *The breast would have to look and feel just right* . . . *Then I've got to reload it:* ibid.; Freeborn interview by Fox.

305 *"they went straight up to each other"* . . . *"Smack their bottoms again"* . . . *"one of these syringes"* . . . *Instead of feeding the babies* . . . *So I had to take them back to my lab* . . . *In the end, there was no need for* any of that work: ibid.

306 *"a bit in awe of"* . . . *"He doesn't need to be, the picture is great"* . . . *Polanski and Richter started riffing* . . . *Polanski asked:* Richter, *Moonwatcher's Memoir,* 140–41.

307 *He urged Kubrick to try it sometime:* Anthony Frewin to author, email, October 15, 2017.

307 *I am struck by the loneliness:* Richter, *Moonwatcher's Memoir,* 141.

307 *"Man emerged from the anthropoid background"* . . . *"And lacking fighting teeth or claws":* "Dawn of Man notes," Stanley Kubrick, May 30, 1967.

308 *"Well, I'll add a little red here"* . . . *"he was going to be learning and crafting and building":* Richter, interview by author.

308 *"I was trying to do as little as possible"* . . . *"I don't want to just go there right away":* ibid.

308 *"That frozen moment at the beginning of history":* Arthur C. Clarke, foreword, in Richter, *Moonwatcher's Memoir,* x.

309 *"Oh, sorry"* . . . *"No, no," said Kubrick. "Use it, it looks good":* Richter, interview by author.

309 *the man-apes stayed in character:* Cantwell, interview by Larson.

310 *"Well, that's all right, I'll make some phony meat":* Freeborn, interview by Fox.

310 *"No, I want real meat":* Shay and Duncan, "2001: A Time Capsule."

310 *"quite rank by the next morning"* . . . *"which all night killed any germs and cured it completely":* Freeborn, interview by Fox.

311 *was suffering from "housemaid's knee"* . . . *"because they thought it may not go down well"* . . . *"I recall, I think, thirty-two retakes":* Woods, interview by Richter.

312 *After a couple deafening blasts, the crows retreated:* Richter, *Moonwatcher's Memoir,* 140.

313 *"The key point in that"* . . . *"one of the great moments of film history":* Richter, interview by author.

313 *"On a field . . . the carpenters build a small rectangular platform"* . . . *"Throw the bones around,*

Dan!" . . . "We shoot and shoot": platform scene draws from Richter, *Moonwatcher's Memoir*, 133–35; Richter, interview by author; Clarke, *Lost Worlds*, 51–52.

314 *"eventually Stanley was satisfied"*: Clarke, *Lost Worlds*, 51–52.

315 *Arthur C. Clarke and I were standing there talking*: LoBrutto, *Stanley Kubrick*, 288–89.

315 *"What was happening with him is he was getting to that cut"* . . . *"and then suddenly we were free of the three million years"*: Richter, interview by author.

316 *"twelve to sixteen, sometimes eighteen hours a day"* . . . *We worked long hours . . . a 20th Century Fox production that had conducted an espionage operation*: Freeborn, interview by Fox.

317 *"I believe the mutual frustration we both felt"*: quoted in Stuart and Kathleen Freeborn, interview by Richter.

318 *"Okay, nobody's going to know about this"* . . . *But the reward came to me after I'd done another picture*: Freeborn, interview by Fox.

CHAPTER 9: END GAME

319 *Man, in a very technical age*: Michel Ciment, "A Propos de Orange Mécanique" ("On the Subject of *A Clockwork Orange*"), *Positif*, June 1972, contained in original manuscript, Agel, ed., *Making of Kubrick's 2001*.

320 *The true total, then, was closer to sixteen thousand steps*: Herb Lightman, "Filming 2001: A Space Odyssey," *American Cinematographer*, June 1968.

320 *"It was so bold"* . . . *"Every day, the amount of information"*: Cantwell, interview by Larson.

322 *"a four-hundred-feet railroad, impaling the studio"*: Clarke to Caras, September 12, 1966.

322 *He had a [shot] where the spaceship* Discovery: Caras, interview by Richter.

323 *"farting its way across the Universe"*: Shay and Duncan, "2001: A Time Capsule."

323 *"looks like shit"*: Pederson, interview by Larson, April 4, 2003.

323 *Ultimately a kind of extended collective improvisation*: ibid.; Shay and Duncan, "2001: A Time Capsule"; Masters, interview by Shay.

323 *"We all designed it"*: Masters, interview by Shay.

324 *"Harry, I don't like it"* . . . *"What the hell is wrong with it?"* . . . *"I want to see* Discovery": Lange, interview by Larson.

326 *The nine-foot-wide space station model*: dimension is from Cantwell, interview by Larson.

327 *"Such was the joy of the unit"*: Birkin, interview by author.

328 *"Bryan, I want you to produce stuff"* . . . *"We've taken the human element out"*: Bryan Loftus, interview by author, September 22, 2016.

328 *"I can't do that"* . . . *"Well, listen, you can read, can't you?"* . . . *"I think at that time, he had a lot more"*: Frewin, interview by author.

329 *"He had a list of ten people"*: Birkin, interview by author.

330 *"fundamentally sound and mutual-interest"*: Kubrick to Caras, August 15, 1966.

330 *"You have to learn to always be shocked"* . . . *You have to learn to bargain*: Kubrick to Caras, December 26, 1966.

331 *"like flu raised to the power of 10"*: Clarke to Kubrick, January 18, 1967.

331 *"We are getting magnificent shots"* . . . *I know you are greatly concerned . . . As a closing thought*: Kubrick to Clarke, January 7, 1967.

332 *"I'm very distressed to learn"*: Tom Craven to Clarke, January 22, 1967.

332 *"Actually, I do not agree with you"*: Clarke to Craven, January 29, 1967.

333 *"Thought you'd be amused"*: Clarke to Kubrick, February 3, 1967.

333 *"pretty disturbed"* by *"Stanley's reticence"*: Dan Terrell to Caras, February 8, 1966.

334 *"Confidential some noses out of joint at MGM"*: Caras to Kubrick, October 10, 1966.

334 *Please don't send me Mike Connolly pieces*: Kubrick to Caras, October 10, 1966.

334 *"I told him I knew nothing"*: Caras to Kubrick, October 10, 1966.

334 *"Please write me today full details"* . . . *Choose your adjectives carefully*: Kubrick to Caras, October 10, 1966.

335 *who he always referred to affectionately as "Big Boy Bob"*: Frewin to author, email, October 15, 2017.

336 *"highly critical of MGM"* . . . *"It was obvious that this was to be a session"* . . . *"Throughout the entire conversation, the humor was good"* . . . *"I and the others involved are professional enough"* . . . *I appreciate the fact that he wants to be advertising manager* . . . *"The whole thing is being treated like"* . . . *"I don't think he's always realistic"* . . . *"Without a doubt, it was O'Brien"* . . . *One last thing*: Caras to Kubrick, November 3, 1966.

338 *"These remarks are so inaccurate and unfriendly"*: Kubrick to Terrell, November 17, 1966.

339 *"Widely rumored, but denied by Metro"*: *Variety*, February 28, 1968.

339 *"Do it right, then do it better"*: Doug Trumbull, "Creating Special Effects for 2001," *American Cinematographer*, June 1968.

340 *"it was just all terrible"*: Trumbull, interview by author.

340 *The director also revived his banana-oil-and-paint*: Pederson, interview by Moritz.

340 *"kind of epiphany"* . . . *"And it was a corridor of light"* . . . *"I think we can make the Star Gate this way"* . . . *"Okay, I get it"* . . . *"'Whatever you need'"*: Trumbull, interview by author.

343 *As a simple explanation of how slit scan works*: Shay and Duncan, "2001: A Time Capsule."

344 *"Doug—also use op-art and electron art"*: notecards, Stanley Kubrick, April 1967.

344 *"elevated the effect from a two-dimensional novelty"*: Shay and Duncan, "2001: A Time Capsule."

344 *"Every day that I was working on that movie"* . . . *"And that was part of the evolutionary transformation"* . . . *"seventeen minutes of uninterrupted"*: Trumbull, interview by author.

346 *suspected some of its authors may have indulged*: Clarke, interview by author, December 19, 2001.

346 *"Were you guys ever"* . . . *"Never, no, not, too busy"* . . . *"Stanley, I've seen friends of mine lose their perspective"* . . . *"I agree with you completely"*: Trumbull, interview by author.

347 *"Stanley was, I think, such a genius-level person"* . . . *"Stanley, you remember when we were on the set"* . . . *"Ya want blue paint?"* . . . *"I don't want to hear"* . . . *"Yeah, absolutely deadly serious"* . . . *"He would constantly be mumbling orders"*: ibid.

349 *"One day in dailies"* . . . *"We all knew he could handle it"* . . . *"On the one hand, there was pride"* . . . *"Who fucked up the shot this time?"* . . . *"You're embarrassing me"*: ibid.

350 *These were then lit from above*: star filming methodologies are from Shay and Duncan, "2001: A Time Capsule."

351 *eight optimal drift directions and speed combinations*: Pederson, interview by Larson.

351 *"And so very early in the production"* . . . *"The waltz rhythm followed the star speed"*: Trumbull, interview by author.

352 *The rotoscopers were in there in the dark all the time*: Shay and Duncan, "2001: A Time Capsule."

353 *Although Clarke spoke about it only to close friends*: Bernstein, interview with author, September 17, 2016. Clarke told Bernstein that Turing should have moved to Ceylon, as he did, to live more freely.

354 *"And so I decided I'm going to purport"* . . . *"Are you gay?"* . . . *"And how's your life?"*: Trumbull, interview by author.

355 *he did find Gustav Mahler's Third Symphony*: David de Wilde, interview by author, November 29, 2016.

355 *Cordell spent a year writing variations on Mahler's Third*: Quincy Jones, quoted in Jon Burlingame, "A Lost Masterpiece Discovered: Alex North's Score for "2001: A Space Odyssey," Dylanna Music, 2012, www.alexnorth2001.com/ABOUT_THE_SCORE.html.

355 *"I know poor Frank had a nervous breakdown"* . . . *Then he did a [recording] session*: de Wilde, interview by author.

356 *"as a model for what he was after"*: Bernstein, "How About a Little Game."

356 *Clarke also cited Mahler and Vaughan Williams*: Clarke to Purdy, September 9, 1968.

356 *his playing Williams's Symphony no. 7 (Antarctic Symphony) while filming Dullea*: Dullea, interview by author.

356 *"ruled out most of the ultramodern composers" . . . "in general . . . film composition tends to lack originality"*: Bernstein, "How About a Little Game."

357 *This provocation was probably a response . . . The director also reportedly consulted*: Paul A. Merkley, "'Stanley Hates This But I Like It': North Vs. Kubrick on the Music for 2001: A Space Odyssey," *Journal of Film Music* 2, no. 1 (Fall 2007); Royal S. Brown, "An Interview with Bernard Herrmann," *Overtones and Undertones: Reading Film Music* (Oakland: University of California Press, 1994), accessed at the Society for the Appreciation of the Music of Bernard Herrmann, www.bernardherrmann.org/articles/an-interview-with-bernard-herrmann.

357 *"Listen, get a couple hundred pounds from the accounts department" . . . "and I don't want to waste time doing this"*: Frewin, interview by author.

357 *"He'd used Mendelssohn's Midsummer Night's Dream"*: Clarke, *Lost Worlds*, 45.

358 *"Mahler's music with that show reel"*: de Wilde, interview by author.

358 *"And we're looking at the wheel"*: Birkin, interview by author.

359 *"Well, you're overruled"*: de Wilde, interview by author.

359 *Ray Lovejoy, the editor*: Jan Harlan to author, email, June 24, 2016.

359 *"Around the lot, there was a strange reaction"*: Cantwell, interview by Larson.

359 *"I also wanted something that wouldn't sound too 'spacey'"*: Clyde Gilmour, "Exclusive: For the First Time, Kubrick Explains His Space Odyssey," *The Toronto Telegram*, May 15, 1968.

360 *"They were just slapping together all kinds of weird creatures"*: Pederson, interview by Larson.

360 *It was a broadcast of "Requiem"*: According to a letter sent by Kubrick secretary Christine Mitchell to Melvyn Bragg of the BBC on August 21, 1967, the broadcast was on the BBC's Third Program, August 18, 1967, 3:00 p.m.

360 *BBC said it didn't have rights to lend it*: Kubrick to Eric Roseberry, August 30, 1967.

360 *MGM's production office proceeded to make a deal . . . "a big mistake"*: Harlan to author, email, June 24, 2017.

360 *"In a letter recently uncovered"*: Alex Ross, "Space Is the Place," *New Yorker*, September 23, 2013.

360 *"The publisher realized"*: Harlan to author, email.

360 *the composer sued Kubrick*: Agel, ed., *Making of Kubrick's 2001*, 12.

360 *denied by the composer . . . "MGM wrote me"*: Michael John White, "Fears of a Clown" (György Ligeti interview) by *The Independent* (UK), October 18, 1989.

362 *"My music, in Kubrick's selection" . . . "It was less wonderful that I was neither asked nor paid"*: quoted in Merkley, "'Stanley Hates This But I Like It.'"

362 *Among the first pieces of music that David de Wilde transferred*: de Wilde, interview by author.

362 *"a big and great piece of majestic music" . . . "He loved already the title"*: Harlan to author, email, June 24, 2017. Note that in this email, Harlan's recollection of the timing of Kubrick's request was revised according to de Wilde's memory, straightening out some prior confusion on the part of researchers as to when Kubrick became aware of a composition so definitive to *2001*'s impact and so resonant with its concept.

363 *I teach you the Superman*: Friedrich Nietzsche, *Thus Spoke Zarathustra: A Book for Everyone and No One*, trans. R. J. Hollingdale (Baltimore: Penguin Books, 1961), 41–42.

363 *Evidence exists that 2001's spectacular opening shot*: Bruce Logan substantiates this supposition in Shay and Duncan, "2001: A Time Capsule."

364 *The new orientation simply looked right . . . "Is this going to work, or is this completely over the*

top?" . . . "I think it's great" . . . "I was really honored to be asked": Trumbull, interview by author.

CHAPTER 10: SYMMETRY AND ABSTRACTION

365 *There are certain thematic ideas:* Gilmour, "For the First Time, Kubrick Explains His Space Odyssey."

365 *"the final stages and trying to bring it together" . . . "He wasn't wasting any time, he was focused" . . . "Everything had to have that excellence"*: Cantwell, interview by Larson.

366 *"fundamentally at right angles" . . . "superbly usable"*: ibid.

366 *"a little kit . . . Just as a piece of rough art"*: Cantwell, interview transcript by Shay.

367 *"There's a level of abstraction" . . . It would not feel like special effects*: Cantwell, interview by Larson.

367 *"If you can describe it, I can film it"*: Clarke, "Son of Dr. Strangelove."

367 *"This is really Stanley Kubrick's movie"*: Arthur C. Clarke talk at MGM promotional event, April 4th, 1968, www.youtube.com/watch?v=HEEtfhxLQbw.

368 *"We chatted, and I suggested"*: Cantwell, interview by Larson.

369 *In each case, the monolith*: descriptions of cutout monoliths from Cantwell, interview transcript by Shay, August 17, 1979.

370 *"a puppet controlled by invisible strings"*: Clarke, *2001: A Space Odyssey*, 12.

370 *"The literal description of these tests"*: Clarke, *Lost Worlds*, 47–48.

370 *"Any sufficiently advanced technology"*: Clarke's laws have been widely promulgated. See "Clark's Three Laws," Wikipedia, https://en.wikipedia.org/wiki/Clarke%27s_three_laws.

370 *"wanted to hint at magic"*: Clarke, interview by Gelmis, *Camera Three*.

371 *in late October, Cantwell conceived of a view*: time frame determined by Stanley Kubrick summary of lunch discussion with the Effects Department, October 20, 1967.

371 *"Once again, we go for the symmetry" . . . "It was just lit and shot"*: Cantwell, interview by Larson.

372 *"a little accelerator motor"*: Shay and Duncan, "2001: A Time Capsule."

372 *which he called "brilliant!"*: Trumbull to author, email, June 26, 2017.

372 *"probably worked close to a hundred hours a week" . . . "vision hacking"*: Cantwell, interview by Larson.

373 *"The lead-in titles were to be" . . . "the Suns of Frasconi's shouldn't be" . . . "He would come in with a new narration" . . . "Arthur was very subtle"*: ibid.

374 *As early as February, he'd asked Caras*: Stanley Kubrick to Caras, February 12, 1967.

374 *"I would describe the quality"*: Stanley Kubrick to Benn Reyes, July 17, 1967.

374 *"perfect, he's got just the right amount"*: Stanley Kubrick to music producer Floyd Peterson, August 21, 1967.

375 *Although he initially found Balsam's rendition "wonderful"*: Stanley Kubrick to Caras, August 18, 1966.

375 *"unctuous, patronizing, neuter quality"*: Agel, ed., *Making of Kubrick's 2001*, 119.

375 *"a charming man"*: "Who's Behind Hal?," *Toronto Telegram*, May 15, 1968.

375 *"I am constantly occupied" . . . "Well, I don't think there is any question"*: *2001: A Space Odyssey*, Kubrick, dir.

375 *Kubrick sat four feet away from the actor*: from Terry O'Reilly, *Voices of Influence*, CBC Radio, January 21, 2012, www.cbc.ca/radio/undertheinfluence/voices-of-influence-1.2801871.

376 *with the director making small revisions*: from Doug Rain Narration folder, November 27, 1967, Kubrick Archive, UAL.

376 *"in order to maintain the required relaxed tone"*: 2010 Odyssey Archive (MMXOA) (blog), http://2010odysseyarchive.blogspot.si/2015/01/douglas-rain-immortal-voice.html.

376 *"a little more concerned"* . . . *"more matter of fact"*: Gerry Flahive to author, email, September 9, 2017; quotes are from his research at Kubrick Archive, UAL.

376 *"I'm sorry about everything"*: Kubrick Archive, UAL.

376 *"difficult to define"* . . . *"never completely freed myself"* . . . *"You're working up your crew psychology report"*: 2001: A Space Odyssey, Stanley Kubrick, dir.

377 *Kubrick had Rain sing, hum, and speak*: transcript of "Kill HAL?" scene, slate 734-1, Kubrick Archive, UAL, kindly provided in an email from Flahive to author, September 15, 2017.

377 *Using a new analog recording device*: Wendy Carlos, "The Eltro Mark II 'Information Rate Changer,'" last modified July 2008, www.wendycarlos.com/other/Eltro-1967.

378 *the scene's cumulative power hit Cantwell*: Cantwell, interview by Larson.

379 *"Death is inflicted like an act outside time"*: Michel Chion, *Kubrick's Cinematic Odyssey*, trans. Claudia Gorbman (London: BFI, 2001), 115.

379 *"I was used to Stanley"* . . . *"Stanley lifted the helmet up"*: de Wilde, interview by author.

380 *"The artist, like the God of creation"*: James Joyce, *A Portrait of the Artist as a Young Man* (New York: B. W. Heubsch, 1916), 252.

380 *"Please advise return cable availability"*: Caras to General Biological Supply House, May 31, 1966.

380 *A cable from Ivor Powell*: Powell to Caras, May 26, 1966.

380 *"Cannot offer quotations"*: General Biological Supply House to Caras, June 2, 1966.

381 *Moore, who'd helped Stuart Freeborn . . . had already made something of a name for herself*: Pam Warner, "Variety Is the Spice of Life to Liz Moore," *Advertiser and Reporter*, June 29, 1967.

381 *"Originally it was going to be much more complex"*: Shay and Duncan, "2001: A Time Capsule."

381 *"If you look at automatons"*: Johnson, interview transcript by Shay, August 3, 1979.

381 *"provided by Geoff Unsworth"*: Trumbull, interview by author.

382 *"with about forty thousand watts of backlight"*: Shay and Duncan, "2001: A Time Capsule."

382 *"Stanley filmed a number of different moves"*: Trumbull, interview by Shay.

383 *"the operator ran away screaming"*: incident is described in Piers Bizony, *2001: Filming the Future* (London: Aurum Press, 1994) as quoted in "Making the Starchild in '2001': A Tribute to Liz Moore," *2001Italia.It* (blog), www.2001italia.it/2013/05/making-starchild-in-2001-tribute-to-liz.html.

383 *"I said, 'Have you ever seen a good alien'"* . . . *"He said, 'I can't come up with anything'"* . . . *"The minute we can imagine it"*: Christiane Kubrick, interview by author, June 5, 2016.

384 *"I've got an idea about this"* . . . *"He'd seen a dotted pattern somewhere"*: Shay and Duncan, "2001: A Time Capsule."

385 *"I want to shoot some footage of aliens"* . . . *"I have come to realize"*: Richter, *Moonwatcher's Memoir* 136–37.

385 *"like a slightly cynical rabbi"*: Clarke, "Son of Dr. Strangelove."

386 *"nice and tight"* . . . *"stamp out perfect rounds"* . . . *And we went all over his body*: Shay and Duncan, "2001: A Time Capsule."

386 *"I think it was too good, in a way"*: Freeborn, interview by Fox.

386 *"And so a flat bunch of lights"* . . . *"I made this really interesting little slit scan machine"* . . . *"'You've just got to stop'"*: Trumbull, interview by author.

388 *"If it's so surreal and so crazy that it's unique"* . . . *"It would be great if I could think of something"* . . . *"I wish I was talented"*: Christiane Kubrick, interview by author, June 5, 2016.

389 *Born in 1904, he emerged from rural poverty in Montana*: Robert O' Brien background is from Anne McGonigle, interview by author, August 30, 2017; Carole Beers, "Robert H. O'Brien; Former Head of MGM Made His Home Here," *Seattle Times*, October 10, 1997; Leslie Eaton, "Robert H. O'Brien, 93, MGM President in '60s," *New York Times*, October 11, 1997.

390 *David Lean's three-hour epic* Doctor Zhivago . . . *No expense was spared . . . With wintry Moscow rising:* Melanie Williams, *David Lean* (Oxford: Oxford University Press, 2016), 184–85.

390 *owned more than five hundred thousand shares . . . two expensive proxy fights:* "Investment: Newest Life of Leo the Lion," *Time,* September 1, 1967; Eaton, "Robert H. O'Brien."

390 *"They were very nervous—quite rightly":* Christiane Kubrick, interview by author, September 22, 2016.

391 *came at MGM's suggestion:* Chion, *Kubrick's Cinematic Odyssey,* 24; "The Odyssey of Alex North," Robert Townson, liner notes for CD *Alex North's 2001: Unused Soundtrack Score,* (Varèse Sarabande VSD 5400), 2003.

391 *"ecstatic at the chance to work with Kubrick":* Merkley, "'Stanley Hates This But I Like It.'"

391 *"couldn't accept the idea of composing part of the score":* Burlingame, "A Lost Masterpiece Rediscovered," www.alexnorth2001.com/ABOUT_THE_SCORE.html.

391 *MGM would pay him $25,000 plus expenses:* Benjamin Melniker to Stanley Kubrick, December 12, 1967.

391 *orchestra was booked . . . scheduled for January 15 and 16:* Philip G. Jones to Stanley Kubrick, December 11, 1967.

391 *who'd orchestrated composer Alec Templeton's "Bach Goes to Town":* Frewin to author, email, October 15, 2017.

392 *"We were given an apartment, a cook, and a car":* Merkley, "Stanley Hates This But I Like It."

392 *"I worked day and night":* Burlingame, "A Lost Masterpiece."

392 *"He came in an ambulance":* de Wilde, interview by author.

392 *"I was present at all the recording sessions":* Merkley, "Stanley Hates This But I Like It."

392 *"He made some very good suggestions":* Schwam and Scorsese, *The Making of 2001,* 130.

392 *"He looked at me, and I looked at him":* de Wilde, interview by author.

393 *"It's a marvelous piece of music" . . . "Stanley hates this, but I like it!":* Merkley, "Stanley Hates This But I Like It." Here I question Merkley's assertion that the piece in question wasn't the one intended to replace *Thus Spoke Zarathustra,* however. In the previous sentence, Brant specifically identifies it as the opening piece, citing kettledrums.

393 *"It's shit":* comment by Con Pederson to Dave Larson after formal interview of Pederson, April 4, 2003; confirmed in Pederson email to author, November 9, 2017.

393 *the rest of the film would "use breathing effects":* Kate McQuiston, *We'll Meet Again: Musical Design in the Films of Stanley Kubrick* (New York: Oxford University Press, 2013), 132.

393 *"Fulfilled my obligation" . . . "clearing rights to temp tracks":* North's notes are at the Margaret Herrick Library of the Academy of Motion Picture Arts and Sciences, Los Angeles.

393 *"But in that instance, as in all others, O'Brien trusted my judgment" . . . Although he and I went over the picture very carefully:* Michel Ciment, *Kubrick* (New York: Holt, Rinehart, and Winston, 1983), 177.

394 *"All along, he was trying to clear the rights":* Merkley, "Stanley Hates This But I Like It."

394 *"something was on his mind" . . . "'What to do, and who to consider'" . . . "For Stanley, that was not something to be devastated by":* Cantwell, interview by Larson.

395 *So far, Gaffney's contribution to 2001 had been . . . "I want scenes in Monument Valley":* LoBrutto, *Stanley Kubrick,* 288–89.

396 *"deader than a doornail" . . . "I knew you were going to get me killed" . . . "You fly, I'll be the radar" . . . "I almost got killed":* ibid., 290–91.

396 *The director replied that in retrospect:* Birkin, interview by author.

397 *"That was as bold as anything else in the film":* Cantwell, interview by Parson.

397 *"we just had a good time" . . . the documentary prologue:* de Wilde, interview by author.

398 *"They were children of the forest"* . . . *"Much of the time you spend"*: Clarke to Kubrick via Reyes, October 31, 1967.

399 *"Since consciousness had first dawned"*: in "Narration" folder, Kubrick Archive, UAL, dated January 22, 1968.

399 *"Narration needs drastically reduced"*: Kubrick to Clarke via Reyes, November 22, 1967.

399 *"Just returned from two-week"*: Clarke to Kubrick, November 23, 1967.

399 *"Sorry about narration"*: Kubrick to Clarke via Reyes, November 23, 1967.

399 *"some new purple prosery"*: Clarke to Kubrick, November 25, 1967.

400 *"I must confess I've been under the impression"*: Kubrick to Clarke, December 27, 1967.

400 *"I don't want any unpleasantness or holdups"*: Clarke to Meredith, December 31, 1967.

400 *"It doesn't look like we'll get to the narration"*: Kubrick to Clarke, January 23, 1968.

400 *"It's impossible to rush"*: Clarke to Kubrick, January 29, 1968.

401 *"pretending to look busy"* . . . *"obviously nervous"* . . . *"Well, Dan, I have to settle your credit"* . . . *"Well, I'll take the starring credit"* . . . *"Can you live with that?"* . . . *"Believe me, Stanley"*: Richter, *Moonwatcher's Memoir*, 142.

401 *"He wouldn't have given it to me"*: Richter, interview by author.

401 *"Everybody was departing"* . . . *"No, I'm not worried because I have Doug"* . . . *"That was one of the biggest challenges"* . . . *"I was extremely averse to this"* . . . *"Hey, Stanley, do whatever you want"* . . . *"Well, I'm going to do it anyway"* . . . *"It was a tough moment"* . . . *"was epic in its complexity"* . . . *"I remember some kind of very uncomfortable"*: Trumbull, interview by author.

403 *"A lot of years have gone by"*: Con Pederson to author, email, September 7, 2017.

403 *"some other hanky-panky"* . . . *"In the last days"* . . . *"And there was always"*: Trumbull, interview by author.

404 *Instead, he had Louis Blau contact*: Blau to Joseph Youngerman, February 27, 1968.

404 *"members of his unit to mark their association"*: "Presentation to Stanley Kubrick by the members of his unit to mark their association on the production of *2001: A Space Odyssey*," February, 1968.

405 *"As you know, everything closed down"*: Kubrick to "Jimmy," March 7, 1968.

CHAPTER 11: RELEASE

406 *I can't do anything about things when they're going well*: Frewin, "Chairman Stanley."

406 *"I had done three years of working for Stanley"* . . . *"Sorry, you can't come in"* . . . *"He couldn't care less"*: de Wilde, interview by author.

407 *When Kubrick finally emerged from Immigration*: Bernstein, interview by author.

408 *interrogated Lange to find out what he knew*: "Memorandum for Fred Ordway File," Stanley Kubrick, September 29, 1967.

408 *claimed damages of $1 million*: "Launch $1-Mil Suit Against Kubrick's '2001,'" *Film and Television Daily*, March 27, 1968.

408 *in a glass-walled dome car*: Caras to Stanley Kubrick, April 5, 1967.

408 *which included converting the mono soundtrack*: de Wilde to author, email, October 5, 2017.

408 *The first time* 2001: A Space Odyssey *was screened at full length*: date is based on a statement in the April 17, 1968, issue of *Variety* to the effect that Kubrick saw the film for the first time only eight days before "the press preview," which was at the Uptown Theater on March 31.

408 *the studio's top brass were present* . . . *"For three years, they had been coming over"* . . . *"they all stood up"* . . . *"Did you work on that?"*: de Wilde, interview by author.

410 *on March 20 Scott Meredith succeeded in extracting*: Meredith to Blau, March 20, 1968.

411 *"When Stanley approved the book"*: Agel, *Making of Kubrick's 2001*, 256.

411 *"Apart from that, Mrs. Lincoln"*: Clarke to Youd, June 14, 1967.

411 *"I'm sure the deal will reemerge"*: Youd to Clarke, July 31, 1967.

411 *"No, I have no part"*: Clarke to Youd, August 28, 1967.

411 *If I have got it straight*: Youd to Clarke, October 2, 1967.

412 *"The book-film situation"*: Clarke to Youd, October 21, 1967.

412 *when the author went for a good-natured verbal duel*: C. S. Lewis to Clarke, February 18, 1954.

412 *My best wishes for every success*: Val Cleaver to Arthur C. Clarke, March 24, 1968.

413 *"Well, today we lost two presidents"*: Ken Kelley, "*Playboy* Interview: Arthur C. Clarke," *Playboy*, July 1986.

413 *"were just streaming out"*: Lockwood, interview by author.

413 *"I have never seen a major film preview"*: Victor Davis, "It's a Fantastic World—Wrapped in Reality," *Daily Express* (UK), April 3, 1968.

414 *"You'll have to see it six times"*: David Lewin, "Exit Hal, the Strangest Star of Them All," *Daily Mail* (UK), April 3, 1968.

414 *"It took four years to complete"*: Donald Zec, "At an Interplanetary First Night in America," *Daily Mirror* (UK), April 3, 1968.

414 *"Despite its marvelous and elaborate sets"*: "Space Film Draws VIPs, Criticism," *Evening Star* (Washington, DC), April 3, 1968.

414 *"one of the sycophants"* . . . *"very nice about it all"*: Lockwood, interview by author.

414 *evidently with some help from Louis Blau* . . . *"George Pal called me"*: Koch to Clarke, April 24, 1968.

415 *Decades later, however, Clarke quietly confirmed*: In Clarke's interview by McAleer, November 25, 1989, he stated flatly, "I was baffled by the ending," and when asked by McAleer if he was speaking for the record, replied, "Put it in quotes"; James Randi, interview by author, March 25, 2016; Michael Moorcock, "Close to Tears, He Left at the Intermission," *New Statesman*, January 8, 2017.

415 *Paul Newman, Joanne Woodward, Gloria Vanderbilt, and Henry Fonda*: Getty Images and *New York Post*, April 4, 1968.

415 *"Lots of alte kackers here"*: Christiane Kubrick, interview by author, June 5, 2016.

416 *"Kubrick trying to prove how boring"* . . . *"were saying things like"* . . . *"He was very upset"*: Randi, interview by author.

416 *a fretful Kubrick had prowled*: Christiane Kubrick, interview by author, June 5, 2016.

417 *"up and down the side aisles"*: Caras, interview by McAleer.

417 *"I've never seen an audience so restless"*: Thomas E. Brown, "*2001*'s Pre- and Post-Premiere Edits," the Kubrick Site, www.visual-memory.co.uk/amk/doc/brown3.html.

417 *"He was suddenly thinking"* . . . *"I didn't know there wasn't any air"* . . . *"They are not going to be the ones"*: Christiane Kubrick, interview by author, June 5, 2016.

417 *left at the intermission*: Moorcock, "Close to Tears."

417 *"Well, that's the end of Stanley Kubrick"*: Tom Mesirow, "Science Fiction and Prophecy: Talking to Arthur C. Clarke," *Los Angeles Review of Books*, July 24, 2013.

417 *By the end, 241 walkouts had been recorded*: figure is from Jack Nicholson in *Stanley Kubrick: A Life in Pictures*, Harlan, dir.

417 *"Whatever it was"* . . . *"Apart from Louie Blau"* . . . *"He looked a little bewildered"*: Bernstein, interview by author.

418 *"It was just a room full of drinks"* . . . *"I've never seen such a full room"* . . . *"schadenfreude and a nasty smile"* . . . *"Terry was trembling"* . . . *"I've always loved him for that"* . . . *"Stanley was tearing himself to shreds"* . . . *"'Oh my God, they really hated it'"* . . . *"What am I going to do?"* . . . *"This is rubbish"* . . . *"Stanley couldn't sleep"* . . . *"'Listen, let's drive'"* . . . *we went to that house*: combines Christiane Kubrick, interviews by author, June 5 and September 22, 2016.

419 *"some sort of great film"*: Penelope Gilliatt, "After Man," *New Yorker*, April 13, 1968.

419 *"somewhere between hypnotic and immensely boring"*: Renata Adler, "2001 Is Up," *New York Times*, April 4, 1968.

419 *"a major disappointment"*: Stanley Kauffmann, "Lost in the Stars," *New Republic*, May 4, 1968.

419 *"a thoroughly uninteresting failure"*: "Review of *2001: A Space Odyssey*," *Village Voice*, April 11, 1968.

419 *he caught wind of her disdain*: Kubrick knew this as early as the night of the April 3 New York premiere, as attested to in Mike Kaplan, "Kubrick: A Marketing Odyssey," *Guardian* (US), November 2, 2007.

420 *"monumentally unimaginative"*: Pauline Kael, "Trash, Art, and the Movies," *Harper's*, February 1969.

420 *she'd left her imprint*: Claudia Luther, "Margaret Booth, 104, Film Editor Had 70-Year Career," *Los Angeles Times*, October 31, 2002.

420 *we can be sure her views*: de Wilde, interview by author.

420 *some going so far as to say*: Neil McAleer, *Arthur C. Clarke: The Authorized Biography* (Chicago: Contemporary Books, 1992), 211.

420 *"at no one's request"*: Agel, *Making of Kubrick's 2001*, 170.

420 *Participants at the April 4 trim meeting*: original manuscript of Agel, *Making of Kubrick's 2001*.

421 *sat up all night working*: Caras, interview by McAleer.

421 *Editing continued across that weekend*: Agel, ed., *Making of Kubrick's 2001*, 170; Brown, "2001's Pre- and Post-Premiere Edits."

421 *Kubrick's edits were artificially constrained*: insight comes from Jon Davison to author, email, July 28, 2017.

421 *"were all present until the decision of the trim"* . . . *The basement in the MGM building* . . . *"He did not seem happy"* . . . *"Not true. No MGM exec ever suggested"* . . . *"It's a lie"*: original manuscript of Agel, *Making of Kubrick's 2001*.

422 *"Stanley will deny anything"*: Caras, interview by McAleer.

422 *made on MGM publicity man Michael Shapiro's Moviola*: David Larson to author, email, July 8, 2017.

423 *"Stanley Kubrick's magnificent work"*: "2001: Why Was it Cut?," letter from Jon F. Davison, *New York Times*, April 28, 1968.

423 *"some wholesale and rather hasty cutting decisions"*: "2001: A Space Odyssey," *Variety*, April 3, 1968.

423 *"Kubrick didn't see a final cut"* . . . *"the world's most extraordinary film"*: "Kubrick Trims '2001' by 19 Mins., Adds Titles to Frame Sequences," *Variety*, April 17, 1968.

424 *Variety was already recording advance ticket sales* . . . *"Because today's filmgoers are predominantly"*: "'Visual' Mod & 'Verbal' Crix: Kubrick's Sure '2001' to Click," *Variety*, April 10, 1968, 5.

425 *By mid-May, Variety was reporting*: "'2001' Draws Repeat and Recant Notices, Also a Quasi-Hippie Public," *Variety*, May 15, 1968, 20.

425 *"The film jumps erratically"*: "2001 Fails Most Gloriously," Joseph Gelmis, *Newsday*, April 4, 1968.

425 *About 100 years ago, Moby-Dick*: Joseph Gelmis, "Another Look at Space Odyssey," *Newsday*, April 20, 1968.

426 *I was fascinated, and most impressed*: Clarke to Gelmis, May 6, 1968.

426 *"The house had a landing with a light"*: Kaplan, "How Stanley Kubrick Kept His Eye on the Budget, Down to the Orange Juice," *Huffington Post*, February 20, 2012.

427 *"It's the only time in my life"* . . . *"There were some really nasty"* . . . *"What kind of skin is*

that' " . . . "I praised myself for one thing": Christiane Kubrick, interviews by author, June 5 and September 22, 2016.

427 *What happens at the end must tap*: A. H. Weiler, "Kazan, Kubrick and Keaton," *New York Times*, April 28, 1968.

428 *"I was so shocked"*: Christiane Kubrick, interview by author, June 5, 2016.

428 *"New York was the only really hostile city"*: Norden, "*Playboy* Interview: Stanley Kubrick," 94.

428 *that summer, Roger Caras and family would regularly drive over*: Caras, interview by Richter.

428 *the two of them blazing away*: Christiane Kubrick, interview by author, September 22, 2016.

CHAPTER 12: AFTERMATH

429 *Two possibilities exist*: Clarke, interview by author, December 2001.

430 *"In those days, you couldn't record television" . . . burst into tears*: Christiane Kubrick, interview by author, June 5, 2016.

430 *"If anyone understands it on the first viewing" . . . "I don't agree with that statement of Arthur's"*: Norden, "*Playboy* Interview: Stanley Kubrick," 92.

430 *"read the book, see the film"*: LoBrutto, *Stanley Kubrick*, 310.

430 *the highest-grossing film of the year*: see "Box Office / Business for *2001: A Space Odyssey* (1968)," IMDb, www.imdb.com/title/tt0062622/business?ref_=ttfc_ql_4; "1968 in Film," Wikipedia, last modified November 17, 2017, https://en.wikipedia.org/wiki/1968_in _film.

431 *"Stanley is now laughing all the way to the bank"*: Clarke to Ray Bradbury, June 6, 1968.

431 *"Stanley and I are laughing"*: Arthur C. Clarke, "The Myth of *2001*," *Cosmos Science-Fantasy Review*, no. 1 (April 1969): 10, reprinted in Clarke, *Report on Planet Three*, 222–24.

431 *"The success of 2001 was a great surprise"*: Clarke interview from *Time out of Mind*, BBC2 series on science fiction, 1979, www.youtube.com/watch?v=b6RCLBHtEhw.

431 *"2001? I see it every week"*: Schwam and Scorsese, *The Making of 2001*, cover.

431 *"Stanley very wisely realized that using narration"*: Clarke, interview by Gelmis, *Camera Three*.

432 *"It's creating more controversy"*: Clarke, interview by Paul Anderson, KPFA-FM, Berkeley, CA, May 20, 1968.

434 *The first one was when I saw 2001*: "A Drive of Titanic Proportions," James Cameron interview with the American Academy of Achievement, June 18, 1999, reprinted in *James Cameron: The Interviews*, ed. Brent Dunham (Jackson: University Press of Mississippi, 2011), 115.

434 *"the big bang"*: quoted in Eric Harrison, "Stanley Kubrick, Film Giant, Dies at 70," *Los Angeles Times*, March 8, 1999.

434 *"ultimate science fiction movie"*: Paul Scanlon, "George Lucas, The Wizard of 'Star Wars,'" *Rolling Stone*, August 25, 1977.

435 *"I'm not sure . . . I would have had the guts"*: in *Standing on the Shoulders of Kubrick: The Legacy of "2001: A Space Odyssey,"* directed by Gary Leva (Burbank, CA: Warner Bros. Home Video, 2007).

435 *In 1967 he informed Clarke*: Clarke to Stanley Kubrick, June 30, 1967.

436 *"I was to go back with my clothes"*: Trumbull, interview by Larson.

436 *"Those effects were not designed and directed" . . . One directs the movie*: Trumbull, interview by author.

437 *"The year was 1968" . . . "Mr. Trumbull was not in charge" . . . "the comparative contributions"*: "An Open Letter from MGM/UA and Stanley Kubrick," *Hollywood Reporter*, August 15, 1984.

437 *"an absolute genius" . . . "Stanley, I'm calling you" . . . "Wow, thanks" . . . "I just want you to know"*: Trumbull, interview by author.

438 *"a running lecture on free will"*: Stanley Kubrick, *Saturday Review*, December 25, 1971.

439 *The results placed* Citizen Kane *first*: the *All Things Considered* poll was conducted on November 14, 1977.

439 *"both an inspiration and cautionary tale"*: Stanley Kubrick's last public statement, a filmed speech to the Directors Guild of America, is available on YouTube, www.youtube.com /watch?v=3p1T3sVX4EY.

440 *By all accounts, the service . . . "I had no idea a week ago" . . . "So we all chuckled" . . . "You know, this is extraordinary" . . . "really one of the most intimate and affecting"*: my account of Stanley Kubrick's funeral is based on Trumbull, interview by author, and also Alexander Walker's moving remembrance from *The Stanley & Us Project*, directed by Mauro di Flaviano, Federico Greco, and Stefano Landini, posted to YouTube on March 8, 2012, available at www .youtube.com/watch?v=k79dtVHD_fk.

441 *"beautiful buffet, drinks, conversation" . . . "Stanley, all this crap that happened"*: Trumbull, interview by author.

442 *"You are deservedly the best-known science fiction writer"*: quoted in "In Memoriam," a new preface to Clarke, *2001: A Space Odyssey,* repr. (1999), v.

442 *subsequent police investigation found the accusation baseless*: "Sci-Fi Novelist Cleared of Sex Charges," BBC News online, April 6, 1998, http://news.bbc.co.uk/2/hi/south_asia/74938 .stm.

442 *The* Mirror *subsequently published an apology*: "Sir Arthur C. Clarke," *Telegraph* (UK) online, March 20, 2008, https://web.archive.org/web/20080326180335/http://www.telegraph.co .uk/news/main.jhtml?xml=%2Fnews%2F2008%2F03%2F19%2Fdb1904.xml.

443 *"I would say 2001 reflects about ninety percent"*: Agel, *Making of Kubrick's 2001,* 136.

443 *"We are privileged to live in the greatest"*: Michael Benson, *Beyond: Visions of the Interplanetary Probes* (New York: Abrams Books, 2003), Arthur C. Clarke, foreword, 9.

443 *"I've had a diverse career"*: Clarke, "Egogram," January 30, 2008.

444 *for about thirty seconds this vast cosmic explosion*: National Aeronautics and Space Administration, "NASA Detects Naked-Eye Explosion Halfway Across Universe," press release, March 20, 2008, www.nasa.gov/centers/goddard/news/topstory/2008/brightest_grb.html.

INDEX

Page numbers in *italics* refer to photographs.

Craven, Thomas, 332–33
Crawford, Joan, 167
Cruikshank, George, 262
Cruise, Tom, 440, 441
Curtis, Paul, 247, 292
cryonics, 115*n*

Dacre, Harry, 210, 377–78
Daily Express, 413
Daily Mirror, 414
"Daisy Bell" (Dacre), 210, 377–78
Dam Busters, The, 91
Dart, Raymond, 10, 72, 261, 266, 267
David and Lisa, 90
Davison, Jon, 423
Davenport, Nigel, 199
Davis, Victor, 413–14
Day the Earth Stood Still, The, 28, 49
de Chirico, Giorgio, 108
Defense, US Department of, 101
Dell Publishing, 236–40, 329, 331, 410, 411
Deneuve, Catherine, 140
Deodato, Eumir, 433
de Silva, Chandra, 14–15
Destination Moon, 23, 28, 49
de Wilde, David, 171, 355, 358, 362, 379–80, 398, 405, 406–10
Dickens, Charles, 262
Dietrich, Marlene, 381
Directors Guild of America (DGA), 404, 439–40
Dirty Dozen, The, 110
Discurio, 357
Disney, Walt, 84
Disney Company, 92, 93, 374
Dixon, Jim, 349–50
Doctor Who, 123
Doctor Zhivago, 85, 336, 390, 424
Douglas, Kirk, 5
Downham, Heather, 190
Drake, Frank, 19
Dr. Strangelove or: How I Learned to Stop Worrying and Love the Bomb, 4, 23–25, 28, 34, 35, 37–39, 41, 43, 49, 58, 86, 87, 93, 118, 139, 170, 186, 202, 233, 242, 315, 337, 355, 356
  Air Force and, 57
  rear projection in, 268
  sci-fi framing device for, 37–38

Sellers in, 41, 43, 262
Southern and, 4, 70, 237, 437
Tony Frewin and, 106–7
2001 and, 145
drugs, hallucinogenic, 98–99, 246, 306, 345–36
Dude, 53
Duggan, Terry, 300–302, *301,* 305
Dullea, Keir, 90, 138, 139, 159–60, 178–79, *179,* 187–88, 190–92, *191,* 194, *195,* 196–98, *198,* 207, 210–11, *211, 217,* 220, 253, 308, 334, 345, 356, 376, 381, 401, 413, 414, 442
Duncan, Jody, 343–44
Dunning, Eric, 252, 254–56, 258
D. W. Griffith Lifetime Achievement Award, 439–40
Dyson, Freeman, 213

Eady Levy, 134–35
Earthrise, 364*n,* 369
Echo satellites, 56–57
Ekanayake, Hector, 30, 53, 59, 78–79, 435
Elizabeth II, Queen, 442
Elstree, *see* Borehamwood Studios
embryos, 145, 380–81
"Encounter in the Dawn" (Clarke), 54–55, 64, 72, 73, 102
Epimenides, 129*n*
Ernst, Max, 107, 108
Esam, John, 246
Ettinger, Robert, 115*n*
Eugene's Flying Ballet, 197, 252
Evening Star, 414
Exploration of Space, The (Clarke), 21
extraterrestrials, 18–21, 27, 40, 63–64, 109, 114, 213, 383–84
  2001 and, 30, 64–66, 83, 92, 108, 115, 383–99
Eyes Wide Shut, 438, 440

Faulkner, William, 420
Fearless Vampire Killers, The, 226, 306
Ferlinghetti, Lawrence, 246
Fine, Donald, 236, 240
Finnegans Wake (Joyce), 3
Finney, Albert, 90
Fleetwood, Ken, 140
Fonda, Henry, 90, 415

Mount Laurel Library
100 Walt Whitman Avenue
Mount Laurel, NJ 08054-9539
856-234-7319
www.mountlaurellibrary.org